Series on Analysis, Applications and Computation – Vol. 11

ISAAC

Hardy Operators on Euclidean Spaces and Related Topics

Series on Analysis, Applications and Computation

More information on this series can be found at http://www.worldscientific.com/series/saac

Series on Analysis, Applications and Computation – Vol. 11

Hardy Operators on Euclidean Spaces and Related Topics

○ Shanzhen Lu
Beijing Normal University, China

○ Zunwei Fu
Linyi University, China

○ Fayou Zhao
Shanghai University, China

○ Shaoguang Shi
Linyi University, China

World Scientific

NEW JERSEY · LONDON · SINGAPORE · BEIJING · SHANGHAI · HONG KONG · TAIPEI · CHENNAI · TOKYO

Published by

World Scientific Publishing Co. Pte. Ltd.

5 Toh Tuck Link, Singapore 596224

USA office: 27 Warren Street, Suite 401-402, Hackensack, NJ 07601

UK office: 57 Shelton Street, Covent Garden, London WC2H 9HE

Library of Congress Cataloging-in-Publication Data

Names: Lu, Shanzhen, 1939–   author. | Fu, Zunwei, author. | Zhao, Fayou, author. |
  Shi, Shaoguang, author.
Title: Hardy operators on Euclidean spaces and related topics / Shanzhen Lu, Zunwei Fu,
  Fayou Zhao, Shaoguang Shi.
Description: 1st edition. | New Jersey : World Scientific, [2023] |
  Series: Series on analysis, applications and computation, 1793-4702 ; vol. 11 |
  Includes bibliographical references and index.
Identifiers: LCCN 2022055992 | ISBN 9789811253676 (hardcover) |
  ISBN 9789811253683 (ebook for institutions) | ISBN 9789811253690 (ebook for individuals)
Subjects: LCSH: Hardy spaces. | Generalized spaces.
Classification: LCC QA331 .L8166 2023 | DDC 515/.9--dc23/eng20230317
LC record available at https://lccn.loc.gov/2022055992

British Library Cataloguing-in-Publication Data
A catalogue record for this book is available from the British Library.

For photocopying of material in this volume, please pay a copying fee through the Copyright Clearance Center, Inc., 222 Rosewood Drive, Danvers, MA 01923, USA. In this case permission to photocopy is not required from the publisher.

For any available supplementary material, please visit
https://www.worldscientific.com/worldscibooks/10.1142/12762#t=suppl

Printed in Singapore

Preface

It is well known that the properties of operators in various function spaces play an essential role in harmonic analysis. Just as Stein [Stein (1993)] stated: average operator, singular integral operator and oscillatory integral operator are the most basic three types of operators in harmonic analysis. The Hardy-Littlewood maximal operator, the Hilbert transform and the Fourier transform, are prime examples of these classes of operators, respectively. Undoubtedly, Hardy operators can be classified as average operators. The study of Hardy operators can be traced back to Hardy's own classic works [Hardy (1920)] and [Hardy (1925)]. Hardy's discrete inequality first promotes the development of number theory. A detailed account of the history of the topic can be found in [Kufner et al. (2007)]. The Hardy operator and the Hardy inequality are of fundamental importance in many branches of mathematical analysis and mathematical physics, and have been intensively studied since their appearances. A rich theory has been developed with the original inequalities on $\mathbb{R}$ extended and refined in many ways, and an extensive literature on them now exists. We refer the reader to the monographs of Opic and Kufner [Opic and Kufner (1990)], Kufner and Persson [Kufner and Persson (2003)], and Kufner, Persson and Samko [Kufner et al. (2017)].

As a higher-dimensional generalization, Faris [Faris (1976)] first defined higher-dimensional Hardy operators during his study of quantum mechanics. Some results on the n-dimensional Hardy operators have been developed with the original inequalities on $\mathbb{R}^n$, such as [Balinsky et al. (2015)], [Opic and Kufner (1990), Chapter 2] and [Kufner et al. (2017)]. Throughout the book, we shall focus on our new properties of the higher-dimensional Hardy operators in the last decade. To be precise, this monograph will focus on higher-dimensional Hardy operators, based on the spherical average form, introduced by Christ and Grafakos [Christ and Grafakos (1995)] (see also [Drábek et al. (1995)]).

The key to this monograph is based on that the Hardy operator is generally smaller than the Hardy-Littlewood maximal operator, which leads to, on the one hand, the operator norm of the Hardy operator itself being smaller than the latter. On the other hand, the former characterizing the weight function class or function spaces is greater than the latter. This is also the main motivation of this monograph.

In Chapter 1, we give the sharp bounds for the n-dimensional (fractional) Hardy operators and the m-linear Hardy operators on the (weighted) Lebesgue spaces, and we also give the sharp inequalities between L^p-norms for the higher-dimensional Hardy operator and its dual. Chapter 2 investigates the sharp bounds for the n-dimensional Hardy type operators in different function spaces, such as (central) Morrey spaces and (central) BMO spaces. We give some new methods to obtain the operator norms. Chapter 3 focuses on the n-dimensional generalization of Hardy operators and their weighted theory. We establish the weighted norm inequalities for the n-dimensional Hardy operators. We also study the propositions on the class of weights which are different with the Muckenhoupt weights. In Chapter 4, using the symmetrical structure of Hardy operators, we give some characterizations of central function spaces via the boundedness and compactness of commutators of n-dimensional Hardy operators. More precisely, we characterize the central $VMO(\mathbb{R}^n)$ space (the central $BMO(\mathbb{R}^n)$ closure of the space of all functions being infinite-times continuously differential with compact support) via commutators of Hardy operators. Chapter 5 extends some of our results to the non Euclidean space. We give the best constants for Hardy operators on the Heisenberg group, and we also establish the sharp estimates of p-adic Hardy operators. The Hausdorff operators can be regarded as a generalization of Hardy operators. In fact, many methods can be transferred from the Hardy operators to the Hausdorff operators. In the last chapter, we give the sharp constants for q-analogue of the Hausdorff inequality in the sense of quantum calculus.

We would like to express our sincere gratitude to Professors Dashan Fan, Loukas Grafakos, Zongguang Liu, Dunyan Yan and Qingyan Wu for their effective cooperates in the study of Hardy operators.

Shanzhen Lu (Beijing Normal University)
Zunwei Fu (Linyi University)
Fayou Zhao (Shanghai University)
Shaoguang Shi (Linyi University)
October, 2022

Contents

Index of Notations

$\mathbb{Z}$ the set of all integers

$\mathbb{Z}_+$ the set of all positive integers $\{1, 2, 3, \ldots\}$

$\mathbb{R}$ the set of real numbers

$\mathbb{R}_+$ the set of positive real numbers

$\mathbb{R}^n$ the Euclidean n-space

$|x|$ $\sqrt{|x_1|^2 + \cdots + |x_n|^2}$ when $x = (x_1, \ldots, x_n) \in \mathbb{R}^n$

$\mathbb{S}^{n-1}$ the unit sphere $\{x \in \mathbb{R}^n : |x| = 1\}$

ω_n the surface area of the unit sphere $\mathbb{S}^{n-1}$

ν_n the volume of the unit ball $\{x \in \mathbb{R}^n : |x| < 1\}$

p' the number $p/(p - 1)$, whenever $0 < p \neq 1 < \infty$

$1'$ the number ∞

∞' the number 1

χ_E the characteristic function of the set E

$A \sim B$ there exist two positive constants C_1 and C_2 such that $C_1 B \leq A \leq C_2 B$

$A \ll B$ A is much smaller than B

Chapter 1

Sharp bounds for Hardy operators on Lebesgue spaces

Hardy in [Hardy (1920)] stated (without proof) that if $a > 0$, $f(x) \geq 0$, $p > 1$ and $\int_a^\infty f^p(x)dx$ is convergent, then

$$\int_a^\infty (Hf(x))^p \, dx \leq \left(\frac{p}{p-1}\right)^p \int_a^\infty f^p(x)dx, \tag{1.1}$$

where H is the classical one-dimensional Hardy operator defined in the form of

$$Hf(x) = \frac{1}{x}\int_0^x f(t)dt, \; x > 0$$

for all nonnegative measurable functions f on $\mathbb{R}_+$. His main aim was to find a new, more elementary proof of Hilbert's inequality for double series, and he showed in [Hardy (1925)] that in fact this inequality follows from the discrete version of (1.1):

$$\sum_{n=1}^\infty \left(\frac{1}{n}\sum_{k=1}^n a_k\right)^p \leq \left(\frac{p}{p-1}\right)^p \sum_{n=1}^\infty a_n^p,$$

where $a_n \geq 0$ and $p > 1$. Hardy in [Hardy (1925)] stated and proved the following inequality: suppose that $f(x) \geq 0$ and $p > 1$; that f is integrable over any finite interval $(0, X)$ and f^p is integrable over $(0, \infty)$. Then

$$\int_0^\infty (Hf(x))^p \, dx \leq \left(\frac{p}{p-1}\right)^p \int_0^\infty f^p(x)dx. \tag{1.2}$$

A natural question is to consider the case of higher dimensions. As we know, the theory in higher dimensions is perceived to be much more difficult and indeed there are significant problems in higher dimensions for which the one-dimensional techniques are not adequate. The higher-dimensional results also provide an elegant approach to study one-dimensional integral operators. As an extension of higher dimensions, we mention the following

1

form. For a locally integral function f on $\mathbb{R}^n$, Christ and Grafakos in [Christ and Grafakos (1995)] introduced the following n-dimensional Hardy operator

$$\mathcal{H}_\beta f(x) = \frac{1}{|B(\mathbf{0}, |x|)|} \int_{B(\mathbf{0}, |x|)} f(y) dy, \quad x \in \mathbb{R}^n \backslash \{0\}, \qquad (1.3)$$

where $B(\mathbf{0}, |x|)$ denotes a ball with center at the origin and radius $|x|$, and $|B(\mathbf{0}, |x|)|$ denotes the volume of the ball $B(\mathbf{0}, |x|)$.

Averaging operators are of fundamental importance in analysis and it is often desirable to obtain sharp norm estimates for them. The study of these operators may not be as delicate as that of maximal operators ([Grafakos and Montgomery-Smith (1997); Melas (2003)]), but still requires the use of certain beautiful and elegant ideas. In the following sections, we shall give the sharp bounds for higher-dimensional Hardy-type operators on Lebesgue spaces.

1.1   Sharp bounds for Hardy operators

The norm of the n-dimensional Hardy operator $\mathcal{H}$ on $L^p(\mathbb{R}^n)$ was evaluated in [Christ and Grafakos (1995)] (see also [Drábek et al. (1995)]) and was found to be equal to that of the one-dimensional Hardy operator H.

Theorem 1.1. *Let $f \in L^p(\mathbb{R}^n)$ and $1 < p < \infty$. Then $\mathcal{H}$ is bounded on $L^p(\mathbb{R}^n)$ with norm equal to the constant $\frac{p}{p-1}$.*

In fact, let $\mathbb{R}_+$ denote the multiplicative group of positive real numbers with Haar measure $\frac{dt}{t}$. The authors in [Christ and Grafakos (1995)] used the group inequality $\|g * K\|_{L^p} \le \|g\|_{L^p} \|K\|_{L^1}$ to prove this theorem.

Fu, Grafakos and Lu et al. in [Fu et al. (2012)] extended the above result to the weighted Lebesgue spaces by using a simpler method.

Theorem 1.2. *Let $\alpha < n(p-1)$ and $1 \le p < \infty$. If $f \in L^p(|x|^\alpha dx)$, then $\mathcal{H}$ maps the weighted Lebesgue space $L^p(|x|^\alpha dx)$ to $L^p(|x|^\alpha dx)$ with norm equal to the constant $\frac{pn}{pn-n-\alpha}$.*

Proof. Let ω_n be the area of the unit sphere $\mathbb{S}^{n-1}$ and $\omega_n = 2\pi^{n/2}/\Gamma(n/2)$. Set

$$g_f(x) = \frac{1}{\omega_n} \int_{|\xi|=1} |f(|x|\xi)| d\xi$$

for $x \in \mathbb{R}^n$. It is easy to see that the operator $\mathcal{H}$ and its restriction to radial functions have the same operator norm on $L^p(|x|^\alpha dx)$. We may therefore assume that f is a nonnegative radial function.

Fix $\alpha < n(p-1)$. Let $B(\mathbf{0}, R)$ be a ball of radius R centered at the origin. By Minkowski's integral inequality, we have

$$\nu_n \|\mathcal{H}f\|_{L^p(|x|^\alpha dx)} = \left(\int_{\mathbb{R}^n} \left| \frac{1}{|x|^n} \int_{B(\mathbf{0},|x|)} f(y)\, dy \right|^p |x|^\alpha\, dx \right)^{1/p}$$

$$= \left(\int_{\mathbb{R}^n} \left| \int_{B(\mathbf{0},1)} f(|x|y)\, dy \right|^p |x|^\alpha\, dx \right)^{1/p}$$

$$\leq \int_{B(\mathbf{0},1)} \left(\int_{\mathbb{R}^n} |f(|y|x)|^p |x|^\alpha\, dx \right)^{1/p} dy$$

$$\leq \int_{B(\mathbf{0},1)} |y|^{-\frac{n-\alpha}{p}}\, dy \|f\|_{L^p(|x|^\alpha dx)}$$

$$= \frac{\nu_n pn}{pn - n - \alpha} \|f\|_{L^p(|x|^\alpha dx)},$$

where $\nu_n = \frac{\pi^{n/2}}{\Gamma(1+n/2)}$ is the volume of the unit ball in $\mathbb{R}^n$. Therefore, we have obtained the upper estimate

$$\|\mathcal{H}\|_{L^p(|x|^\alpha dx) \to L^p(|x|^\alpha dx)} \leq \frac{pn}{pn - n - \alpha}.$$

Conversely, for $0 < \varepsilon < \min\{1, n/p' - \alpha/p\}$, we take

$$f_\varepsilon(x) = |x|^{-\frac{n+\alpha}{p} - \varepsilon} \chi_{\{|x|>1\}}(x).$$

By a simple calculation, we see that

$$\|f_\varepsilon\|^p_{L^p(|x|^\alpha dx)} = \omega_n (p\varepsilon)^{-1}.$$

It follows that

$$\mathcal{H}(f_\varepsilon)(x) = \nu_n^{-1} |x|^{-\frac{n+\alpha}{p} - \varepsilon} \int_{|x|^{-1} < |y| < 1} |y|^{-\frac{n}{p} - \frac{\alpha}{p} - \varepsilon}\, dy \chi_{\{|x|>1\}}(x).$$

Then we have that $\|\mathcal{H}(f_\varepsilon)\|_{L^p(|x|^\alpha dx)}$ is equal to

$$\frac{1}{\nu_n} \left(\int_{|x|>1} \left(|x|^{-\frac{n+\alpha}{p} - \varepsilon} \int_{|x|^{-1} < |y| < 1} |y|^{-\frac{n+\alpha}{p} - \varepsilon}\, dy \right)^p |x|^\alpha dx \right)^{1/p}$$

$$\geq \frac{1}{\nu_n} \left(\int_{|x|>\varepsilon^{-1}} \left(|x|^{-\frac{n+\alpha}{p} - \varepsilon} \int_{\varepsilon < |y| < 1} |y|^{-\frac{n+\alpha}{p} - \varepsilon}\, dy \right)^p |x|^\alpha dx \right)^{1/p}$$

$$= \frac{\omega_n}{\nu_n} \cdot \frac{1 - \varepsilon^{n - \frac{n}{p} - \frac{\alpha}{p} - \varepsilon}}{n - \frac{n}{p} - \frac{\alpha}{p} - \varepsilon} \cdot \varepsilon^\varepsilon \|f_\varepsilon\|_{L^p(|x|^\alpha dx)}.$$

Consequently, we obtain

$$\|\mathcal{H}\|_{L^p(|x|^\alpha dx) \to L^p(|x|^\alpha dx)} \geq \frac{\omega_n}{\nu_n} \cdot \frac{1 - \varepsilon^{n - \frac{n}{p} - \frac{\alpha}{p} - \varepsilon}}{n - \frac{n}{p} - \frac{\alpha}{p} - \varepsilon} \cdot \varepsilon^\varepsilon.$$

We deduce that

$$\|\mathcal{H}\|_{L^p(|x|^\alpha dx)\to L^p(|x|^\alpha dx)} \geq \frac{pn}{pn-n-\alpha}$$

by letting $\varepsilon \to 0^+$ (using the fact that $\varepsilon^\varepsilon \to 1$). □

Remark 1.1. The norm of the operator $\mathcal{H}$ from $L^p(|x|^\alpha dx)$ to itself is independent of n exactly when $\alpha = 0$.

1.2 Sharp bounds for fractional Hardy operators

Recall that, for a locally integrable function f on $\mathbb{R}^n$, the n-dimensional fractional Hardy operator $\mathcal{H}_\beta$ is defined by

$$\mathcal{H}_\beta f(x) = \frac{1}{|B(0,|x|)|^{1-\frac{\beta}{n}}} \int_{|y|<|x|} f(y)dy, \quad x \in \mathbb{R}^n\backslash\{0\}, \qquad (1.4)$$

where $0 \leq \beta < n$ (see also [Fu et al. (2007)]). Obviously, $\mathcal{H}_0$ will be reduced to the n-dimensional Hardy operator $\mathcal{H}$ when $\beta = 0$. For $0 \leq \beta < n$, the fractional Hardy operator is closely related to the fractional Hardy-Littlewood maximal operator M_β which is defined by

$$M_\beta f(x) = \sup_{r>0} \frac{1}{|B(x,r)|^{1-\frac{\beta}{n}}} \int_{|y-x|<r} |f(y)|dy, \quad x \in \mathbb{R}^n,$$

and to the Riesz potential I_β defined by

$$I_\beta f(x) = \int_{\mathbb{R}^n} \frac{f(y)}{|x-y|^{n-\beta}}, \quad x \in \mathbb{R}^n.$$

It is then easy to see that

$$M_\beta f(x) = \sup_{y\in\mathbb{R}^n} \mathcal{H}_\beta(|f(\cdot+x)|)(y), \quad x \in \mathbb{R}^n$$

and

$$\mathcal{H}_\beta(|f|)(x) \leq 2^{n-\beta} M_\beta f(x) \leq 2^{n-\beta}|B(0,1)|^{\frac{\beta}{n}-1} I_\beta(|f|)(x), \quad x \in \mathbb{R}^n\backslash\{0\}.$$

In addition, for $0 \leq \beta < n$, $1 < p \leq n/\beta$ and $1/p - 1/q = \beta/n$, the following two statements are well-known (see [Lu et al. (2007)]):

(a) If $f \in L^p(\mathbb{R}^n)$, then

$$\|M_\beta f\|_{L^q(\mathbb{R}^n)} \leq C\|f\|_{L^p(\mathbb{R}^n)};$$

(b) If $f \in L^1(\mathbb{R}^n)$, then for any $\lambda > 0$,

$$|\{x \in \mathbb{R}^n : M_\beta f(x) > \lambda\}| \leq \left(\frac{C}{\lambda}\|f\|_{L^1(\mathbb{R}^n)}\right)^{\frac{n}{n-\beta}}.$$

However the constants C, the bounds of operator M_β, are not given explicit expression of depending on the parameters p, q and β.

In what follows, $B(\cdot, \cdot)$ is the usual beta function defined by $B(z, w) = \int_0^1 t^{z-1}(1-t)^{w-1}dt$, where z and w are complex numbers with positive real parts. The following relationship between the gamma and beta functions is valid: $B(z,w)\Gamma(z+w) = \Gamma(z)\Gamma(w)$.

Let us recall some results of the fractional Hardy operator. For the one-dimensional case, Bliss in [Bliss (1930)] worked out the best possible constant C_0 in the inequality

$$\|\mathcal{H}_\beta f\|_{L^q(\mathbb{R}_+)} \leq C_0 \|f\|_{L^p(\mathbb{R}_+)},$$

where $0 < \beta < 1$, $1 < p < \infty$, $1/p - 1/q = \beta$ and

$$C_0 = (p'/q)^{1/q} \left(\frac{1}{q\beta} \cdot B\left(\frac{1}{q\beta}, \frac{1}{q'\beta}\right) \right)^{-\beta}.$$

For more information about the fractional Hardy operator, we refer to ([Burenkov (2011)], [Boyd (1971)], [Kufner and Persson (2003)], [Manakov (1992)]) and references therein.

In this section, the bounds of the operator $\mathcal{H}_\beta$ from $L^p(\mathbb{R}^n)$ to $L^q(\mathbb{R}^n)$ and from $L^1(\mathbb{R}^n)$ to $L^{\frac{n}{n-\beta}, \infty}(\mathbb{R}^n)$ are explicitly worked out. Furthermore, we shall show that the constant 1 is the bound of operator $\mathcal{H}_\beta$ from $L^1(\mathbb{R}^n)$ to $L^{\frac{n}{n-\beta}, \infty}(\mathbb{R}^n)$ and is the best possible, that is,

$$\|\mathcal{H}_\beta\|_{L^1(\mathbb{R}^n) \to L^{\frac{n}{n-\beta}, \infty}(\mathbb{R}^n)} = 1.$$

Theorem 1.3. *Suppose that $0 < \beta < n$, $1 < p < q < \infty$ and $1/p - 1/q = \beta/n$. If $f \in L^p(\mathbb{R}^n)$, then we have*

$$\|\mathcal{H}_\beta f\|_{L^q(\mathbb{R}^n)} \leq A\|f\|_{L^p(\mathbb{R}^n)}. \tag{1.5}$$

Moreover,

$$\|\mathcal{H}_\beta\|_{L^p(\mathbb{R}^n) \to L^q(\mathbb{R}^n)} = A,$$

where

$$A = \left(\frac{p'}{q}\right)^{1/q} \left(\frac{n}{q\beta} \cdot B\left(\frac{n}{q\beta}, \frac{n}{q'\beta}\right) \right)^{-\beta/n}.$$

It is worth mentioning that the proof in [Fu et al. (2012)] is not applicable to the fractional Hardy operator. Although our idea partly come from [Bliss (1930)], there are some essential difficulties producing in the higher-dimensional case. The first difficulty is how to deal with the high-dimensional case. We shall use the rotation method as in [Fu et al. (2012)]

to reduce the n-dimensional case to the one-dimensional case. The second difficulty here is how to reconstruct some auxiliary functions to achieve the sharp bound, which is quite different from that of [Bliss (1930)].

In order to reduce the dimension of function spaces, we need the following lemma which was obtained in [Lu et al. (2013), p. 7].

Lemma 1.1. *For a function $f \in L^p(\mathbb{R}^n)$, let*

$$g_f(y) = \frac{1}{\omega_n} \int_{|\xi|=1} |f(|y|\xi)| \, d\xi, \; y \in \mathbb{R}^n.$$

Then

$$\mathcal{H}_\beta(|f|)(x) = \mathcal{H}_\beta(g_f)(x)$$

and

$$\|g_f\|_{L^p(\mathbb{R}^n)} \leq \|f\|_{L^p(\mathbb{R}^n)}.$$

Remark 1.2. It follows from the above lemma that

$$\frac{\|\mathcal{H}_\beta f\|_{L^q(\mathbb{R}^n)}}{\|f\|_{L^p(\mathbb{R}^n)}} \leq \frac{\|\mathcal{H}_\beta(g_f)\|_{L^q(\mathbb{R}^n)}}{\|g_f\|_{L^p(\mathbb{R}^n)}}.$$

Therefore, the norm of the operator $\mathcal{H}_\beta$ from $L^p(\mathbb{R}^n)$ to $L^q(\mathbb{R}^n)$ is equal to the norm that $\mathcal{H}_\beta$ restricts to radial and nonnegative functions.

In order to prove Theorem 1.3, we need to construct an auxiliary function $W(x, y, z)$ and to study the continuity and differentiability of the function $W(x, y, z)$. In order to show the properties of $W(x, y, z)$, we firstly introduce another auxiliary function $\phi(u)$, which is closely related with the function $W(x, y, z)$.

Assume that f is a nonnegative continuous function on $\mathbb{R}_+$. By Hölder's inequality,

$$\int_0^s f(r) r^{n-1} dr \leq \left(\int_0^s f^p(r) r^{n-1} dr \right)^{1/p} s^{\frac{n(p-1)}{p}} n^{(1-p)/p}.$$

If we denote by $y = n \int_0^s f(r) r^{n-1} dr$ and $z = n \int_s^\infty f^p(r) r^{n-1} dr$, then

$$\lim_{s \to 0} s^{-\frac{n(p-1)}{p}} y = 0. \tag{1.6}$$

Let $\phi(u)$ be the function defined by the equation

$$\phi(u) = u^p \int_0^1 \frac{\eta^{q-2}}{\left(1 - u + u\eta^{q\beta/n}\right)^{n/\beta}} d\eta$$

$$= \frac{1}{[(1-u)u^{p-1}]^{\frac{n}{q\beta}}} \int_0^U \frac{\zeta^{q-2}}{\left(1 + \zeta^{q\beta/n}\right)^{n/\beta}} d\zeta, \tag{1.7}$$

where $U = [u/(1-u)]^{n/q\beta}$. In order to easily describe the point in the three-dimensional rectangular coordinate system, we shall use x instead of s. For each point (x, y, z) in the octant R of xyz-space where $x > 0$, $y > 0$ and $z > 0$, the equation

$$\phi(u) = x^{n(p-1)}y^{-p}z \tag{1.8}$$

has a unique solution $u(x, y, z)$, since

$$\phi(0) = 0, \quad \phi(1) = \infty, \quad \phi'(u) > 0 \text{ for } 0 < u < 1. \tag{1.9}$$

If (x, y, z) approaches a point $(x_1, 0, 1)$, where $x_1 > 0$, along a continuous curve in R, then by (1.9), we know that $u(x, y, z)$ approaches 1. It follows from the second form of $\phi(u)$, (1.6), (1.8) and (1.9) that

$$\lim_{x \to x_1} \frac{x^{(p-1)n}}{y^p(1-u)^{-\frac{n}{q\beta}}} = \int_0^\infty \frac{\zeta^{q-2}}{(1+\zeta^{\frac{q\beta}{n}})^{n/\beta}} d\zeta$$

$$= \frac{n}{q\beta} \cdot \frac{\Gamma(\frac{n}{q\beta})\Gamma((1-\frac{1}{q})\frac{n}{\beta})}{\Gamma(\frac{n}{\beta})}. \tag{1.10}$$

If (x, y, z) approaches a point $(x_2, y_2, 0)$, where $x_2 > 0$, $y_2 > 0$, along a continuous curve in R. It follows from the first form of $\phi(u)$, (1.8) and (1.9) that $u(x, y, z) \to 0$ and

$$\lim_{x \to x_2} \frac{u^p}{z} = (q-1)\frac{x_2^{(p-1)n}}{y_2^p}. \tag{1.11}$$

Let $W(x, y, z)$ be the function defined by the equation

$W(x, y, z)$

$$= \frac{n}{q(\beta-n)+n}\left(\frac{1}{1-u} \cdot \frac{y^q}{x^{q(n-\beta)-n}} + \frac{z}{(1-u)u^{p-1}} \cdot \frac{y^{q-p}}{x^{n(q-p)-q\beta}}\right),$$

where the function $W(x, y, z)$ is determined when u is replaced by the function $u(x, y, z)$ in (1.7). In order to calculate the derivatives of W with respect to x, y and z, we need to give some necessary estimates. It follows from the second form for $\phi(u)$ that

$$\phi'(u) = \frac{n}{q\beta}\left(\frac{1+pu-p}{u(1-u)} \cdot \frac{x^{n(p-1)}z}{y^p} + \frac{u^{p-1}}{1-u}\right).$$

When W is treated as a function of x, y, z, u, the partial derivative of W with respect to u is

$$\frac{\partial W}{\partial u} = \frac{n}{q(\beta-n)+n}\left(\frac{1}{(1-u)^2} \cdot \frac{y^q}{x^{q(n-\beta)-n}}\right.$$

$$+\frac{y^{q-p}z}{x^{n(q-p)-q\beta}}\left(\frac{1}{(1-u)^2u^{p-1}}-\frac{(p-1)u^{p-2}}{(1-u)u^{2(p-1)}}\right)\right)$$

$$=\frac{n}{q(\beta-n)+n}\left(\frac{1}{(1-u)^2}\cdot\frac{y^q}{x^{q(n-\beta)-n}}\right.$$

$$+\frac{y^{q-p}z}{x^{n(q-p)-q\beta}}\left(\frac{1}{1-u}-\frac{p-1}{u}\right)\frac{1}{(1-u)u^{p-1}}\right)$$

$$=\frac{q\beta}{q(\beta-n)+n}\cdot\frac{1}{(1-u)u^{p-1}}\cdot\frac{y^q}{x^{q(n-\beta)-n}}$$

$$\times\left(\frac{u^{p-1}}{1-u}\frac{x^{n(p-1)}}{y^p}\cdot z\cdot\frac{1+pu-p}{u(1-u)}\right)\frac{n}{q\beta}$$

$$=\frac{q\beta}{q(\beta-n)+n}\cdot\frac{1}{(1-u)u^{p-1}}\cdot\frac{y^q}{x^{q(n-\beta)-n}}\phi'(u).$$

In view of (1.8), we have

$$x=\left(y^p z^{-1}\phi(u)\right)^{\frac{1}{n(p-1)}},\quad y=x^{\frac{n(p-1)}{p}}z^{\frac{1}{p}}\phi(u)^{-\frac{1}{p}},\quad z=x^{-n(p-1)}y^p\phi(u).$$

Then

$$\frac{\partial x}{\partial u}=\frac{1}{n(p-1)}y^{\frac{p}{n(p-1)}}z^{-\frac{1}{n(p-1)}}\left(\phi(u)\right)^{\frac{1}{n(p-1)}-1}\phi'(u),$$

$$\frac{\partial y}{\partial u}=-\frac{1}{p}x^{\frac{n(p-1)}{p}}z^{\frac{1}{p}}\left(\phi(u)\right)^{-\frac{1}{p}-1}\phi'(u),\quad \frac{\partial z}{\partial u}=x^{-n(p-1)}y^p\phi'(u).$$

Using the above calculations and (1.8), we can get the derivatives of W with respect to x, y and z as follows:

$$W_x=\frac{\partial W}{\partial x}+\frac{\partial W}{\partial u}\frac{\partial u}{\partial x}$$

$$=\frac{n}{q(\beta-n)+n}\left((n-q(n-\beta))\frac{y^q}{1-u}x^{n-q(n-\beta)-1}\right.$$

$$+\frac{q\beta-n(q-p)}{(1-u)u^{p-1}}\cdot\frac{y^{q-p}z}{x^{n(q-p)-q\beta+1}}\right)$$

$$-\frac{p\beta}{(1-u)u^{p-1}}\cdot\frac{y^{q-p}z}{x^{n(q-p)-q\beta+1}}$$

$$=\frac{ny^q}{1-u}\cdot\frac{1}{x^{q(n-\beta)-n+1}},$$

and analogously,

$$W_y=\frac{p}{1-p}\cdot\frac{y^{q-1}}{(1-u)x^{q(n-\beta)-n}},\quad W_z=\frac{1}{1-p}\cdot\frac{1}{(1-u)u^{p-1}}\cdot\frac{y^{q-p}}{x^{q(n-\beta)-np}}.$$

Let

$$\lambda = \frac{1}{1-p} \cdot \frac{1}{(1-u)u^{p-1}} \cdot \frac{y^{q-p}}{x^{q(n-\beta)-np}}$$

and $g = yux^{-n}$. Then W_x, W_y, W_z can be rewritten as

$$W_x = nx^{n-q(n-\beta)-1}y^q - n(p-1)\lambda g^p x^{n-1}, \quad W_y = p\lambda g^{p-1}, \quad W_z = \lambda.$$

When (x, y, z) approaches $(x_1, 0, 1)$ or $(x_2, y_2, 0)$, the expressions (1.10) and (1.11) show that

$$\lim_{x \to x_1} W = -A, \quad \lim_{x \to x_2} W = \frac{n}{q(\beta - n) + n} \cdot \frac{y_2^q}{x_2^{q(n-\beta)-n}}. \tag{1.12}$$

Proof of Theorem 1.3. It follows from Lemma 1.1 that the norm of the operator $\mathcal{H}_\beta$ from $L^p(\mathbb{R}^n)$ to $L^q(\mathbb{R}^n)$ is equal to the norm that $\mathcal{H}_\beta$ restricts to radial functions. Consequently, without loss of generality, it suffices to carry out the proof of the theorem by assuming that f is a nonnegative, radial, smooth function with compact support on $\mathbb{R}^n$.

Using the polar coordinate transformation, we can rewrite (1.5) as

$$n \int_0^\infty \left(n \int_0^s f(r) r^{n-1} dr \right)^q s^{q(\beta-n)+n-1} ds$$

$$\leq A^q \left(n \int_0^\infty f^p(r) r^{n-1} dr \right)^{q/p}. \tag{1.13}$$

Then it reduces to prove that (1.13) holds.

Without loss of generality, we may assume that

$$n \int_0^\infty f^p(r) r^{n-1} dr = 1.$$

If

$$n \int_0^\infty f^p(r) r^{n-1} dr \neq 1,$$

then we can replace f by

$$\left(n \int_0^\infty f^p(r) r^{n-1} dr \right)^{-1} f.$$

Consider the curve C in xyz-space defined by the equations

$$y = n \int_0^x f(r) r^{n-1} dr, \quad z = n \int_x^\infty f^p(r) r^{n-1} dr$$

and $z(0) = 1$. For such a curve there exists a smallest interval (x_1, x_2) such that

$$0 \leq x_1 < x_2 \leq \infty,$$

$$y(x) \equiv 0, \ z(x) \equiv 1 \ \text{ on } 0 \le x \le x_1,$$

$$y(x) \equiv \text{const.}, \ z(x) \equiv 0 \text{ on } x_2 \le x < \infty.$$

Considering the function $W(x, y(x), z(x))$, we get the derivative of W with respect to x as

$$W'(x) = nx^{n-1} \left(x^{-q(n-\beta)} y^q - \lambda \left((p-1)h^p - ph^{p-1}f + f^p \right) \right).$$

Let $\psi(h, f) = (p-1)h^p - ph^{p-1}f + f^p$. Since $h \ge 0$ and $f \ge 0$, then $\psi(h, f)$ is always positive except at its root $h = f$. And $W'(x)$ can be rewritten as

$$W'(x) = nx^{n-1}x^{-q(n-\beta)}y^q - nx^{n-1}\lambda\psi(h, f).$$

Let

$$I = n \int_0^\infty \left(\int_0^x f(r)r^{n-1}dr \right)^q x^{q(\beta-n)+n-1}dx$$

and $x_2 < \infty$, then on the sub-curves $C_{01}, C_{12}, C_{2\infty}$ of C corresponding to the intervals $(0, x_1), (x_1, x_2), (x_2, \infty)$,

$$I(C_{01}) = 0, \ I(C_{2\infty}) = -\frac{n}{q(\beta - n) + n} \cdot \frac{y_2^q}{x_2^{q(n-\beta)-n}}.$$

In view of (1.12), we get

$$I(C_{12}) = \int_{x_1}^{x_2} dW + \lambda \int_{x_1}^{x_2} nx^{n-1}\psi(h, f)dx$$

$$= W(x_2) - W(x_1) + \lambda \int_{x_1}^{x_2} nx^{n-1}\psi(h, f)dx$$

$$= A + \frac{n}{q(\beta - n) + n} \cdot \frac{y_2^q}{x_2^{q(n-\beta)-n}} + \lambda \int_{x_1}^{x_2} nx^{n-1}\psi(h, f)dx,$$

where $y_2 = y(x_2)$. It follows from $\lambda < 0$ and $\psi(h, f) \ge 0$ that

$$I(C) = I(C_{12}) + I(C_{2\infty}) \le A.$$

When $x_2 = \infty$, the value of $I(C)$ can be calculated by taking the limit. We omit the details.

Therefore, by the equivalence of (1.5) and (1.13), we obtain that

$$\|\mathcal{H}_\beta\|_{L^p(\mathbb{R}^n) \to L^q(\mathbb{R}^n)} = \sup_{\|f\|_{L^p(\mathbb{R}^n)} \ne 0} \frac{\|\mathcal{H}_\beta f\|_{L^q(\mathbb{R}^n)}}{\|f\|_{L^p(\mathbb{R}^n)}} \le A.$$

On the other hand, take $\widetilde{f}(x) = \dfrac{1}{(1+|x|^{q\beta})^{1+\frac{n}{q\beta}}}$. It follows from

$$n \int_0^s \widetilde{f}(r)r^{n-1}dr = \frac{s^n}{(1 + s^{q\beta})^{\frac{n}{q\beta}}}$$

that the left-hand side of (1.13) is $\frac{n}{q\beta} \cdot B\left(\frac{n}{q\beta} + 1, \frac{n}{q'\beta} - 1\right)$. It is easy to verify that

$$n \int_0^\infty \widetilde{f}^p(r) r^{n-1} dr = \frac{n}{q\beta} \cdot B\left(\frac{n}{q\beta}, \frac{n}{q'\beta}\right).$$

Therefore,

$$\|\mathcal{H}_\beta\|_{L^p(\mathbb{R}^n) \to L^q(\mathbb{R}^n)} = \sup_{\|f\|_{L^p(\mathbb{R}^n)} \neq 0} \frac{\|\mathcal{H}_\beta f\|_{L^q(\mathbb{R}^n)}}{\|f\|_{L^p(\mathbb{R}^n)}} \geq \frac{\|\mathcal{H}_\beta \widetilde{f}\|_{L^q(\mathbb{R}^n)}}{\|\widetilde{f}\|_{L^p(\mathbb{R}^n)}} = A,$$

which completes the proof. $\square$

The next result gives the sharp weak bound for the fractional Hardy operator from $L^1(\mathbb{R}^n)$ to $L^{\frac{n}{n-\beta}, \infty}(\mathbb{R}^n)$.

Theorem 1.4. *Let* $0 < \beta < n$ *and* $f \in L^1(\mathbb{R}^n)$. *Then*

$$\left|\{x \in \mathbb{R}^n : \mathcal{H}_\beta f(x) > \lambda\}\right| \leq \left(\frac{1}{\lambda} \|f\|_{L^1(\mathbb{R}^n)}\right)^{\frac{n}{n-\beta}}$$

holds for any $\lambda > 0$. *Moreover, the constant 1 is sharp.*

Proof. Since

$$\mathcal{H}_\beta f(x) = \frac{1}{(\nu_n |x|^n)^{1-\frac{\beta}{n}}} \int_{|y|<|x|} |f(y)| dy \leq \frac{1}{(\nu_n |x|^n)^{1-\frac{\beta}{n}}} \|f\|_{L^1(\mathbb{R}^n)},$$

we have

$$\left|\{x \in \mathbb{R}^n : |\mathcal{H}_\beta f(x)| > \lambda\}\right| \leq \left|\left\{x \in \mathbb{R}^n : \frac{1}{(\nu_n |x|^n)^{1-\frac{\beta}{n}}} \|f\|_{L^1(\mathbb{R}^n)} > \lambda\right\}\right|$$

$$= \left|\left\{x \in \mathbb{R}^n : |x| < \left(\frac{\|f\|_{L^1(\mathbb{R}^n)}}{\lambda \nu_n^{1-\frac{\beta}{n}}}\right)^{\frac{1}{n-\beta}}\right\}\right|$$

$$= \left(\frac{\|f\|_{L^1(\mathbb{R}^n)}}{\lambda}\right)^{\frac{n}{n-\beta}}.$$

Next we shall show that the constant 1 is a sharp bound by constructing a suitable function. In fact, set

$$g(x) = \chi_{\{x \in \mathbb{R}^n : |x|<1\}}(x).$$

Then we have

$$\|g\|_{L^1(\mathbb{R}^n)} = \nu_n.$$

It follows that

$$\mathcal{H}_\beta g(x) = \frac{1}{(\nu_n |x|^n)^{1-\frac{\beta}{n}}} \int_{|y|<|x|} \chi_{\{y\in\mathbb{R}^n:\, |y|<1\}}(y) dy$$

$$= \frac{|\{y \in \mathbb{R}^n : |y| < \min\{|x|,1\}\}|}{(\nu_n |x|^n)^{1-\frac{\beta}{n}}}.$$

We assert that $\mathcal{H}_\beta g(x) \le \nu_n^{\frac{\beta}{n}}$ for all $x \in \mathbb{R}^n$ and rewrite

$$\left\{ x \in \mathbb{R}^n : \mathcal{H}_\beta g(x) > \nu_n^{\frac{\beta}{n}} \right\} := E_1 \bigcup E_2,$$

where

$$E_1 := B(\mathbf{0}, 1) \bigcap \left\{ x \in \mathbb{R}^n : \mathcal{H}_\beta g(x) > \nu_n^{\frac{\beta}{n}} \right\},$$

$$E_2 := (B(\mathbf{0}, 1))^c \bigcap \left\{ x \in \mathbb{R}^n : \mathcal{H}_\beta g(x) > \nu_n^{\frac{\beta}{n}} \right\}$$

and $(B(\mathbf{0}, |x|))^c = \mathbb{R}^n \backslash B(\mathbf{0}, |x|)$. Next we estimate E_1 and E_2, respectively. If $x \in E_1$, then $|x| < 1$. It follows that

$$\mathcal{H}_\beta g(x) \le \frac{|\{y \in \mathbb{R}^n : |y| < |x|\}|}{(\nu_n |x|^n)^{1-\frac{\beta}{n}}} = (\nu_n |x|^n)^{\frac{\beta}{n}} < \nu_n^{\frac{\beta}{n}}.$$

If $x \in E_2$, then $|x| \ge 1$. It follows that

$$\mathcal{H}_\beta g(x) \le \frac{|\{y \in \mathbb{R}^n : |y| < 1\}|}{(\nu_n |x|^n)^{1-\frac{\beta}{n}}} \le \nu_n^{\frac{\beta}{n}}.$$

Consequently, we have $E_1 = E_2 = \varnothing$. This in turn implies that $\mathcal{H}_\beta g(x) \le \nu_n^{\frac{\beta}{n}}$.

For any $0 < \lambda < \nu_n^{\frac{\beta}{n}}$, we conclude that

$$|\{x \in \mathbb{R}^n : |\mathcal{H}_\beta g(x)| > \lambda\}| \qquad\qquad (1.14)$$

$$= \left| B(\mathbf{0}, 1) \bigcap \left\{ x \in \mathbb{R}^n : \nu_n^{\frac{\beta}{n}} |x|^\beta > \lambda \right\} \right|$$

$$+ \left| (B(\mathbf{0}, 1))^c \bigcap \left\{ x \in \mathbb{R}^n : \frac{\nu_n^{\frac{\beta}{n}}}{|x|^{n-\beta}} > \lambda \right\} \right|$$

$$= \left| \left\{ x \in \mathbb{R}^n : \frac{\lambda^{1/\beta}}{\nu_n^{1/n}} < |x| < 1 \right\} \right| + \left| \left\{ x \in \mathbb{R}^n : 1 \le |x| < \left(\frac{\nu_n^{\frac{\beta}{n}}}{\lambda} \right)^{\frac{1}{n-\beta}} \right\} \right|$$

$$= \left(\frac{\nu_n}{\lambda} \right)^{\frac{n}{n-\beta}} - \lambda^{\frac{n}{\beta}}.$$

If there exists a constant C such that

$$\left|\{x \in \mathbb{R}^n : |\mathcal{H}_\beta f(x)| > \lambda\}\right| \leq \left(\frac{C\|f\|_{L^1}}{\lambda}\right)^{\frac{n}{n-\beta}}$$

for all $f \in L^1(\mathbb{R}^n)$. Then we can choose that

$$f(x) = \chi_{\{x \in \mathbb{R}^n : |x| < 1\}}(x).$$

It follows from equality (1.14) that

$$\left(\frac{\nu_n}{\lambda}\right)^{\frac{n}{n-\beta}} - \lambda^{\frac{n}{\beta}} \leq \left(\frac{C\nu_n}{\lambda}\right)^{\frac{n}{n-\beta}}$$

always holds for every $0 < \lambda < \nu_n^{\frac{\beta}{n}}$. Letting $\lambda \to 0^+$, this forces that $C \geq 1$. This means that the constant 1 is sharp. $\qquad\square$

By using the similar method, we can obtain the weak-type (p, p) bounds of n-dimensional Hardy operators for $1 \leq p \leq \infty$.

Theorem 1.5. [Zhao et al. (2012)] *For $1 \leq p \leq \infty$, the inequality*

$$\|\mathcal{H}f\|_{L^{p,\infty}(\mathbb{R}^n)} \leq 1 \cdot \|f\|_{L^p(\mathbb{R}^n)}$$

holds. Moreover,

$$\|\mathcal{H}\|_{L^p(\mathbb{R}^n) \to L^{p,\infty}(\mathbb{R}^n)} = 1.$$

Obviously, the best constant in the weak-type $(1, 1)$ inequality for $\mathcal{H}$ is independent of n.

1.3   Sharp inequalities for dual Hardy operators

The dual (or adjoint) operator H^* of the one-dimensional Hardy operator is

$$H^* f(x) = \int_x^\infty \frac{f(t)}{t} dt, \quad x > 0$$

for all nonnegative measurable functions f on $\mathbb{R}_+$.

Now, let us briefly present some background to the problems we are about to investigate. It is known from [Hardy et al. (1952), (9.9.1) and (9.9.2), p. 244] that, for $1 < p < \infty$, the operators H and H^* satisfy the inequalities

$$\frac{p-1}{p}\|Hf\|_{L^p(\mathbb{R}_+)} \leq \|H^* f\|_{L^p(\mathbb{R}_+)} \leq p\|Hf\|_{L^p(\mathbb{R}_+)}, \tag{1.15}$$

(see also [Komori (2003)] for more details). These estimates are not optimal and the constants in (1.15) are further improved by Kolyada in [Komori (2003)] who arrived at the following sharp conclusion.

Theorem A [Kolyada (2014), Theorem 1.1]   Let f be a nonnegative measurable function on $\mathbb{R}_+$. Then

$$(p-1)\|Hf\|_{L^p(\mathbb{R}_+)} \leq \|H^*f\|_{L^p(\mathbb{R}_+)} \leq (p-1)^{1/p}\|Hf\|_{L^p(\mathbb{R}_+)}$$

if $1 < p \leq 2$, and

$$(p-1)^{1/p}\|Hf\|_{L^p(\mathbb{R}_+)} \leq \|H^*f\|_{L^p(\mathbb{R}_+)} \leq (p-1)\|Hf\|_{L^p(\mathbb{R}_+)}$$

if $2 \leq p < \infty$. All constants are the best possible.

It is interesting to generalize the one-dimensional theory to the higher-dimensional setting. The aim of the following work is to prove that Theorem A holds for the Euclidean space $\mathbb{R}^n$ (instead of $\mathbb{R}_+$).

Similar to the one-dimensional case, the dual operator of $\mathcal{H}$ is given by

$$\mathcal{H}^*f(x) = \int_{|y|>|x|} \frac{f(y)}{|B(\mathbf{0},|y|)|} dy,$$

where f is a nonnegative measurable function on $\mathbb{R}^n$.

As we know, a linear operator has the same norm as the norm of its adjoint. So $\mathcal{H}^*$ maps $L^{p'}(\mathbb{R}^n)$ into $L^{p'}(\mathbb{R}^n)$ with the operator norm $p/(p-1)$ if and only if $\mathcal{H}$ maps $L^p(\mathbb{R}^n)$ into $L^p(\mathbb{R}^n)$ with the operator norm $p/(p-1)$ for $1 < p < \infty$, where p' is the conjugate index of p. Applying Fubini's theorem, a straightforward computation then shows that

$$\mathcal{H}f(x) = \frac{1}{|B(\mathbf{0},|x|)|} \int_{B(\mathbf{0},|x|)} \left(\int_{\{z:\, |y|<|z|<|x|\}} \frac{f(z)}{|B(\mathbf{0},|z|)|} dz \right) dy$$

$$\leq \frac{1}{|B(\mathbf{0},|x|)|} \int_{B(\mathbf{0},|x|)} \mathcal{H}^*f(y)dy,$$

and

$$\mathcal{H}^*f(x) = \int_{\{z:\, |z|>|x|\}} \left(\frac{1}{|B(\mathbf{0},|z|)|^2} \int_{\{y:\, |x|<|y|<|z|\}} f(y)dy \right) dz$$

$$\leq \int_{\{z:\, |z|>|x|\}} \frac{\mathcal{H}f(z)}{|B(\mathbf{0},|z|)|} dz.$$

With these estimates, the fairly rough inequalities similar to the one-dimensional case will be established as follows

$$\frac{p-1}{p} \|\mathcal{H}f\|_{L^p(\mathbb{R}^n)} \leq \|\mathcal{H}^*f\|_{L^p(\mathbb{R}^n)} \leq p \|\mathcal{H}f\|_{L^p(\mathbb{R}^n)}$$

for any $1 < p < \infty$.

We shall establish an analogous result of Theorem A which in fact is motivated by the interesting result given by Kolyada in [Kolyada (2014)]. We are going to give the relation between L^p norms of the higher-dimensional Hardy operator and its dual operator, and to establish best constants in inequalities. It is interesting to find that sharp constants are dimension free, which means that they do not depend on the dimension of the underlying space.

Our results can be stated as follows.

Theorem 1.6. *Let f be a nonnegative measurable function on $\mathbb{R}^n$. Then*

$$(p-1)\|\mathcal{H}f\|_{L^p(\mathbb{R}^n)} \leq \|\mathcal{H}^*f\|_{L^p(\mathbb{R}^n)} \leq (p-1)^{1/p}\|\mathcal{H}f\|_{L^p(\mathbb{R}^n)} \qquad (1.16)$$

if $1 < p \leq 2$, and

$$(p-1)^{1/p}\|\mathcal{H}f\|_{L^p(\mathbb{R}^n)} \leq \|\mathcal{H}^*f\|_{L^p(\mathbb{R}^n)} \leq (p-1)\|\mathcal{H}f\|_{L^p(\mathbb{R}^n)} \qquad (1.17)$$

if $2 \leq p < \infty$. All constants in (1.16) and (1.17) are sharp.

Our method is quite elementary by use of the rotation method. As we said, the main results show that the n-dimensional inequalities are equivalent to one-dimensional ones with the identical constants. We must show that there is an optimizer for the inequality, i.e., that there is a function f that actually gives equality in Theorems 1.6 with the sharp constant. Indeed, the problem of finding the best constant is reduced to test equality over the class of radial functions.

We shall need the following lemma which is analogous to the one-dimensional case on $\mathbb{R}$ presented in [Komori (2003)] in which the author did not provide the proof.

Lemma 1.2. *Let $1 < p < \infty$. Assume that f is a nonnegative measurable function on $\mathbb{R}^n$. If $\mathcal{H}f \in L^p(\mathbb{R}^n)$, then*

$$|x|^{n/p}\mathcal{H}f(x) \to 0 \text{ as } |x| \to 0^+ \text{ or } |x| \to \infty.$$

Proof. For any $x \in \mathbb{R}^n\backslash\{0\}$, there exists $j \in \mathbb{Z}$ such that $2^j < |x| \leq 2^{j+1}$. Since f is nonnegative, and by the definition of $\mathcal{H}$, we have

$$2^{-n}\mathcal{H}f(2^j) \leq \mathcal{H}f(x) \leq 2^n\mathcal{H}f(2^{j+1}). \qquad (1.18)$$

On the one hand, we find that

$$\|\mathcal{H}f\|_{L^p(\mathbb{R}^n)}^p \geq \int_{\{x:\, |x|>1\}} \left(\frac{1}{\nu_n|x|^n}\int_{B(\mathbf{0},|x|)} f(y)dy\right)^p dx$$

$$= \sum_{j=0}^{\infty} \int_{\{x:\ 2^j < |x| \le 2^{j+1}\}} \left(\frac{1}{\nu_n |x|^n} \int_{B(\mathbf{0},|x|)} f(y) dy \right)^p dx$$

$$\ge \sum_{j=0}^{\infty} \int_{\{x:\ 2^j < |x| \le 2^{j+1}\}} \left(2^{-n} \mathcal{H} f(2^j) \right)^p dx$$

$$= \frac{\nu_n (2^n - 1)}{2^{np}} \sum_{j=0}^{\infty} 2^{jn} \left(\mathcal{H} f(2^j) \right)^p .$$

Noting that $\mathcal{H} f \in L^p(\mathbb{R}^n)$, it implies that

$$\lim_{j \to \infty} 2^{jn} \left(\mathcal{H} f(2^j) \right)^p = 0.$$

It follows from (1.18) that for $2^j < |x| \le 2^{j+1}$,

$$2^{-np} 2^{jn} \left(\mathcal{H} f(2^j) \right)^p \le |x|^n \left(\mathcal{H} f(x) \right)^p \le 2^{np} 2^{(j+1)n} \left(\mathcal{H} f(2^{j+1}) \right)^p ,$$

which clearly yields

$$\lim_{|x| \to \infty} |x|^n \left(\mathcal{H} f(x) \right)^p = 0.$$

On the other hand, we have

$$\|\mathcal{H} f\|_{L^p(\mathbb{R}^n)}^p \ge \int_{\{x:\ 0 < |x| \le 1\}} \left(\frac{1}{\nu_n |x|^n} \int_{B(\mathbf{0},|x|)} f(y) dy \right)^p dx$$

$$= \sum_{j=-\infty}^{0} \int_{\{x:\ 2^{j-1} < |x| \le 2^j\}} \left(\frac{1}{\nu_n |x|^n} \int_{B(\mathbf{0},|x|)} f(y) dy \right)^p dx$$

$$\ge \sum_{j=-\infty}^{0} \int_{\{x:\ 2^{j-1} < |x| \le 2^j\}} \left(2^{-n} \mathcal{H} f(2^{j-1}) \right)^p dx$$

$$= \frac{\nu_n (2^n - 1)}{2^{np}} \sum_{j=-\infty}^{0} 2^{(j-1)n} \left(\mathcal{H} f(2^{j-1}) \right)^p .$$

Using that $\mathcal{H} f \in L^p(\mathbb{R}^n)$, it implies that

$$\lim_{j \to -\infty} 2^{(j-1)n} \left(\mathcal{H} f(2^{j-1}) \right)^p = 0,$$

and then we must have

$$\lim_{|x| \to 0^+} |x|^n \left(\mathcal{H} f(x) \right)^p = 0,$$

which completes the proof.   □

Proof of Theorem 1.6. Assume that $\mathcal{H}f$ and $\mathcal{H}^*f$ belong to $L^p(\mathbb{R}^n)$. It is easy to see that $\mathcal{H}$ and $\mathcal{H}^*$ are radial operators. Write

$$F_n(s) = \int_{\mathbb{S}^{n-1}} s^{n-1} f(\delta_s y') d\sigma(y').$$

The norm of $\|\mathcal{H}f\|_{L^p(\mathbb{R}^n)}^p$ can be written in the polar coordinates form as follows:

$$
\begin{aligned}
\|\mathcal{H}f\|_{L^p(\mathbb{R}^n)}^p &= \frac{\omega_n}{\nu_n^p} \int_0^\infty t^{n-1-np} \left(\int_0^t \int_{\mathbb{S}^{n-1}} s^{n-1} f(\delta_s y') d\sigma(y') ds \right)^p dt \\
&= \frac{\omega_n}{\nu_n^p} \int_0^\infty t^{n-1-np} \left(\int_0^t F_n(s) ds \right)^p dt.
\end{aligned}
\tag{1.19}
$$

Similarly, we may rewrite $\|\mathcal{H}^*f\|_{L^p(\mathbb{R}^n)}^p$ as

$$
\|\mathcal{H}^*f\|_{L^p(\mathbb{R}^n)}^p = \frac{\omega_n}{\nu_n^p} \int_0^\infty s^{n-1} \left(\int_s^\infty \frac{F_n(t)}{t^n} dt \right)^p ds.
\tag{1.20}
$$

Applying integration by parts and using Lemma 1.2, (1.19) can be estimated as follows:

$$
\begin{aligned}
\|\mathcal{H}f\|_{L^p(\mathbb{R}^n)}^p &= \frac{\omega_n}{\nu_n^p} \left(\frac{1}{n-np} t^{n-np} \left(\int_0^t F_n(s) ds \right)^p \Bigg|_{t=0}^\infty \right. \\
&\quad \left. - \frac{p}{n-np} \int_0^\infty t^{n-np} F_n(t) \left(\int_0^t F_n(s) ds \right)^{p-1} dt \right) \\
&= \frac{1}{\nu_n^{p-1}} \cdot \frac{p}{p-1} \int_0^\infty t^{n-np} F_n(t) \left(\int_0^t F_n(s) ds \right)^{p-1} dt.
\end{aligned}
\tag{1.21}
$$

Next we shall prove the theorem in three steps.

STEP 1. We shall show that

$$\|\mathcal{H}^*f\|_{L^p(\mathbb{R}^n)} \le (p-1)^{1/p} \|\mathcal{H}f\|_{L^p(\mathbb{R}^n)}$$

if $1 < p \le 2$, and

$$(p-1)^{1/p} \|\mathcal{H}f\|_{L^p(\mathbb{R}^n)} \le \|\mathcal{H}^*f\|_{L^p(\mathbb{R}^n)}$$

if $2 \le p < \infty$.

For $t > 0$, let

$$\Phi_n(s,t) = \int_s^t \frac{F_n(u)}{u^n} du, \quad 0 < s \le t < \infty.$$

Then an easy calculation shows that

$$n \int_0^t s^{n-1} \Phi_n(s,t) ds = \int_0^t F_n(u) du,
\tag{1.22}$$

and also for any $q > 0$

$$\left(\int_s^\infty \frac{F_n(t)}{t^n} dt\right)^q = \int_s^\infty \frac{\partial}{\partial t} \left(\int_s^t \frac{F_n(u)}{u^n} du\right)^q dt$$

$$= q \int_s^\infty \frac{F_n(t)}{t^n} \Phi_n(s,t)^{q-1} dt. \qquad (1.23)$$

It follows from (1.23) with $q = p$ and the Fubini theorem that (1.20) can be written as

$$\|\mathcal{H}^* f\|_{L^p(\mathbb{R}^n)}^p = \frac{\omega_n}{\nu_n^p} \cdot p \int_0^\infty s^{n-1} \int_s^\infty \frac{F_n(t)}{t^n} \Phi_n(s,t)^{p-1} dt ds$$

$$= \frac{\omega_n}{\nu_n^p} \cdot p \int_0^\infty \frac{F_n(t)}{t^n} \int_0^t s^{n-1} \Phi_n(s,t)^{p-1} ds dt. \qquad (1.24)$$

From (1.21) and (1.22), we have the identity

$$\|\mathcal{H} f\|_{L^p(\mathbb{R}^n)}^p = \left(\frac{n}{\nu_n}\right)^{p-1} \cdot \frac{p}{p-1}$$

$$\times \int_0^\infty t^{n(1-p)} F_n(t) \left(\int_0^t s^{n-1} \Phi_n(s,t) ds\right)^{p-1} dt. \qquad (1.25)$$

Obviously,

$$\|\mathcal{H}^* f\|_{L^2(\mathbb{R}^n)}^2 = \|\mathcal{H} f\|_{L^2(\mathbb{R}^n)}^2.$$

Thus we shall only deal with the case $p \neq 2$. We now divide p into two cases: $1 < p < 2$ and $2 < p < \infty$.

CASE I. $1 < p < 2$. By Hölder's inequality with exponents $(p-1)^{-1}$ and $(2-p)^{-1}$, we have that

$$\int_0^t s^{n-1} \Phi_n(s,t)^{p-1} ds \leq \left(\int_0^t s^{n-1} \Phi_n(s,t) ds\right)^{p-1} \left(\int_0^t s^{n-1} ds\right)^{2-p}$$

$$= n^{p-2} t^{n(2-p)} \left(\int_0^t s^{n-1} \Phi_n(s,t) ds\right)^{p-1}.$$

The above estimate together with (1.24) and (1.25) gives that

$$\|\mathcal{H}^* f\|_{L^p(\mathbb{R}^n)}^p \leq \left(\frac{n}{\nu_n}\right)^{p-1} \cdot p \int_0^\infty F_n(t) t^{n-np} \left(\int_0^t s^{n-1} \Phi_n(s,t) ds\right)^{p-1} dt$$

$$= (p-1) \|\mathcal{H} f\|_{L^p(\mathbb{R}^n)}^p.$$

CASE II. $p > 2$. Using Hölder's inequality with exponents $p-1$ and $(p-1)/(p-2)$, we have

$$\left(\int_0^t s^{n-1} \Phi_n(s,t) ds\right)^{p-1} \leq \int_0^t s^{n-1} \Phi_n(s,t)^{p-1} ds \left(\int_0^t s^{n-1} ds\right)^{p-2}$$

$$= n^{2-p} t^{n(p-2)} \int_0^t s^{n-1} \Phi_n(s,t)^{p-1} ds.$$

With the help of (1.24) and (1.25), we have, as a consequence, that

$$\|\mathcal{H}f\|_{L^p(\mathbb{R}^n)}^p \leq \frac{\omega_n}{\nu_n^p} \cdot \frac{p}{p-1} \int_0^\infty \frac{F_n(t)}{t^n} \int_0^t s^{n-1} \Phi_n(s,t)^{p-1} ds dt$$

$$= \frac{1}{p-1} \|\mathcal{H}^*f\|_{L^p(\mathbb{R}^n)}^p.$$

Thus, we finish the proof of Step 1.

STEP 2. We shall prove that

$$(p-1)\|\mathcal{H}f\|_{L^p(\mathbb{R}^n)} \leq \|\mathcal{H}^*f\|_{L^p(\mathbb{R}^n)}, \text{ if } 1 < p < 2$$

and

$$\|\mathcal{H}^*f\|_{L^p(\mathbb{R}^n)} \leq (p-1)\|\mathcal{H}f\|_{L^p(\mathbb{R}^n)}, \text{ if } 2 < p < \infty.$$

Noting that for any $q > 0$, we have

$$\left(\int_s^\infty \frac{F_n(t)}{t^n} dt\right)^q = q \int_s^\infty \frac{F_n(t)}{t^n} \left(\int_t^\infty \frac{F_n(u)}{u^n} du\right)^{q-1} dt. \qquad (1.26)$$

Applying (1.26) with the exponent $q = p$, (1.20) can be written as

$$\|\mathcal{H}^*f\|_{L^p(\mathbb{R}^n)}^p = \frac{\omega_n}{\nu_n^p} \int_0^\infty s^{n-1} \left(\int_s^\infty \frac{F_n(t)}{t^n} dt\right)^p ds$$

$$= \frac{\omega_n}{\nu_n^p} \cdot p \int_0^\infty s^{n-1} \int_s^\infty \frac{F_n(t)}{t^n} \left(\int_t^\infty \frac{F_n(u)}{u^n} du\right)^{p-1} dt ds$$

$$= p\nu_n^{1-p} \int_0^\infty F_n(t) \left(\int_t^\infty \frac{F_n(u)}{u^n} du\right)^{p-1} dt. \qquad (1.27)$$

Once again, we use (1.26) with the exponent $q = p - 1$ and switch the order of integration. This gives

$$\|\mathcal{H}^*f\|_{L^p(\mathbb{R}^n)}^p$$

$$= \nu_n^{1-p} \cdot p(p-1) \int_0^\infty F_n(t) \left(\int_t^\infty \frac{F_n(u)}{u^n} \left(\int_u^\infty \frac{F_n(v)}{v^n} dv\right)^{p-2} du\right) dt$$

$$= \nu_n^{1-p} \cdot p(p-1) \int_0^\infty \frac{F_n(u)}{u^n} \left(\int_u^\infty \frac{F_n(v)}{v^n} dv\right)^{p-2} \left(\int_0^u F_n(t) dt\right) du.$$

Define now the following functions

$$\varphi_n(u) = \frac{F_n^{1/(p-1)}(u)}{u^n} \int_0^u F_n(t) dt$$

and

$$\psi_n(u) = F_n(u)^{(p-2)/(p-1)} \left(\int_u^\infty \frac{F_n(v)}{v^n} dv\right)^{p-2}.$$

Then we have

$$\|\mathcal{H}^* f\|_{L^p(\mathbb{R}^n)}^p = \nu_n^{1-p} \cdot p(p-1) \int_0^\infty \varphi_n(u)\psi_n(u)du. \qquad (1.28)$$

With the help of expressions of $\|\mathcal{H}f\|_{L^p(\mathbb{R}^n)}^p$ in (1.21) and $\|\mathcal{H}^* f\|_{L^p(\mathbb{R}^n)}^p$ in (1.27), we have that

$$\int_0^\infty \varphi_n(u)^{p-1}du = \nu_n^{p-1} \cdot \frac{p-1}{p} \|\mathcal{H}f\|_{L^p(\mathbb{R}^n)}^p \qquad (1.29)$$

and

$$\int_0^\infty \psi_n(u)^{(p-1)/(p-2)}du = \nu_n^{p-1} \cdot \frac{1}{p} \|\mathcal{H}^* f\|_{L^p(\mathbb{R}^n)}^p \qquad (1.30)$$

for any $p > 1$. We now also divide p into two cases: $1 < p < 2$ and $2 < p < \infty$.

CASE I. $1 < p < 2$. Using in (1.28) the Hölder inequality with exponents $p-1$ and $(p-1)(p-2)^{-1}$ (see [Hardy et al. (1952), Theorem 189, p. 140]), and by (1.29) and (1.30), we have

$$\|\mathcal{H}^* f\|_{L^p(\mathbb{R}^n)}^p \geq \nu_n^{1-p} \cdot p(p-1) \cdot \left(\nu_n^{p-1} \cdot \frac{p-1}{p} \|\mathcal{H}f\|_{L^p(\mathbb{R}^n)}^p\right)^{1/(p-1)}$$

$$\times \left(\nu_n^{p-1} \cdot \frac{1}{p} \|\mathcal{H}^* f\|_{L^p(\mathbb{R}^n)}^p\right)^{(p-2)/(p-1)}$$

$$= (p-1)^{\frac{p}{p-1}} \|\mathcal{H}f\|_{L^p(\mathbb{R}^n)}^{\frac{p}{p-1}} \|\mathcal{H}^* f\|_{L^p(\mathbb{R}^n)}^{\frac{p(p-2)}{p-1}}.$$

Hence

$$\|\mathcal{H}^* f\|_{L^p(\mathbb{R}^n)} \geq (p-1)\|\mathcal{H}f\|_{L^p(\mathbb{R}^n)}$$

in the case $1 < p < 2$.

CASE II. $p > 2$. Using in (1.28) the Hölder inequality with exponents $p-1$ and $(p-1)(p-2)^{-1}$ together with equalities (1.29) and (1.30), we have

$$\|\mathcal{H}^* f\|_{L^p(\mathbb{R}^n)}^p \leq (p-1)^{\frac{p}{p-1}} \cdot \|\mathcal{H}f\|_{L^p(\mathbb{R}^n)}^{\frac{p}{p-1}} \|\mathcal{H}^* f\|_{L^p(\mathbb{R}^n)}^{\frac{p(p-2)}{p-1}}.$$

Hence

$$\|\mathcal{H}^* f\|_{L^p(\mathbb{R}^n)} \leq (p-1)\|\mathcal{H}f\|_{L^p(\mathbb{R}^n)}$$

in the case $p > 2$.

STEP 3. In order to prove that the constants in (1.16) and (1.17) are sharp, we are going now to construct some suitable functions. We shall prove it by choosing three classes of functions.

CASE I. We shall prove that the constant $(1/(p-1))^{1/p}$ in the right-hand side of (1.16) and the left-hand side of (1.17) is sharp.

For any $\varepsilon > 0$, take $f_\varepsilon(x) = \chi_{\{x:\ 1\leq|x|\leq1+\varepsilon\}}(x)$. On the one hand, we have

$$\|\mathcal{H}f_\varepsilon\|_{L^p(\mathbb{R}^n)}^p = \int_{\mathbb{R}^n}\left(\frac{1}{|B(\mathbf{0},|x|)|}\int_{B(\mathbf{0},|x|)}\chi_{\{y:\ 1\leq|y|\leq1+\varepsilon\}}(y)dy\right)^p dx$$

$$= \int_{\{x:\ 1<|x|<1+\varepsilon\}}\left(\frac{1}{|B(\mathbf{0},|x|)|}\int_{\{y:\ 1\leq|y|<|x|\}}dy\right)^p dx$$

$$+ \int_{\{x:\ |x|\geq1+\varepsilon\}}\left(\frac{1}{|B(\mathbf{0},|x|)|}\int_{\{y:\ 1\leq|y|\leq1+\varepsilon\}}dy\right)^p dx$$

$$=: I_1 + I_2.$$

By simple calculations, we conclude that

$$I_1 = \omega_n\int_1^{1+\varepsilon}t^{n-1-np}(t^n-1)^p dt$$

$$\leq \nu_n\cdot\frac{((1+\varepsilon)^n-1)^p}{p-1}\cdot\left(1-(1+\varepsilon)^{n(1-p)}\right)$$

and

$$I_2 = \nu_n\cdot\frac{((1+\varepsilon)^n-1)^p}{p-1}\cdot(1+\varepsilon)^{n(1-p)}.$$

With the above estimates, we obtain that

$$\frac{((1+\varepsilon)^n-1)^p(1+\varepsilon)^{n(1-p)}}{p-1}\leq\nu_n^{-1}\|\mathcal{H}f_\varepsilon\|_p^p\leq\frac{((1+\varepsilon)^n-1)^p}{p-1}.$$

On the other hand, for $\mathcal{H}^*f_\varepsilon$, we have

$$\|\mathcal{H}^*f_\varepsilon\|_{L^p(\mathbb{R}^n)}^p = \int_{\mathbb{R}^n}\left(\int_{\{y:\ |y|>|x|\}}\frac{\chi_{\{y:\ 1\leq|y|\leq1+\varepsilon\}}(y)}{\nu_n|y|^n}dy\right)^p dx$$

$$= \int_{\{x:\ |x|\leq1\}}\left(\int_{\{y:\ |y|>|x|\}}\frac{\chi_{\{y:\ 1\leq|y|\leq1+\varepsilon\}}(y)}{\nu_n|y|^n}dy\right)^p dx$$

$$+ \int_{\{x:\ |x|>1\}}\left(\int_{\{y:\ |y|>|x|\}}\frac{\chi_{\{y:\ 1\leq|y|\leq1+\varepsilon\}}(y)}{\nu_n|y|^n}dy\right)^p dx$$

$$=: I_3 + I_4.$$

It is easy to calculate that

$$I_3 = \nu_n\cdot n^p\cdot(\ln(1+\varepsilon))^p$$

and

$$I_4 = n^p \cdot \omega_n \int_1^{1+\varepsilon} t^{n-1} \left(\ln \left(\frac{1+\varepsilon}{t} \right) \right)^p dt \leq \nu_n \cdot n^p \cdot (\ln(1+\varepsilon))^p \left((1+\varepsilon)^n - 1 \right).$$

Consequently,

$$n^p (\ln(1+\varepsilon))^p \leq \nu_n^{-1} \| \mathcal{H}^* f_\varepsilon \|_{L^p(\mathbb{R}^n)}^p \leq n^p (1+\varepsilon)^n (\ln(1+\varepsilon))^p.$$

Applying the asymptotic relations $(1+x)^\alpha - 1 \sim \alpha x$ and $\ln(1+x) \sim x$ as $x \to 0$ for any $\alpha \in \mathbb{R}$, we obtain that

$$\lim_{\varepsilon \to 0^+} \frac{\| \mathcal{H} f_\varepsilon \|_{L^p(\mathbb{R}^n)}}{\| \mathcal{H}^* f_\varepsilon \|_{L^p(\mathbb{R}^n)}} = \left(\frac{1}{p-1} \right)^{1/p}.$$

CASE II. We shall prove that the constant $p-1$ in the left-hand side of (1.16) and the right-hand side of (1.17) is sharp. To this end, we need to choose different functions according to the range of p.

Let $1 < p < 2$. For any ε such that $0 < \varepsilon < n/p$, define

$$f_\varepsilon(x) = |x|^{\varepsilon - n/p} \chi_{\{x:\, 0<|x|\leq 1\}}(x).$$

Then we have

$$\| \mathcal{H} f_\varepsilon \|_{L^p(\mathbb{R}^n)}^p = \frac{1}{\nu_n^p} \int_{\mathbb{R}^n} \left(\frac{1}{|x|^n} \int_{B(0,|x|)} |y|^{\varepsilon - n/p} \chi_{\{y:\, 0<|y|\leq 1\}}(y) dy \right)^p dx$$

$$\geq \frac{1}{\nu_n^p} \int_{\{x:\, |x|<1\}} \frac{1}{|x|^{np}} \left(\int_{\{y:\, 0<|y|<|x|\}} |y|^{\varepsilon - n/p} dy \right)^p dx$$

$$= n^p \cdot \omega_n \cdot \frac{1}{p\varepsilon} \cdot \left(\frac{1}{\varepsilon + n(1 - 1/p)} \right)^p.$$

To estimate the $L^p(\mathbb{R}^n)$ norm of $\mathcal{H}^* f_\varepsilon$, by using polar coordinates, we conclude that

$$\| \mathcal{H}^* f_\varepsilon \|_{L^p(\mathbb{R}^n)}^p = \int_{\mathbb{R}^n} \left(\int_{\{y:\, |y|>|x|\}} \frac{|y|^{\varepsilon - n/p} \chi_{\{y:\, 0<|y|\leq 1\}}(y)}{\nu_n |y|^n} dy \right)^p dx$$

$$= \int_{\{x:\, |x|<1\}} \left(\int_{\{y:\, |y|>|x|\}} \frac{|y|^{\varepsilon - n/p} \chi_{\{y:\, 0<|y|\leq 1\}}(y)}{\nu_n |y|^n} dy \right)^p dx$$

$$= \frac{\omega_n^{p+1}}{\nu_n^p} \int_0^1 \left(\int_t^1 r^{\varepsilon - n/p - 1} dr \right)^p t^{n-1} dt$$

$$= \frac{\omega_n^{p+1}}{\nu_n^p} \cdot \frac{1}{(n/p - \varepsilon)^p} \int_0^1 \left(t^{\varepsilon - n/p} - 1 \right)^p t^{n-1} dt$$

$$\leq \frac{\omega_n^{p+1}}{\nu_n^p} \cdot \frac{1}{(n/p - \varepsilon)^p} \int_0^1 t^{p\varepsilon - 1} dt$$

$$= n^p \cdot \omega_n \cdot \frac{1}{p\varepsilon} \cdot \frac{1}{(n/p - \varepsilon)^p}.$$

Therefore, we obtain

$$\lim_{\varepsilon \to 0^+} \frac{\|\mathcal{H}f_\varepsilon\|_{L^p(\mathbb{R}^n)}}{\|\mathcal{H}^* f_\varepsilon\|_{L^p(\mathbb{R}^n)}} \geq \frac{1}{p-1}.$$

It implies that the constant $1/(p-1)$ in the left-hand side of (1.16) is the best possible.

At last we now deal with the case of $p > 2$. For any ε satisfying $0 < \varepsilon < n(1 - 1/p)$, take $f_\varepsilon(x) = |x|^{-\varepsilon - n/p} \chi_{\{x:\ |x| \geq 1\}}(x)$. Then we have

$$
\begin{aligned}
\|\mathcal{H}f_\varepsilon\|_{L^p(\mathbb{R}^n)}^p &= \frac{1}{\nu_n^p} \int_{\mathbb{R}^n} \left(\frac{1}{|x|^n} \int_{B(0,|x|)} |y|^{-\varepsilon - n/p} \chi_{\{y:\ |y| \geq 1\}}(y) dy \right)^p dx \\
&= \frac{\omega_n^{p+1}}{\nu_n^p} \int_1^\infty \left(\int_1^t r^{-\varepsilon + n(1 - 1/p) - 1} dr \right)^p t^{n(1-p)-1} dt \\
&= \frac{\omega_n^{p+1}}{\nu_n^p} \cdot \frac{1}{(n - n/p - \varepsilon)^p} \int_1^\infty \left(t^{n - n/p - \varepsilon} - 1 \right)^p t^{n(1-p)-1} dt \\
&\leq \frac{\omega_n^{p+1}}{\nu_n^p} \cdot \frac{1}{(n(1 - 1/p) - \varepsilon)^p} \int_1^\infty t^{-p\varepsilon - 1} dt \\
&= n^p \cdot \omega_n \cdot \frac{1}{p\varepsilon} \cdot \frac{1}{(n(1 - 1/p) - \varepsilon)^p}.
\end{aligned}
$$

In the same way as above, we have

$$
\begin{aligned}
\|\mathcal{H}^* f_\varepsilon\|_{L^p(\mathbb{R}^n)}^p &= \int_{\mathbb{R}^n} \left(\int_{\{y:\ |y| > |x|\}} \frac{|y|^{-\varepsilon - n/p} \chi_{\{y:\ |y| \geq 1\}}(y)}{\nu_n |y|^n} dy \right)^p dx \\
&\geq \int_{\{x:\ |x| \geq 1\}} \left(\int_{\{y:\ |y| > |x|\}} \frac{|y|^{-\varepsilon - n/p}}{\nu_n |y|^n} dy \right)^p dx \\
&= n^p \cdot \omega_n \cdot \frac{1}{p\varepsilon} \cdot \frac{1}{(n/p + \varepsilon)^p}.
\end{aligned}
$$

Letting $\varepsilon \to 0^+$, we obtain that

$$\lim_{\varepsilon \to 0^+} \frac{\|\mathcal{H}^* f_\varepsilon\|_{L^p(\mathbb{R}^n)}}{\|\mathcal{H}f_\varepsilon\|_{L^p(\mathbb{R}^n)}} \geq p - 1.$$

The sharpness assertion in the left-hand side of (1.17) is thereby established. Combining all estimates together, we finish the proof. $\qquad \square$

Remark 1.3. It is worth noting that we do not need to assume $f \in L^p(\mathbb{R}^n)$ in Theorem 1.6. We can easily see that the condition $\mathcal{H}f \in L^p(\mathbb{R}^n)$ does not imply that $f \in L^p(\mathbb{R}^n)$. Indeed, set $f(x) = (|x| - 1)^{-1/p} \chi_{\{x:\ 1 \leq |x| \leq 2\}}(x)$, $p > 1$. Obviously,

$$\mathcal{H}f(x) = 0 \quad \text{for} \quad 0 < |x| \leq 1.$$

So we have

$$\|\mathcal{H}f\|^p_{L^p(\mathbb{R}^n)} = \int_{\{x:\ |x| \geq 1\}} (\mathcal{H}f(x))^p \, dx$$

$$= \int_{\{x:\ |x| \geq 1\}} \left(\frac{1}{|B(\mathbf{0}, |x|)|} \int_{B(\mathbf{0}, |x|)} (|y| - 1)^{-1/p} \chi_{\{y:\ 1 \leq |y| \leq 2\}}(y) dy \right)^p dx$$

$$\leq \int_{\{x:\ |x| \geq 1\}} \left(\frac{2^{n-1}\omega_n}{\nu_n} \cdot \frac{p}{p-1} \cdot \frac{1}{|x|^n} \right)^p dx$$

$$= \left(\frac{2^{n-1}\omega_n}{\nu_n} \cdot \frac{p}{p-1} \right)^p \cdot \frac{\nu_n}{p-1}.$$

Hence, $\mathcal{H}f \in L^p(\mathbb{R}^n)$. However, a simple estimate shows that

$$\|f\|^p_{L^p(\mathbb{R}^n)} = \int_{\mathbb{R}^n} ((|x| - 1)^{-1/p} \chi_{\{x:\ 1 \leq |x| \leq 2\}}(x))^p dx$$

$$= \omega_n \int_1^2 t^{n-1}(t-1)^{-1} dt$$

$$\geq \omega_n \int_1^2 (t-1)^{-1} dt$$

$$= \infty,$$

which implies $f \notin L^p(\mathbb{R}^n)$.

1.4   Sharp bounds for m-linear Hardy and Hilbert operators

We now study m-linear averaging operators analogous to the operator in (1.3). The study of multilinear averaging operators is a byproduct of the recent interest in multilinear singular integral operator theory; this subject was founded in the 1970s by Coifman and Meyer [Coifman and Meyer (1975)] in their comprehensive study of many singular integral operators of multiparameter function input, such as the Calderón commutators, paraproducts, and pseudodifferential operators.

In the sequel we use the following notation: for $1 \leq i \leq m$, $y_i = (y_{1i}, y_{2i}, \ldots, y_{ni})$ will denote elements of $\mathbb{R}^n$. The Euclidean norms of each y_i is $|y_i| = \sqrt{\sum_{j=1}^n |y_{ji}|^2}$ and of the m-tuple $(y_1, y_2, \ldots, y_m)$ is

$$|(y_1, y_2, \ldots, y_m)| = \sqrt{\sum_{i=1}^m |y_i|^2}.$$

We use this notation in the following definition of the m-linear Hardy operator.

Definition 1.1. Let $m \in \mathbb{N}$, $f_1, f_2, \ldots, f_m$ be locally integrable functions on $\mathbb{R}^n$. The m-linear Hardy operator is defined by

$$
\mathcal{H}^m(f_1, \ldots, f_m)(x)
$$
$$
= \frac{1}{\nu_{mn}} \cdot \frac{1}{|x|^{mn}} \int_{|(y_1, \ldots, y_m)| < |x|} f_1(y_1) \cdots f_m(y_m) \, dy_1 \cdots dy_m, \qquad (1.31)
$$

where $x \in \mathbb{R}^n \backslash \{0\}$. The 2-linear operator will be referred to as bilinear.

Two other variants of m-linear Hardy operators (of one-dimensional nature) were introduced and studied by Bényi and Oh [Bényi and Oh (2006)]. Our approach is simpler than that in both [Bényi and Oh (2006)] and [Christ and Grafakos (1995)] and easily adapts to the multilinear setting. It relies on the method of rotations and on the principle that many positive averaging operators attain their (weighted or unweighted) Lebesgue space operator norms on the subspace of radial functions.

The main result is the following:

Theorem 1.7. Let $m \in \mathbb{N}$, $f_i \in L^{p_i}(|x|^{\frac{\alpha_i p_i}{p}} dx)$, $1 < p_i < \infty$, $1 \le p < \infty$, $i = 1, \ldots, m$, $1/p = 1/p_1 + \cdots + 1/p_m$, $\alpha_i < pn(1 - 1/p_i)$, and $\alpha = \alpha_1 + \cdots + \alpha_m$. Then the m-linear Hardy operator $\mathcal{H}^m$ defined in (1.31) maps the product of weighted Lebesgue spaces $L^{p_1}(|x|^{\frac{\alpha_1 p_1}{p}} dx) \times \cdots \times L^{p_m}(|x|^{\frac{\alpha_m p_m}{p}} dx)$ to $L^p(|x|^\alpha dx)$ with norm equal to the constant

$$
\frac{\nu_n^m}{\nu_{mn}} \cdot \frac{pmn}{pmn - n - \alpha} \cdot \frac{1}{2^{m-1}} \cdot \frac{\prod_{i=1}^m \Gamma(\frac{n}{2}(1 - \frac{1}{p_i} - \frac{\alpha_i}{pn}))}{\Gamma(\frac{n}{2}(m - \frac{1}{p} - \frac{\alpha}{pn}))}.
$$

Proof. We first give the proof of unweighted case when $m = 2$.
Set as before $\omega_n = 2\pi^{\frac{n}{2}}/\Gamma(\frac{n}{2})$. For $i = 1, 2$, let

$$
g_{f_i}(y_i) = \omega_n^{-1} \int_{|\xi_i| = 1} |f_i(|y_i|\xi_i)| \, d\xi_i, \quad y_i \in \mathbb{R}^n.
$$

Obviously, $g_{f_1}(y_1)$ and $g_{f_2}(y_2)$ are nonnegative radial functions and

$$
\nu_{2n} \mathcal{H}^2(g_{f_1}, g_{f_2})(x) = \frac{1}{|x|^{2n}} \int_{|(y_1, y_2)| < |x|} g_{f_1}(y_1) g_{f_2}(y_2) \, dy_1 dy_2
$$
$$
= \frac{1}{|x|^{2n}} \int_{|(y_1, y_2)| < |x|} \prod_{i=1}^2 \left(\frac{1}{\nu_n} \int_{|\xi_i| = 1} |f_i(|y_i|\xi_i)| d\xi_i \right) dy_1 dy_2
$$

$$= \frac{1}{\nu_n^2 |x|^{2n}} \int_{|\xi_1|=1} \int_{|\xi_2|=1} \left(\int_{|(y_1,y_2)|<|x|} \prod_{i=1}^{2} |f_i(|y_i|\xi_i)| dy_1 dy_2 \right) d\xi_1 d\xi_2$$

$$= \frac{1}{|x|^{2n}} \int_{|(y_1,y_2)|<|x|} |f_1(y_1)f_2(y_2)| \, dy_1 dy_2 = \nu_{2n} \mathcal{H}^2(|f_1|,|f_2|)(x).$$

Also by Minkowski's inequality, we have

$$\|g_{f_1}\|_{L^{p_1}(\mathbb{R}^n)} = \left(\int_{\mathbb{R}^n} |g_{f_1}(x_1)|^{p_1} dx_1 \right)^{1/p_1}$$

$$\leq \frac{1}{\omega_n} \int_{|\xi_1|=1} \left(\int_{\mathbb{R}^n} |f_1(|x_1|\xi_1)|^{p_1} dx_1 \right)^{1/p_1} d\xi_1$$

$$= \frac{1}{\omega_n} \int_{|\xi_1|=1} \left(\int_{\mathbb{R}^n} |f_1(x_1)|^{p_1} dx_1 \right)^{1/p_1} d\xi_1$$

$$= \|f_1\|_{L^{p_1}(\mathbb{R}^n)}.$$

Similarly, for g_{f_2}, we obtain

$$\|g_{f_2}\|_{L^{p_2}(\mathbb{R}^n)} \leq \|f_2\|_{L^{p_2}(\mathbb{R}^n)}.$$

Therefore one has that

$$\frac{\|\mathcal{H}^2(f_1,f_2)\|_{L^p(\mathbb{R}^n)}}{\|f_1\|_{L^{p_1}(\mathbb{R}^n)}\|f_2\|_{L^{p_2}(\mathbb{R}^n)}} \leq \frac{\|\mathcal{H}^2(|f_1|,|f_2|)\|_{L^p(\mathbb{R}^n)}}{\|f_1\|_{L^{p_1}(\mathbb{R}^n)}\|f_2\|_{L^{p_2}(\mathbb{R}^n)}}$$

$$\leq \frac{\|\mathcal{H}^2(g_{f_1},g_{f_2})\|_{L^p(\mathbb{R}^n)}}{\|g_{f_1}\|_{L^{p_1}(\mathbb{R}^n)}\|g_{f_2}\|_{L^{p_2}(\mathbb{R}^n)}}.$$

This implies that the operator $\mathcal{H}^2$ and its restriction to radial functions have the same operator norm in $L^p(\mathbb{R}^n)$. So, without loss of generality, we assume that f_i, $i = 1, 2$ are radial functions in the rest of the proof.

By Minkowski's integral inequality and Hölder's inequality, we have

$$\nu_{2n} \|\mathcal{H}^2(f_1,f_2)\|_{L^p(\mathbb{R}^n)}$$

$$= \left(\int_{\mathbb{R}^n} \left| \frac{1}{|x|^{2n}} \int_{|(y_1,y_2)|<|x|} f_1(y_1)f_2(y_2) \, dy_1 dy_2 \right|^p dx \right)^{1/p}$$

$$= \left(\int_{\mathbb{R}^n} \left| \int_{|(z_1,z_2)|<1} f_1(|x|z_1)f_2(|x|z_2) \, dz_1 dz_2 \right|^p dx \right)^{1/p}$$

$$= \left(\int_{\mathbb{R}^n} \left| \int_{|(z_1,z_2)|<1} f_1(|z_1|x)f_2(|z_2|x) \, dz_1 dz_2 \right|^p dx \right)^{1/p}$$

$$\leq \int_{|(z_1,z_2)|<1} \left(\int_{\mathbb{R}^n} |f_1(|z_1|x) f_2(|z_2|x)|^p \, dx \right)^{1/p} dz_1 dz_2$$

$$\leq \int_{|(z_1,z_2)|<1} \prod_{i=1}^{2} \left(\int_{\mathbb{R}^n} |f_i(x)|^{p_i} \, dx \right)^{1/p_i} |z_1|^{-\frac{n}{p_1}} |z_2|^{-\frac{n}{p_2}} \, dz_1 dz_2$$

$$= C_1 \prod_{i=1}^{2} \|f_i\|_{L^{p_i}(\mathbb{R}^n)},$$

where

$$C_1 = \int_{|(z_1,z_2)|<1} |z_1|^{-\frac{n}{p_1}} |z_2|^{-\frac{n}{p_2}} \, dz_1 dz_2$$

$$= \int_{\mathbb{S}^{n-1}} \int_{\mathbb{S}^{n-1}} \int_{\rho_1^2+\rho_2^2 \leq 1, \rho_1 \geq 0, \rho_2 \geq 0} \rho_1^{-\frac{n}{p_1}} \rho_2^{-\frac{n}{p_2}} \rho_1^{n-1} \rho_2^{n-1} \, d\rho_1 d\rho_2 \, dz_1' dz_2'$$

$$= \omega_n^2 \int_0^{\frac{\pi}{2}} \int_0^1 r^{n-\frac{n}{p_1}-1} r^{n-\frac{n}{p_2}-1} (\cos\theta)^{n-\frac{n}{p_1}-1} (\sin\theta)^{n-\frac{n}{p_2}-1} \, r dr d\theta$$

$$= \frac{\omega_n^2}{n(2-\frac{1}{p})} \int_0^{\frac{\pi}{2}} (\cos\theta)^{n-\frac{n}{p_1}-1} (\sin\theta)^{n-\frac{n}{p_2}-1} \, d\theta$$

$$= \frac{\omega_n^2}{n(2-\frac{1}{p})} \int_0^1 (1-t^2)^{\frac{1}{2}(n-\frac{n}{p_1}-1)} t^{n-\frac{n}{p_2}-1} (1-t^2)^{-\frac{1}{2}} dt \quad (t = \sin\theta)$$

$$= \frac{\omega_n^2}{n} \cdot \frac{p}{2p-1} \cdot \frac{1}{2} \int_0^1 (1-x)^{\frac{n}{2}-\frac{n}{2p_1}-1} x^{\frac{n}{2}-\frac{n}{2p_2}-1} dx \quad (x = t^2)$$

$$= \frac{\omega_n^2}{2n} \cdot \frac{p}{2p-1} \cdot B\left(\frac{n}{2} - \frac{n}{2p_1}, \frac{n}{2} - \frac{n}{2p_2} \right).$$

We write above $z_i = \rho_i z_i'$, where $z_i' \in \mathbb{S}^{n-1}$ and $\rho_i > 0$. Therefore, it follows that

$$\|\mathcal{H}^2\|_{L^{p_1}(\mathbb{R}^n) \times L^{p_2}(\mathbb{R}^n) \to L^p(\mathbb{R}^n)} \leq \frac{\omega_n^2}{\omega_{2n}} \cdot \frac{p}{2p-1} \cdot B\left(\frac{n}{2} - \frac{n}{2p_1}, \frac{n}{2} - \frac{n}{2p_2} \right).$$

Now for $0 < \varepsilon < \min\{1, \frac{(p_1-1)n}{p_2}, \frac{n}{p_2'}\}$, we define

$$f_1^\varepsilon(x_1) = \begin{cases} 0 & |x_1| \leq \frac{\sqrt{2}}{2}, \\ |x_1|^{-\frac{n}{p_1} - \frac{p_2\varepsilon}{p_1}} & |x_1| > \frac{\sqrt{2}}{2}; \end{cases} \quad f_2^\varepsilon(x_2) = \begin{cases} 0 & |x_2| \leq \frac{\sqrt{2}}{2}, \\ |x_2|^{-\frac{n}{p_2} - \varepsilon} & |x_2| > \frac{\sqrt{2}}{2}. \end{cases}$$

By an elementary calculation, we obtain that

$$\|f_1^\varepsilon\|_{L^{p_1}(\mathbb{R}^n)}^{p_1} = \|f_2^\varepsilon\|_{L^{p_2}(\mathbb{R}^n)}^{p_2} = \frac{\omega_n}{p_2\varepsilon} (\sqrt{2})^{p_2\varepsilon}.$$

Consequently, we have that $\mathcal{H}^2(f_1^\varepsilon, f_2^\varepsilon)(x) = 0$ when $|x| \leq 1$ and that $\mathcal{H}^2(f_1^\varepsilon, f_2^\varepsilon)(x)$ is equal to

$$\nu_{2n}^{-1} |x|^{-\frac{n+p_2\varepsilon}{p}} \int_{|(y_1,y_2)|<1; |y_1| > \frac{\sqrt{2}}{2|x|}; |y_2| > \frac{\sqrt{2}}{2|x|}} |y_1|^{-\frac{n+p_2\varepsilon}{p_1}} |y_2|^{-\frac{n}{p_2}-\varepsilon} dy_1 dy_2$$

when $|x| > 1$. And $\nu_{2n}^p \, \|\mathcal{H}^2(f_1^\varepsilon, f_2^\varepsilon)\|_{L^p}^p$ is equal to

$$\int_{|x|>1}\left[|x|^{-\frac{n}{p}-\frac{p_2\varepsilon}{p}} \int_{|(y_1,y_2)|<1;\, |y_1|>\frac{\sqrt{2}}{2|x|};\, |y_2|>\frac{\sqrt{2}}{2|x|}} \frac{|y_1|^{-\frac{n}{p_1}-\frac{p_2\varepsilon}{p_1}}}{|y_2|^{\frac{n}{p_2}+\varepsilon}} dy_1 dy_2 \right]^p dx$$

$$\geq \int_{|x|>\frac{1}{\varepsilon}}\left[|x|^{-\frac{n}{p}-\frac{p_2\varepsilon}{p}} \int_{|(y_1,y_2)|<1;\, |y_1|>\frac{\sqrt{2}}{2/\varepsilon};\, |y_2|>\frac{\sqrt{2}}{2/\varepsilon}} \frac{|y_1|^{-\frac{n}{p_1}-\frac{p_2\varepsilon}{p_1}}}{|y_2|^{\frac{n}{p_2}+\varepsilon}} dy_1 dy_2 \right]^p dx$$

$$= \int_{|x|>\frac{1}{\varepsilon}} |x|^{-n-p_2\varepsilon} dx \left(\int_{|(y_1,y_2)|<1;\, |y_1|>\frac{\sqrt{2}}{2/\varepsilon};\, |y_2|>\frac{\sqrt{2}}{2/\varepsilon}} \frac{|y_1|^{-\frac{n+p_2\varepsilon}{p_1}}}{|y_2|^{\frac{n}{p_2}+\varepsilon}} dy_1 dy_2 \right)^p$$

$$= (C_2\, C_3)^p \, ,$$

where C_2 and C_3 are the second and third factors in the last term, respectively. We now calculate the values of the constants C_2 and C_3. Writing $y_i = \rho_i z_i'$, we have

$$C_3 = \int_{|(y_1,y_2)|<1;\, |y_1|>\frac{\sqrt{2}}{2/\varepsilon};\, |y_2|>\frac{\sqrt{2}}{2/\varepsilon}} |y_1|^{-\frac{n}{p_1}-\frac{p_2\varepsilon}{p_1}} |y_2|^{-\frac{n}{p_2}-\varepsilon} dy_1 dy_2$$

$$= \int_{\mathbb{S}^{n-1}} \int_{\mathbb{S}^{n-1}} \int_{\rho_1^2+\rho_2^2<1;\, \rho_1>\frac{\sqrt{2}}{2/\varepsilon};\, \rho_2>\frac{\sqrt{2}}{2/\varepsilon}} \rho_1^{n-\frac{n+p_2\varepsilon}{p_1}-1} \rho_2^{n-\frac{n}{p_2}-\varepsilon-1} d\rho_1 d\rho_2 dz_1' dz_2'$$

$$= \omega_n^2 \int_{\rho_1=\frac{\sqrt{2}}{2/\varepsilon}}^{1} \int_{\rho_2=\frac{\sqrt{2}}{2/\varepsilon}}^{\sqrt{1-\rho_1^2}} \rho_1^{n-\frac{n}{p_1}-\frac{p_2\varepsilon}{p_1}-1} \rho_2^{n-\frac{n}{p_2}-\varepsilon-1} d\rho_2 d\rho_1$$

$$= \frac{\omega_n^2}{n-\frac{n}{p_2}-\varepsilon} \left(\frac{1}{2} \int_{\frac{1}{2\varepsilon-2}}^{1} (1-t)^{\frac{1}{2}(n-\frac{n}{p_2}-\varepsilon)} t^{\frac{1}{2}(n-\frac{n}{p_1}-\frac{p_2\varepsilon}{p_1})-1} dt \right.$$

$$\left. - \left(\frac{\sqrt{2}}{2/\varepsilon}\right)^{n-\frac{n}{p_2}-\varepsilon} \frac{1}{n-\frac{n}{p_1}-\frac{p_2\varepsilon}{p_1}} \left(1 - \left(\frac{\sqrt{2}}{2/\varepsilon}\right)^{n-\frac{n}{p_1}-\frac{p_2\varepsilon}{p_1}}\right) \right)$$

$$= \frac{\omega_n^2}{n-\frac{n}{p_2}-\varepsilon} \left(\frac{1}{2}\left(\int_0^1 - \int_0^{\frac{\varepsilon^2}{2}} \right)(1-t)^{\frac{1}{2}(n-\frac{n}{p_2}-\varepsilon)} t^{\frac{1}{2}(n-\frac{n}{p_1}-\frac{p_2\varepsilon}{p_1})-1} dt \right.$$

$$\left. - \left(\frac{\sqrt{2}\varepsilon}{2}\right)^{n-\frac{n}{p_2}} \frac{(\sqrt{2})^\varepsilon}{\varepsilon^\varepsilon} \frac{1}{n-\frac{n+p_2\varepsilon}{p_1}} \left(1 - \left(\frac{\sqrt{2}\varepsilon}{2}\right)^{n-\frac{n}{p_1}} (\sqrt{2})^{\frac{p_2\varepsilon}{p_1}} \frac{1}{\varepsilon^{p_2\varepsilon/p_1}}\right) \right)$$

$$= \frac{\omega_n^2}{n-\frac{n}{p_2}-\varepsilon} \left(\frac{1}{2} B\left(\frac{n}{2}-\frac{n}{2p_1}-\frac{p_2\varepsilon}{2p_1}, \frac{n-\varepsilon}{2}-\frac{n}{2p_2}+1\right) \right.$$

$$- \frac{1}{2} \int_0^{\frac{\varepsilon^2}{2}} (1-t)^{\frac{1}{2}(n-\frac{n}{p_2}-\varepsilon)} t^{\frac{1}{2}(n-\frac{n}{p_1}-\frac{p_2\varepsilon}{p_1})-1} dt$$

$$\left. - \left(\frac{\sqrt{2}\varepsilon}{2}\right)^{n-\frac{n}{p_2}} \frac{(\sqrt{2})^\varepsilon}{\varepsilon^\varepsilon} \frac{1}{n-\frac{n+p_2\varepsilon}{p_1}} \left(1 - \left(\frac{\sqrt{2}\varepsilon}{2}\right)^{n-\frac{n}{p_1}} (\sqrt{2})^{\frac{p_2\varepsilon}{p_1}} \frac{1}{\varepsilon^{p_2\varepsilon/p_1}}\right) \right).$$

For C_2 we have

$$C_2 = \left(\int_{|x|>1/\varepsilon} |x|^{-n-p_2\varepsilon} dx \right)^{1/p} = \omega_n^{\frac{1}{p}} \left(\int_{1/\varepsilon}^{\infty} r^{-n-p_2\varepsilon} r^{n-1} dr \right)^{1/p}$$

$$= (\varepsilon^\varepsilon)^{\frac{p_2}{p}} \left(\sqrt{2}/2 \right)^{\frac{p_2\varepsilon}{p}} \frac{p_2^{1/p_1} p_2^{1/p_2}}{p_2^{1/p}} \prod_{j=1}^{2} \|f_j^\varepsilon\|_{L^{p_j}(\mathbb{R}^n)}$$

$$= (\varepsilon^\varepsilon)^{\frac{p_2}{p}} \left(\sqrt{2}/2 \right)^{\frac{p_2\varepsilon}{p}} \prod_{j=1}^{2} \|f_j^\varepsilon\|_{L^{p_j}(\mathbb{R}^n)}.$$

Let $\varepsilon \to 0^+$. By the beta-function property $B(p, q+1) = \frac{q}{p+q} B(p,q)$, $p, q > 0$, we obtain

$$\|\mathcal{H}^2\|_{L^{p_1}(\mathbb{R}^n) \times L^{p_2}(\mathbb{R}^n) \to L^p(\mathbb{R}^n)} \geq \frac{\omega_n^2}{\nu_{2n}} \cdot \frac{p}{2p-1} \cdot B\left(\frac{n}{2} - \frac{n}{2p_1}, \frac{n}{2} - \frac{n}{2p_2} \right).$$

For the weighted case when $m = 2$. The proof of the upper bound in this case is similar to that of the previous case. For the proof of the lower bound, for a sufficiently small $\varepsilon \in (0,1)$, we take

$$f_1^\varepsilon(x_1) = \begin{cases} 0 & |x_1| \leq \frac{\sqrt{2}}{2}, \\ |x_1|^{-\frac{n+p_2\varepsilon}{p_1} - \frac{\alpha_1}{p}} & |x_1| > \frac{\sqrt{2}}{2}; \end{cases}$$

$$f_2^\varepsilon(x_2) = \begin{cases} 0 & |x_2| \leq \frac{\sqrt{2}}{2}, \\ |x_2|^{-\frac{n}{p_2} - \frac{\alpha_2}{p} - \varepsilon} & |x_2| > \frac{\sqrt{2}}{2}. \end{cases}$$

These functions show that the claimed norm of the operator on weighted Lebesgue spaces is indeed obtained as $\varepsilon \to 0^+$.

Since the proof of the weighted case when $m \geq 3$ is similar to that of case $m = 2$, we omit the details. $\qquad \square$

In the following we focus on the positive real numbers $(0, \infty)$ with the usual Lebesgue measure. Let T^m be the m-linear Hilbert operator

$$T^m(f_1, \ldots, f_m)(x) = \int_0^\infty \cdots \int_0^\infty \frac{f_1(x_1) \cdots f_m(x_m)}{(x + x_1 + \cdots + x_m)^m} dx_1 \cdots dx_m,$$

(1.32)

where $x > 0$. The following is a known sharp estimate

$$\int_0^\infty T^1(f)(x) g(x) \, dx \leq \frac{\pi}{\sin(\pi/p)} \|f\|_{L^p(0,\infty)} \|g\|_{L^{p'}(0,\infty)},$$

where $1/p+1/p' = 1, 1 < p < \infty$; see for instance [Grafakos (2008)], [Hardy et al. (1952)], and [Schur (1911)]. We have the following result concerning T^1.

Proposition 1.1. *Let* $1 < p < \infty$ *and* $-1 < \alpha < p - 1$. *Then for any function* f *in* $L^p(x^\alpha dx)$, *we have*

$$\|T^1(f)\|_{L^p(x^\alpha dx)} \leq \frac{\pi}{\sin(\pi(\alpha + 1)/p)}\|f\|_{L^p(x^\alpha dx)}.$$

Moreover,

$$\|T^1\|_{L^p(x^\alpha dx) \to L^p(x^\alpha dx)} = \frac{\pi}{\sin(\pi(\alpha + 1)/p)}.$$

Proof. By the Minkowski integral inequality, we have

$$\|T^1(f)\|_{L^p(x^\alpha dx)} = \left(\int_0^\infty \left|\int_0^\infty \frac{f(y)}{x + y}dy\right|^p x^\alpha dx\right)^{1/p}$$

$$\leq \left(\int_0^\infty \left(\int_0^\infty \frac{|f(yx)|}{1 + y}dy\right)^p x^\alpha dx\right)^{1/p}$$

$$\leq \int_0^\infty \left(\int_0^\infty |f(yx)|^p x^\alpha dx\right)^{1/p} \frac{1}{1 + y}dy$$

$$= \int_0^\infty \left(\int_0^\infty |f(x)|^p x^\alpha dx\right)^{1/p} \frac{y^{-\frac{\alpha+1}{p}}}{1 + y}dy$$

$$= B\left(1 - \frac{\alpha + 1}{p}, \frac{\alpha + 1}{p}\right)\|f\|_{L^p(x^\alpha dx)}.$$

Thus, one deduces the estimate

$$\|T^1\|_{L^p(x^\alpha dx) \to L^p(x^\alpha dx)} \leq B\left(1 - \frac{\alpha + 1}{p}, \frac{\alpha + 1}{p}\right).$$

To obtain a lower bound for the operator norm, we take $0 < \varepsilon < \min\{1, p - \alpha - 1\}$, and define $f_\varepsilon(x) = x^{-\frac{1}{p} - \frac{\alpha}{p} - \frac{\varepsilon}{p}}\chi_{\{|x|>1\}}$. A calculation yields that

$$\|f_\varepsilon\|_{L^p(x^\alpha dx)}^p = 1/\varepsilon.$$

We have

$$\|T^1(f_\varepsilon)\|_{L^p(x^\alpha dx)} = \left(\int_0^\infty \left(\int_1^\infty \frac{y^{-\frac{1+\alpha+\varepsilon}{p}}}{x + y}dy\right)^p x^\alpha dx\right)^{1/p}$$

$$\geq \left(\int_1^\infty \left(\int_1^\infty \frac{y^{-\frac{1+\alpha+\varepsilon}{p}}}{x + y}dy\right)^p x^\alpha dx\right)^{1/p}$$

$$= \left(\int_1^\infty \left(\int_{\frac{1}{x}}^\infty \frac{y^{-\frac{1+\alpha+\varepsilon}{p}}}{1+y} dy \right)^p x^{-1-\varepsilon} dx \right)^{1/p}.$$

Note that

$$\int_{\frac{1}{x}}^\infty \frac{y^{-\frac{1+\alpha+\varepsilon}{p}}}{1+y} dy = \int_0^\infty \frac{y^{-\frac{1+\alpha+\varepsilon}{p}}}{1+y} dy - \int_0^{\frac{1}{x}} \frac{y^{-\frac{1+\alpha+\varepsilon}{p}}}{1+y} dy$$

$$\geq \int_0^\infty \frac{y^{-\frac{1+\alpha+\varepsilon}{p}}}{1+y} dy - \int_0^{\frac{1}{x}} y^{-\frac{1+\alpha+\varepsilon}{p}} dy$$

$$= B\left(1 - \frac{1+\alpha+\varepsilon}{p}, \frac{1+\alpha+\varepsilon}{p}\right) - \frac{x^{-1+\frac{1+\alpha+\varepsilon}{p}}}{1 - \frac{1+\alpha+\varepsilon}{p}}.$$

It follows that $\|T^1(f_\varepsilon)\|_{L^p(x^\alpha dx)}$ is more than

$$\left(\int_1^\infty x^{-1-\varepsilon} dx \right)^{1/p} B\left(1 - \frac{1+\alpha+\varepsilon}{p}, \frac{1+\alpha+\varepsilon}{p}\right)$$

$$- \frac{p}{p-1-\alpha-\varepsilon} \left(\int_1^\infty x^{(-1+\frac{1+\alpha+\varepsilon}{p})p-1-\varepsilon} dx \right)^{1/p}$$

$$= \frac{1}{\varepsilon^{1/p}} B\left(1 - \frac{1+\alpha+\varepsilon}{p}, \frac{1+\alpha+\varepsilon}{p}\right) - \frac{p}{p-1-\alpha-\varepsilon} \cdot \frac{1}{(p-\alpha-1)^{1/p}}.$$

Letting $\varepsilon \to 0^+$, we deduce that

$$\lim_{\varepsilon \to 0^+} \frac{\|T^1(f_\varepsilon)\|_{L^p(x^\alpha dx)}}{\|f_\varepsilon\|_{L^p(x^\alpha dx)}} \geq B\left(1 - \frac{\alpha+1}{p}, \frac{1+\alpha}{p}\right).$$

Thus we have

$$\|T^1\|_{L^p(x^\alpha dx) \to L^p(x^\alpha dx)} \geq B\left(1 - \frac{\alpha+1}{p}, \frac{1+\alpha}{p}\right) = \frac{\pi}{\sin(\pi(\alpha+1)/p)}$$

and this concludes the proof of the proposition. □

Next we recall the following result from Bényi and Oh [Bényi and Oh (2006)].

Theorem 1.8. *Let* $m \geq 2$, $f_i \in L^{p_i}(0,\infty)$, $i = 1,\ldots,m$, $1 < p_i < \infty$, $1 < p < \infty$, *and* $1/p = 1/p_1 + \cdots + 1/p_m$. *Then*

$$\|T^m\|_{L^{p_1}(0,\infty) \times \cdots \times L^{p_m}(0,\infty) \to L^p(0,\infty)} = \prod_{i=1}^m \Gamma\left(\frac{1}{p_i'}\right) \Gamma\left(\frac{1}{p}\right) \Gamma(m)^{-1}.$$

We provide the following weighted extension of this result.

Theorem 1.9. *Let* $m \in \mathbb{N}$, f_i *be in* $L^{p_i}\left(x^{\frac{\alpha_i p_i}{p}}\,dx\right)$, $i = 1,\ldots,m$, $1 < p_i < \infty$, $1 \le p < \infty$, $1/p = 1/p_1 + \cdots + 1/p_m$, $-(1+1/p_i)p < \alpha_i < p(1 - 1/p_i)$ *and* $\alpha = \alpha_1 + \cdots + \alpha_m$. *Then*

$$\|T^m\|_{L^{p_1}\left(x^{\frac{\alpha_1 p_1}{p}}\,dx\right) \times \cdots \times L^{p_m}\left(x^{\frac{\alpha_m p_m}{p}}\,dx\right) \to L^p(x^\alpha dx)} = \frac{\prod_{i=1}^m \Gamma(\frac{1}{p_i'} - \frac{\alpha_i}{p})\Gamma(\frac{1+\alpha}{p})}{\Gamma(m)}.$$

Proof. For simplicity we only provide the proof for $m = 2$. (The case $m = 1$ is the essence of Proposition 1.1.) The idea of the proof of Theorem 1.2 yields the stated upper bound. It therefore suffices to show that the constant obtained in this way is also a lower bound. For

$$0 < \varepsilon < \min\left\{1, \frac{p_1}{p_2}\left(\frac{1}{p_1'} - \frac{\alpha_1}{p}\right), \frac{1}{p_2'} - \frac{\alpha_2}{p}\right\},$$

we take

$$f_1^\varepsilon(x_1) = x_1^{-\frac{1}{p_1} - \frac{\alpha_1}{p} - \frac{p_2\varepsilon}{p_1}} \chi_{\{x_1 > 1\}}(x_1)$$

and

$$f_2^\varepsilon(x_2) = x_2^{-\frac{1}{p_2} - \frac{\alpha_2}{p} - \varepsilon} \chi_{\{x_2 > 1\}}(x_2).$$

We have that

$$\|f_1^\varepsilon\|^{p_1}_{L^{p_1}\left(x_1^{\frac{\alpha_1 p_1}{p}}\,dx_1\right)} = \|f_2^\varepsilon\|^{p_2}_{L^{p_2}\left(x_2^{\frac{\alpha_2 p_2}{p}}\,dx_2\right)} = \frac{1}{p_2\varepsilon}.$$

Therefore $\|T^2(f_1^\varepsilon, f_2^\varepsilon)\|_{L^p(x^\alpha dx)}$ is

$$\left(\int_0^\infty \left(\int_0^\infty \int_0^\infty \frac{f_1^\varepsilon(x_1)f_2^\varepsilon(x_2)}{(x + x_1 + x_2)^2}\,dx_1 dx_2\right)^p x^\alpha\,dx\right)^{1/p}$$

$$\ge \left(\int_1^\infty \left(\int_1^\infty \int_1^\infty \frac{x_1^{-\frac{1}{p_1} - \frac{\alpha_1}{p} - \frac{p_2\varepsilon}{p_1}} x_2^{-\frac{1}{p_2} - \frac{\alpha_2}{p} - \varepsilon}}{(x + x_1 + x_2)^2}\,dx_1 dx_2\right)^p x^\alpha\,dx\right)^{1/p}$$

$$= \left(\int_1^\infty x^{-1-p_2\varepsilon}\left(\int_{1/x}^\infty \int_{1/x}^\infty \frac{x_1^{-\frac{1}{p_1} - \frac{\alpha_1}{p} - \frac{p_2\varepsilon}{p_1}} x_2^{-\frac{1}{p_2} - \frac{\alpha_2}{p} - \varepsilon}}{(1 + x_1 + x_2)^2}\,dx_1 dx_2\right)^p dx\right)^{1/p}$$

$$\ge \left(\int_{1/\varepsilon}^\infty x^{-1-p_2\varepsilon}\left(\int_\varepsilon^\infty \int_\varepsilon^\infty \frac{x_1^{-\frac{1}{p_1} - \frac{\alpha_1}{p} - \frac{p_2\varepsilon}{p_1}} x_2^{-\frac{1}{p_2} - \frac{\alpha_2}{p} - \varepsilon}}{(1 + x_1 + x_2)^2}\,dx_1 dx_2\right)^p dx\right)^{1/p}$$

$$= \left(\frac{\varepsilon^{\varepsilon p_2}}{p_2\varepsilon}\right)^{1/p} \int_\varepsilon^\infty \int_\varepsilon^\infty \frac{x_1^{-1/p_1 - \alpha_1/p - p_2\varepsilon/p_1} x_2^{-1/p_2 - \alpha_2/p - \varepsilon}}{(1 + x_1 + x_2)^2}\,dx_1 dx_2.$$

Next, we write

$$\int_\varepsilon^\infty \int_\varepsilon^\infty \frac{x_1^{-1/p_1-\alpha_1/p-p_2\varepsilon/p_1} x_2^{-1/p_2-\alpha_2/p-\varepsilon}}{(1+x_1+x_2)^2} \, dx_1 dx_2 = I_1 - I_2,$$

where

$$I_1 = \int_0^\infty \int_\varepsilon^\infty \frac{x_1^{-1/p_1-\alpha_1/p-p_2\varepsilon/p_1} x_2^{-1/p_2-\alpha_2/p-\varepsilon}}{(1+x_1+x_2)^2} \, dx_1 dx_2$$

and

$$I_2 = \int_0^\varepsilon \int_\varepsilon^\infty \frac{x_1^{-1/p_1-\alpha_1/p-p_2\varepsilon/p_1} x_2^{-1/p_2-\alpha_2/p-\varepsilon}}{(1+x_1+x_2)^2} \, dx_1 dx_2 \,.$$

I_1 can be written as

$$
\begin{aligned}
I_1 &= \int_0^\infty \int_\varepsilon^\infty \frac{x_1^{-1/p_1-\alpha_1/p-p_2\varepsilon/p_1} x_2^{-1/p_2-\alpha_2/p-\varepsilon}}{(1+x_1+x_2)^2} \, dx_1 dx_2 \\
&= \int_0^\infty \int_0^\infty \frac{x_1^{-1/p_1-\alpha_1/p-p_2\varepsilon/p_1} x_2^{-1/p_2-\alpha_2/p-\varepsilon}}{(1+x_1+x_2)^2} \, dx_1 dx_2 \\
&\quad - \int_0^\infty \int_0^\varepsilon \frac{x_1^{-1/p_1-\alpha_1/p-p_2\varepsilon/p_1} x_2^{-1/p_2-\alpha_2/p-\varepsilon}}{(1+x_1+x_2)^2} \, dx_1 dx_2 \\
&= B\Big(\frac{1}{p_1'} - \frac{\alpha_1}{p} - \frac{p_2\varepsilon}{p_1}, 1 + \frac{1}{p_1} + \frac{\alpha_1}{p} + \frac{p_2\varepsilon}{p_1}\Big) B\Big(\frac{1}{p_2'} - \frac{\alpha_2}{p} - \varepsilon, \frac{\alpha+1+p_2\varepsilon}{p}\Big) \\
&\quad - \int_0^\varepsilon \int_0^\infty \frac{x_1^{-1/p_1-\alpha_1/p-p_2\varepsilon/p_1} x_2^{-1/p_2-\alpha_2/p-\varepsilon}}{(1+x_1+x_2)^2} \, dx_2 dx_1 \\
&= B\Big(\frac{1}{p_1'} - \frac{\alpha_1}{p} - \frac{p_2\varepsilon}{p_1}, 1 + \frac{1}{p_1} + \frac{\alpha_1}{p} + \frac{p_2\varepsilon}{p_1}\Big) B\Big(\frac{1}{p_2'} - \frac{\alpha_2}{p} - \varepsilon, \frac{\alpha+1+p_2\varepsilon}{p}\Big) \\
&\quad - B\Big(\frac{1}{p_2'} - \frac{\alpha_2}{p_2} - \varepsilon, 1 + \frac{1}{p_2} + \frac{\alpha_2}{p_2} + \varepsilon\Big) \int_0^\varepsilon \frac{x_1^{-\frac{1}{p_1}-\frac{\alpha_1}{p}-\frac{p_2\varepsilon}{p_1}}}{(1+x_1)^{1+\frac{1}{p_2}+\frac{\alpha_2}{p}+\varepsilon}} \, dx_1 \\
&\geq B\Big(\frac{1}{p_1'} - \frac{\alpha_1}{p} - \frac{p_2\varepsilon}{p_1}, 1 + \frac{1}{p_1} + \frac{\alpha_1}{p} + \frac{p_2\varepsilon}{p_1}\Big) B\Big(\frac{1}{p_2'} - \frac{\alpha_2}{p} - \varepsilon, \frac{\alpha+1+p_2\varepsilon}{p}\Big) \\
&\quad - B\Big(\frac{1}{p_2'} - \frac{\alpha_2}{p_2} - \varepsilon, 1 + \frac{1}{p_2} + \frac{\alpha_2}{p_2} + \varepsilon\Big) \frac{\varepsilon^{1-\frac{1}{p_1}-\frac{\alpha_1}{p}-\frac{p_2\varepsilon}{p_1}}}{1 - \frac{1}{p_1} - \frac{\alpha_1}{p} - \frac{p_2\varepsilon}{p_1}} \,.
\end{aligned}
$$

We estimate I_2 in the following way:

$$I_2 = \int_0^\varepsilon \int_\varepsilon^\infty \frac{x_1^{-1/p_1-\alpha_1/p-p_2\varepsilon/p_1} x_2^{-1/p_2-\alpha_2/p-\varepsilon}}{(1+x_1+x_2)^2} \, dx_1 dx_2$$

$$= \int_0^\varepsilon \int_0^\infty \frac{x_1^{-1/p_1 - \alpha_1/p - p_2\varepsilon/p_1} x_2^{-1/p_2 - \alpha_2/p - \varepsilon}}{(1 + x_1 + x_2)^2} \, dx_1 dx_2$$

$$- \int_0^\varepsilon \int_0^\varepsilon \frac{x_1^{-1/p_1 - \alpha_1/p - p_2\varepsilon/p_1} x_2^{-1/p_2 - \alpha_2/p - \varepsilon}}{(1 + x_1 + x_2)^2} \, dx_1 dx_2$$

$$= B\left(\frac{1}{p_1'} - \frac{\alpha_1}{p} - \frac{p_2\varepsilon}{p_1}, 1 + \frac{1}{p_1} + \frac{\alpha_1}{p} + \frac{p_2\varepsilon}{p_1}\right) \int_0^\varepsilon \frac{x_2^{-\frac{1}{p_2} - \frac{\alpha_2}{p} - \varepsilon}}{(1 + x_2)^{1 + \frac{1}{p_1} + \frac{\alpha_1}{p} + \frac{p_2\varepsilon}{p_1}}} \, dx_2$$

$$- \int_0^\varepsilon \int_0^\varepsilon \frac{x_1^{-1/p_1 - \alpha_1/p - p_2\varepsilon/p_1} x_2^{-1/p_2 - \alpha_2/p - \varepsilon}}{(1 + x_1 + x_2)^2} \, dx_1 dx_2$$

$$\leq B\left(\frac{1}{p_1'} - \frac{\alpha_1}{p} - \frac{p_2\varepsilon}{p_1}, 1 + \frac{1}{p_1} + \frac{\alpha_1}{p} + \frac{p_2\varepsilon}{p_1}\right) \frac{\varepsilon^{1 - \frac{1}{p_2} - \frac{\alpha_2}{p} - \varepsilon}}{1 - \frac{1}{p_2} - \frac{\alpha_2}{p} - \varepsilon}$$

$$- \int_0^\varepsilon \int_0^\varepsilon \frac{x_1^{-1/p_1 - \alpha_1/p - p_2\varepsilon/p_1} x_2^{-1/p_2 - \alpha_2/p - \varepsilon}}{(1 + x_1 + x_2)^2} \, dx_1 dx_2.$$

The previous expressions for I_1 and I_2 allow us to easily compute their limits as $\varepsilon \to 0^+$. Indeed, letting $\varepsilon \to 0^+$, we obtain

$$\lim_{\varepsilon \to 0^+} \frac{\|T^2(f_1^\varepsilon, f_2^\varepsilon)\|_{L^p(x^\alpha dx)}}{\|f_1^\varepsilon\|_{L^{p_1}\left(x^{\frac{\alpha_1 p_1}{p}} dx\right)} \|f_2^\varepsilon\|_{L^{p_2}\left(x^{\frac{\alpha_2 p_2}{p}} dx\right)}}$$

$$\geq B\left(1 - \frac{\alpha_1}{p} - \frac{1}{p_1}, 1 + \frac{\alpha_1}{p} + \frac{1}{p_1}\right) B\left(1 - \frac{\alpha_2}{p} - \frac{1}{p_2}, \frac{1 + \alpha}{p}\right)$$

$$= \Gamma\left(1 - \frac{\alpha_1}{p} - \frac{1}{p_1}\right) \Gamma\left(1 - \frac{\alpha_2}{p} - \frac{1}{p_2}\right) \Gamma\left(\frac{1 + \alpha}{p}\right).$$

This estimate provides the reverse norm inequality and finishes the proof of Theorem 1.9. □

Obviously, both the m-linear Hardy operator (1.31) and the m-linear Hilbert operator (1.32) map $L^1 \times \cdots \times L^1$ to weak $L^{1/m}$. It follows by interpolation that they map $L^{p_1} \times \cdots \times L^{p_m}$ to L^p when $1/p_1 + \cdots + 1/p_m = 1/p > 1$. We do not know anything at the moment what the norm of these operators are on these spaces when $p < 1$.

We provide some remarks related to the case $m = 2$ and $1/2 \leq p < 1$. These easily extend to general $m \geq 2$.

Proposition 1.2. *Let f_i be in $L^{p_i}(\mathbb{R}^n)$, $i = 1, 2$, $1 < p_i < \infty$, $1/2 < p < 1$, and $1/p = 1/p_1 + 1/p_2$. Then*

$$\|\mathcal{H}^2(f_1, f_2)\|_{L^p(\mathbb{R}^n)} \leq C'_{p, p_1, p_2, n} \|f_1\|_{L^{p_1}(\mathbb{R}^n)} \|f_2\|_{L^{p_2}(\mathbb{R}^n)},$$

where

$$\frac{\omega_n^2}{\omega_{2n}} \cdot \frac{p}{2p-1} \cdot B\left(\frac{n}{2} - \frac{n}{2p_1}, \frac{n}{2} - \frac{n}{2p_2}\right) \leq C'_{p,p_1,p_2,n} \leq \frac{\nu_n^2}{\nu_{2n}} \cdot \frac{p_1}{p_1 - 1} \cdot \frac{p_2}{p_2 - 1}.$$

Proof. The idea of the proof of Theorem 1.2 yields a lower bound. For the upper bound, since the condition $|(y_1, y_2)| < |x|$ implies that $|y_1| < |x|$ and $|y_2| < |x|$, we obtain the estimate

$$|\mathcal{H}^2(f_1, f_2)(x)| = \frac{1}{\nu_{2n}} \cdot \frac{1}{|x|^{2n}} \left| \int_{|(y_1, y_2)| < |x|} f_1(y_1) f_2(y_2) \, dy_1 dy_2 \right|$$

$$\leq \frac{1}{\nu_{2n}} \cdot \frac{1}{|x|^{2n}} \int_{|y_1| < |x|} |f_1(y_1)| \, dy_1 \int_{|y_2| < |x|} |f_2(y_2)| \, dy_2$$

$$\leq \frac{\nu_n^2}{\nu_{2n}} \mathcal{H}(f_1)(x) \mathcal{H}(f_2)(x).$$

By Hölder's inequality ($\frac{1}{p_1/p} + \frac{1}{p_2/p} = 1$) and Theorem 1 in [Christ and Grafakos (1995)], we obtain

$$\|\mathcal{H}^2(f_1, f_2)\|_{L^p(\mathbb{R}^n)} \leq \frac{\nu_n^2}{\nu_{2n}} \cdot \frac{p_1}{p_1 - 1} \cdot \frac{p_2}{p_2 - 1} \|f_1\|_{L^{p_1}(\mathbb{R}^n)} \|f_2\|_{L^{p_2}(\mathbb{R}^n)}.$$

This proves the claimed estimate.   □

We have an analogous proposition for the bilinear Hilbert operator T^2.

Proposition 1.3. *Let f_i be in $L^{p_i}(\mathbb{R}_+)$, $i = 1, 2$, $1 < p_i < \infty$, $1/2 < p < 1$, and $1/p = 1/p_1 + 1/p_2$. Then we have*

$$\|T^2(f_1, f_2)\|_{L^p(\mathbb{R}_+)} \leq C''_{p,p_1,p_2,n} \|f_1\|_{L^{p_1}(\mathbb{R}_+)} \|f_2\|_{L^{p_2}(\mathbb{R}_+)},$$

where

$$\Gamma(1/p'_1)\Gamma(1/p'_2)\Gamma(1/p) \leq C''_{p,p_1,p_2} \leq \frac{\pi}{\sin(\pi/p_1)} \cdot \frac{\pi}{\sin(\pi/p_2)}.$$

At last, our guess for the sharp bounds in the case $1/m < p < 1$ are the constants obtained by the radial counterexamples.

1.5   Sharp bounds for Hardy type operators on product spaces

For the multi-dimensional case $n \geq 2$, Pachpatte [Pachpatte (1992)] introduced the multivariate Hardy operator which can be regarded as the rectangle averaging form:

$$P^n(f)(x_1, \ldots, x_n) = \frac{1}{x_1 \cdots x_n} \int_0^{x_1} \cdots \int_0^{x_n} f(t_1, \ldots, t_n) dt_1 \cdots dt_n, \quad (1.33)$$

where the function f is a nonnegative measurable function on $\mathbb{R}_+^n = (0, \infty) \times \cdots \times (0, \infty)$ and $x_i > 0$, $i = 1, 2, \ldots, n$, and proved that for $1 < p < \infty$,

$$\int_{\mathbb{R}_+^n} (P^n(f)(x))^p \, dx \leq \left(\frac{p}{p-1} \right)^{np} \int_{\mathbb{R}_+^n} f^p(x) dx.$$

For $1 < p < \infty$, the norm of P^n is $\|P^n\|_{L^p(\mathbb{R}_+^n) \to L^p(\mathbb{R}_+^n)} = (\frac{p}{p-1})^n$ which depends on dimensions of the underlying space. However, $\|\mathcal{H}\|_{L^p(\mathbb{R}^n) \to L^p(\mathbb{R}^n)}$, the norm of $\mathcal{H}$, is still $\frac{p}{p-1}$, and does not depend on the dimensions of the space. The reason of generating the difference between these two types of operators is, roughly speaking, that each variable can independently dilate by itself in the operator P^n, nevertheless, for the operator $\mathcal{H}$, all variables dilate by the same scale simultaneously. Generally speaking, the spherical averaging operator has better properties than the rectangle averaging operator does, such as the Hardy-Littlewood maximal function.

For the operator P^n, we note that every variable is defined on the 1-dimensional space. In this section, we shall extend the definition of P^n so that every variable is spherical average defined on a higher-dimensional space.

Next we shall give the definition of Hardy type operator on higher-dimensional product spaces as follows and discuss the corresponding properties.

Definition 1.2. Let $m \in \mathbb{N}$, $n_i \in \mathbb{N}$, $x_i \in \mathbb{R}^{n_i}$, $1 \leq i \leq m$, and f be a integrable function on $\mathbb{R}^{n_1} \times \mathbb{R}^{n_2} \times \cdots \times \mathbb{R}^{n_m}$. The Hardy type operator is defined by

$$\mathcal{H}_m f(x) = \left(\prod_{i=1}^m \frac{1}{|B(0, |x_i|)|} \right) \int_{|y_1| < |x_1|} \cdots \int_{|y_m| < |x_m|} |f(y_1, \ldots, y_m)| dy_1 \cdots dy_m,$$

$$(1.34)$$

where $x = (x_1, x_2, \ldots, x_m) \in \mathbb{R}^{n_1} \times \mathbb{R}^{n_2} \times \cdots \times \mathbb{R}^{n_m}$ with $\prod_{i=1}^m |x_i| \neq 0$.

We remark that if $m = 1$, then the operator $\mathcal{H}_m$ will become $\mathcal{H}$ defined by (1.3); if $n_1 = n_2 = \cdots = n_m = 1$, then $\mathcal{H}_m$ will become P^m defined by (1.33). Consequently, the operator $\mathcal{H}_m$ includes both P^m and $\mathcal{H}$. It is much significant to discuss the properties of $\mathcal{H}_m$.

We now formulate our main theorem for $\mathcal{H}_m$ as follows.

Theorem 1.10. *Let* $1 < p < \infty$, $m \in \mathbb{N}$, $n_i \in \mathbb{N}$, $x_i \in \mathbb{R}^{n_i}$, $i = 1, \ldots, m$. *If* $f \in L^p(\mathbb{R}^{n_1} \times \mathbb{R}^{n_2} \times \cdots \times \mathbb{R}^{n_m}, |x|^{\vec{\alpha}} dx)$, *where* $|x|^{\vec{\alpha}} := |x_1|^{\alpha_1} |x_2|^{\alpha_2} \cdots |x_m|^{\alpha_m}$

and $\alpha_j < (p-1)n_j$, then the Hardy type operator $\mathcal{H}_m$ defined in (1.34) is bounded on $L^p(\mathbb{R}^{n_1} \times \mathbb{R}^{n_2} \times \cdots \times \mathbb{R}^{n_m}, |x|^{\vec{\alpha}}dx)$, moreover, the norm of $\mathcal{H}_m$ can be obtained as follows,

$$\|\mathcal{H}_m\|_{L^p(|x|^{\vec{\alpha}}dx)\to L^p(|x|^{\vec{\alpha}}dx)} = \prod_{j=1}^{m}\left(\frac{p}{p-1-\frac{\alpha_j}{n_j}}\right).$$

Proof. We merely give the proof with the case $m = 2$ for the sake of clarity in writing, and the same is true for the general case $m > 2$. We adapt some ideas and methods used in Section 1.4. Set

$$g_f(x_1, x_2) = \frac{1}{\omega_{n_1}} \cdot \frac{1}{\omega_{n_2}} \int_{\mathbb{S}^{n_1-1}} \int_{\mathbb{S}^{n_2-1}} |f(|x_1|\xi_1, |x_2|\xi_2)|d\sigma(\xi_1)d\sigma(\xi_2),$$

where $\omega_{n_i} = \frac{2\pi^{\frac{n_i}{2}}}{\Gamma(\frac{n_i}{2})}$ and $x_i \in \mathbb{R}^{n_i}$, $i = 1, 2$. Obviously, g_f is a nonnegative radial function with respect to the variables x_1 and x_2, respectively.

It follows that $\mathcal{H}_2(g_f)(x_1, x_2)$ is equal to

$$\frac{1}{\omega_{n_1}\omega_{n_2}} \cdot \frac{1}{|B(\mathbf{0}, |x_1|)|} \cdot \frac{1}{|B(\mathbf{0}, |x_2|)|} \int_{B(\mathbf{0}, |x_1|)} \int_{B(\mathbf{0}, |x_2|)}$$

$$\int_{\mathbb{S}^{n_1-1}} \int_{\mathbb{S}^{n_2-1}} |f(|y_1|\xi_1, |y_2|\xi_2)|d\sigma(\xi_1)d\sigma(\xi_2)dy_1dy_2$$

$$= \frac{1}{\omega_{n_1}\omega_{n_2}} \int_{\mathbb{S}^{n_1-1}} \int_{\mathbb{S}^{n_2-1}} \frac{1}{|B(\mathbf{0}, |x_1|)|} \cdot \frac{1}{|B(\mathbf{0}, |x_2|)|}$$

$$\times \int_{B(\mathbf{0}, |x_1|)} \int_{B(\mathbf{0}, |x_2|)} |f(|y_1|\xi_1, |y_2|\xi_2)|dy_1dy_2d\sigma(\xi_1)d\sigma(\xi_2)$$

$$= \int_{\mathbb{S}^{n_1-1}} \int_{\mathbb{S}^{n_2-1}} \frac{1}{|B(\mathbf{0}, |x_1|)|} \cdot \frac{1}{|B(\mathbf{0}, |x_2|)|}$$

$$\times \int_0^{|x_1|} \int_0^{|x_2|} |f(r_1\xi_1, r_2\xi_2)|r_1^{n_1-1}r_2^{n_2-1}dr_1dr_2d\sigma(\xi_1)d\sigma(\xi_2)$$

$$= \frac{1}{|B(\mathbf{0}, |x_1|)|} \cdot \frac{1}{|B(\mathbf{0}, |x_2|)|} \int_{B(\mathbf{0}, |x_1|)} \int_{B(\mathbf{0}, |x_2|)} |f(y_1, y_2)|dy_1dy_2$$

$$= \mathcal{H}_2(f)(x_1, x_2).$$

Using the generalized Minkowski inequality and the Hölder inequality, we conclude that $\|g_f\|_{L^p(|x|^{\vec{\alpha}})}$ is equal to

$$\frac{1}{\omega_{n_1}\omega_{n_2}} \left(\int_{\mathbb{R}^{n_1}} \int_{\mathbb{R}^{n_2}}\right.$$

$$\left.\left(\int_{\mathbb{S}^{n_1-1}} \int_{\mathbb{S}^{n_2-1}} |f(|x_1|\xi_1, |x_2|\xi_2)|d\sigma(\xi_1)d\sigma(\xi_2)\right)^p |x|^{\vec{\alpha}}dx_1dx_2\right)^{1/p}$$

$$\leq \frac{1}{\omega_{n_1}\omega_{n_2}} \int_{\mathbb{S}^{n_1-1}} \int_{\mathbb{S}^{n_2-1}}$$

$$\times \left(\int_{\mathbb{R}^{n_1}} \int_{\mathbb{R}^{n_2}} |f(|x_1|\xi_1, |x_2|\xi_2)|^p |x_1|^{\alpha_1}|x_2|^{\alpha_2} dx_1 dx_2 \right)^{1/p} d\sigma(\xi_1) d\sigma(\xi_2)$$

$$\leq \left(\frac{1}{\omega_{n_1}\omega_{n_2}} \int_{\mathbb{S}^{n_1-1}} \int_{\mathbb{S}^{n_2-1}} \int_{\mathbb{R}^{n_1}} \int_{\mathbb{R}^{n_2}} |f(|x_1|\xi_1, |x_2|\xi_2)|^p \right.$$

$$\left. \times |x_1|^{\alpha_1}|x_2|^{\alpha_2} dx_1 dx_2 d\sigma(\xi_1) d\sigma(\xi_2) \right)^{1/p}$$

$$= \left(\int_{\mathbb{S}^{n_1-1}} \int_{\mathbb{S}^{n_2-1}} \int_0^\infty \int_0^\infty |f(r_1\xi_1, r_2\xi_2)|^p \right.$$

$$\left. \times r_1^{\alpha_1+n_1-1} r_2^{\alpha_2+n_2-1} dr_1 dr_2 d\sigma(\xi_1) d\sigma(\xi_2) \right)^{1/p}$$

$$= \|f\|_{L^p(|x|^{\vec{\alpha}} dx)}.$$

Thus we conclude that the following inequality

$$\frac{\|\mathcal{H}_2 f\|_{L^p(|x|^{\vec{\alpha}} dx)}}{\|f\|_{L^p(|x|^{\vec{\alpha}} dx)}} \leq \frac{\|\mathcal{H}_2(g_f)\|_{L^p(|x|^{\vec{\alpha}} dx)}}{\|g_f\|_{L^p(|x|^{\vec{\alpha}} dx)}}$$

holds provided that $\|f\|_{L^p(|x|^{\vec{\alpha}} dx)} \neq 0$. In addition, clearly if f is a nonnegative radial function, then we have $g_f = f$. This means that the norm of the operator $\mathcal{H}_2$ is equal to the norm that $\mathcal{H}_2$ restricts to the set of nonnegative radial functions. Consequently, without loss of generality, it suffices to fulfil the proof of the theorem by assuming that f is a nonnegative radial function.

Substituting the variables $z_1 = y_1/|x_1|$ and $z_2 = y_2/|x_2|$, we have that

$$\|\mathcal{H}_2 f\|_{L^p(|x|^{\vec{\alpha}} dx)}$$

$$= \left(\int_{\mathbb{R}^{n_1}} \int_{\mathbb{R}^{n_2}} \left(\frac{1}{|B(\mathbf{0}, |x_1|)||B(\mathbf{0}, |x_2|)|} \right)^p \right.$$

$$\left. \times \left(\int_{|y_1|<|x_1|} \int_{|y_2|<|x_2|} f(y_1, y_2) dy_1 dy_2 \right)^p |x_1|^{\alpha_1}|x_2|^{\alpha_2} dx_1 dx_2 \right)^{1/p}$$

$$= \frac{1}{\nu_{n_1}\nu_{n_2}} \left(\int_{\mathbb{R}^{n_1}} \int_{\mathbb{R}^{n_2}} \left(\int_{|z_1|<1} \int_{|z_2|<1} f(z_1|x_1|, z_2|x_2|) dz_1 dz_2 \right)^p \right.$$

$$\left. \times |x_1|^{\alpha_1}|x_2|^{\alpha_2} dx_1 dx_2 \right)^{1/p}.$$

Using the generalized Minkowski inequality again and noting that f is a radial function with respect to the first variable and the second variable, respectively, we have that $\|\mathcal{H}_2(f)\|_{L^p(|x|^{\vec{\alpha}}dx)}$ is not greater than

$$\frac{1}{\nu_{n_1}\nu_{n_2}}\int_{|z_1|<1}\int_{|z_2|<1}$$

$$\left(\int_{\mathbb{R}^{n_1}}\int_{\mathbb{R}^{n_2}}(f(z_1|x_1|,z_2|x_2|))^p|x_1|^{\alpha_1}|x_2|^{\alpha_2}dx_1dx_2\right)^{1/p}dz_1dz_2$$

$$=\frac{1}{\nu_{n_1}\nu_{n_2}}\int_{|z_1|<1}\int_{|z_2|<1}\left(\int_{\mathbb{R}^{n_1}}\int_{\mathbb{R}^{n_2}}(f(x_1,x_2))^p\right.$$

$$\left.\times\left(\frac{|x_1|}{|z_1|}\right)^{\alpha_1}\left(\frac{|x_2|}{|z_2|}\right)^{\alpha_2}dx_1dx_2\right)^{1/p}|z_1|^{-\frac{n_1}{p}}|z_2|^{-\frac{n_2}{p}}dz_1dz_2$$

$$=\frac{1}{\nu_{n_1}\nu_{n_2}}\int_{|z_1|<1}\int_{|z_2|<1}|z_1|^{-\frac{n_1+\alpha_1}{p}}|z_2|^{-\frac{n_2+\alpha_2}{p}}dz_1dz_2\|f\|_{L^p(|x|^{\vec{\alpha}}dx)}$$

$$=\prod_{j=1}^{2}\left(\frac{p}{p-1-\alpha_j/n_j}\right)\|f\|_{L^p(|x|^{\vec{\alpha}}dx)},$$

where $\nu_{n_i}=\frac{\pi^{n_i/2}}{\Gamma(1+n_i/2)}$ is the volume of the unit ball in $\mathbb{R}^{n_i}$, $i=1,2$.

Therefore, it implies that

$$\|\mathcal{H}_2\|_{L^p(|x|^{\vec{\alpha}}dx)\to L^p(|x|^{\vec{\alpha}}dx)}\leq\prod_{j=1}^{2}\left(\frac{p}{p-1-\alpha_j/n_j}\right).$$

Next we need to prove the converse inequality. For the purpose of getting the sharp bound, we set

$$0<\varepsilon<\min\left\{1,\frac{(p_1-1)n}{p_2},\frac{(p_2-1)n}{p_2}\right\},$$

and define

$$f_\varepsilon(x_1,x_2)=|x_1|^{-\frac{n_1+\alpha_1}{p}+\varepsilon}|x_2|^{-\frac{n_2+\alpha_2}{p}+\varepsilon}\chi_{\{|x_1|<1,\,|x_2|<1\}}(x_1,x_2).$$

It follows from the elementary calculation that $\|f_\varepsilon\|_{L^p(|x|^{\vec{\alpha}})}$ is

$$\left(\int_{|x_1|<1}\int_{|x_2|<1}|x_1|^{(\varepsilon-\frac{n_1+\alpha_1}{p})p}|x_2|^{(\varepsilon-\frac{n_2+\alpha_2}{p})p}|x|^{\vec{\alpha}}dx_2dx_1\right)^{1/p}$$

$$=\left(\int_{|x_1|<1}|x_1|^{(-\frac{n_1}{p}+\varepsilon)p}dx_1\right)^{1/p}\left(\int_{|x_2|<1}|x_2|^{(-\frac{n_2}{p}+\varepsilon)p}dx_2\right)^{1/p}$$

$$= \left(\frac{\omega_{n_1}}{p\varepsilon}\right)^{1/p} \left(\frac{\omega_{n_2}}{p\varepsilon}\right)^{1/p}.$$

We rewrite $\mathcal{H}_2(f_\varepsilon)$ as follows

$$\mathcal{H}_2(f_\varepsilon)(x_1, x_2)$$

$$= \frac{1}{|B(\mathbf{0}, |x_1|)||B(\mathbf{0}, |x_2|)|} \int_{|y_1|<|x_1|} \int_{|y_2|<|x_2|} f_\varepsilon(y_1, y_2) dy_1 dy_2$$

$$= \frac{1}{|B(\mathbf{0}, 1)||B(\mathbf{0}, 1)|} \int_{|z_1|<1} \int_{|z_2|<1} f_\varepsilon(z_1|x_1|, z_2|x_2|) dz_1 dz_2$$

$$= \frac{|x_1|^{-\frac{n_1+\alpha_1}{p}+\varepsilon}|x_2|^{-\frac{n_2+\alpha_2}{p}+\varepsilon}}{|B(\mathbf{0}, 1)||B(\mathbf{0}, 1)|} \int_{\{|z_1|<1, |z_1|<1/|x_1|\}}$$

$$\int_{\{|z_2|<1, |z_2|<1/|x_2|\}} |z_1|^{-\frac{n_1+\alpha_1}{p}+\varepsilon}|z_2|^{-\frac{n_2+\alpha_2}{p}+\varepsilon} dz_2 dz_1.$$

Thus we have

$$\|\mathcal{H}_2(f_\varepsilon)\|_{L^p(|x|^{\tilde{\alpha}}dx)}^p$$

$$= \frac{1}{(\nu_{n_1}\nu_{n_2})^p} \int_{\mathbb{R}^{n_1}} \int_{\mathbb{R}^{n_2}} \left| \int_{\{|z_1|<1, |z_1|<\frac{1}{|x_1|}\}} \int_{\{|z_2|<1, |z_2|<\frac{1}{|x_2|}\}} \right.$$

$$\left. |z_1|^{-\frac{n_1+\alpha_1}{p}+\varepsilon}|z_2|^{-\frac{n_2+\alpha_1}{p}+\varepsilon} dz_2 dz_1 \right|^p |x_1|^{p\varepsilon-n_1-\alpha_1}|x_2|^{p\varepsilon-n_2-\alpha_2} |x|^{\tilde{\alpha}} dx_2 dx_1$$

$$\geq \frac{1}{(\nu_{n_1}\nu_{n_2})^p} \int_{|x_1|<1} \int_{|x_2|<1} \left| \int_{|z_1|<1} \int_{|z_2|<1} |z_1|^{-\frac{n_1+\alpha_1}{p}+\varepsilon} \right.$$

$$\left. \times |z_2|^{-\frac{n_2+\alpha_2}{p}+\varepsilon} dz_2 dz_1 \right|^p |x_1|^{p\varepsilon-n_1}|x_2|^{p\varepsilon-n_2} dx_2 dx_1$$

$$= \frac{1}{(\nu_{n_1}\nu_{n_2})^p} \cdot \frac{\nu_{n_1}}{p\varepsilon} \cdot \frac{\nu_{n_2}}{p\varepsilon} \left(\int_{|z_1|<1} \int_{|z_2|<1} |z_1|^{-\frac{n_1+\alpha_1}{p}+\varepsilon}|z_2|^{-\frac{n_2+\alpha_2}{p}+\varepsilon} dz_2 dz_1 \right)^p$$

$$= \left(\frac{p}{p-1-\alpha_1/n+p\varepsilon/n_1} \cdot \frac{p}{p-1-\alpha_2/n+p\varepsilon/n_2} \right)^p \|f_\varepsilon\|_{L^p(|x|^{\tilde{\alpha}}dx)}^p.$$

Therefore,

$$\frac{\|\mathcal{H}_2(f_\varepsilon)\|_{L^p(|x|^{\tilde{\alpha}}dx)}}{\|f_\varepsilon\|_{L^p(|x|^{\tilde{\alpha}}dx)}} \geq \frac{p}{p-1-\alpha_1/n+p\varepsilon/n_1} \cdot \frac{p}{p-1-\alpha_2/n+p\varepsilon/n_2}.$$

Consequently, using the definition of the norm of the operator and letting $\varepsilon \to 0^+$, we conclude that

$$\|\mathcal{H}_2\|_{L^p(|x|^{\tilde{\alpha}}dx)\to L^p(|x|^{\tilde{\alpha}}dx)} \geq \prod_{j=1}^{2} \left(\frac{p}{p-1-\alpha_j/n_j} \right).$$

This finishes the proof. $\qquad\qquad\qquad\qquad\qquad\qquad\qquad\quad \square$

1.6 Notes

Theorem 1.2 was given by Fu, Grafakos and Lu et al. in [Fu et al. (2012)] for $n \geq 2$, and by Landau, Schur and Hardy in [Landau et al. (1926)] for $n = 1$ (see also [Hardy et al. (1952), Theorem 330]). Theorem 1.3 was originated by the work of Bliss in [Bliss (1930)] for one-dimension and proved by Zhao and Lu in [Zhao and Lu (2015)] for the higher dimensions. Theorems 1.4 and 1.10 were due to Lu, Yan and Zhao in [Lu et al. (2013)]. Theorem 1.6 was obtained by Zhao and Liu in [Zhao and Liu (2021)] for $n \geq 2$ and by Kolyada in [Komori (2003)] for $n = 1$. In fact, Theorem 1.6 is still holds when the underlying space $\mathbb{R}^n$ is replaced by the Heisenberg group, see [Zhao and Liu (2021)]. The authors in [Fu et al. (2012)] gave the sharp estimate for the m-linear Hilbert operator T^m on the power weighted Lebesgue spaces, see Theorem 1.9. The unweighted case with the sharp bound was proved in [Bényi and Oh (2006)].

Chapter 2

Hardy operators on other function spaces

2.1 Hardy operators on Campanato spaces and Morrey-Herz spaces

In what follows, denote by $Q(x, R)$ the cube centered at x with sides (parallel to the axes) length R, $|Q(x, R)|$ the Lebesgue measure of $Q(x, R)$ and

$$f_{Q(x, R)} = \frac{1}{|Q(x, R)|} \int_{Q(x, R)} f(y) dy.$$

To study the local behavior of solutions to second order elliptic partial differential equations, Morrey [Morrey (1938)] introduced the classical Morrey spaces $L^{q, \lambda}(\mathbb{R}^n)$.

Definition 2.1. Let $1 \leq q < \infty$ and $-1/q \leq \lambda$. The Morrey space $L^{q, \lambda}(\mathbb{R}^n)$ is defined by

$$L^{q, \lambda}(\mathbb{R}^n) = \left\{ f \in L^q_{\mathrm{loc}}(\mathbb{R}^n) : \|f\|_{L^{q, \lambda}(\mathbb{R}^n)} < \infty \right\},$$

where

$$\|f\|_{L^{q, \lambda}(\mathbb{R}^n)} := \sup_{a \in \mathbb{R}^n, R > 0} \left(\frac{1}{|Q(a, R)|^{1+\lambda q}} \int_{Q(a, R)} |f(x)|^q dx \right)^{1/q}.$$

Obviously, $L^{q, \lambda}(\mathbb{R}^n)$ is a Banach space. We also replace cubes $Q(a, R)$ by balls $B(a, R)$ centered at the point a with radius R. Their corresponding norms are equivalent.

Note that $L^{q, -1/q}(\mathbb{R}^n) = L^q(\mathbb{R}^n)$, $L^{q, 0}(\mathbb{R}^n) = L^\infty(\mathbb{R}^n)$. If $\lambda > 0$, then $L^{q, \lambda}(\mathbb{R}^n) = \{0\}$. Therefore we only consider the case $-1/q < \lambda < 0$ below.

Definition 2.2. Let $b \in L_{\mathrm{loc}}(\mathbb{R}^n)$. We say that $b \in BMO(\mathbb{R}^n)$ if and only if

$$\sup_{Q \subset \mathbb{R}^n} \frac{1}{|Q|} \int_Q |b(x) - b_Q| dx < \infty.$$

The BMO norm of b is defined by

$$\|b\|_* := \sup_{Q \subset \mathbb{R}^n} \frac{1}{|Q|} \int_Q |b(x) - b_Q| dx.$$

If one regards two functions whose difference is a constant as one, then the space BMO is a Banach space.

Remark 2.1. The John-Nirenberg inequality implies that

$$\|b\|_* \sim \sup_{Q \subset \mathbb{R}^n} \left\{ \frac{1}{|Q|} \int_Q |b(x) - b_Q|^p dx \right\}^{1/p},$$

where $1 < p < \infty$.

Definition 2.3. The Lipschitz space $Lip_\beta(\mathbb{R}^n)$ is the space of functions f satisfying

$$\|f\|_{Lip_\beta(\mathbb{R}^n)} := \sup_{x,h \in \mathbb{R}^n, h \neq 0} \frac{|f(x+h) - f(x)|}{|h|^\beta} < \infty,$$

where $0 < \beta \leq 1$.

Remark 2.2. When $0 < \beta < 1$, $Lip_\beta(\mathbb{R}^n) = \dot{\Lambda}_\beta(\mathbb{R}^n)$, where $\dot{\Lambda}_\beta(\mathbb{R}^n)$ is the homogeneous Besov-Lipschitz space.

Definition 2.4. Let $1 \leq p < \infty$ and $-n/p \leq \alpha < 1$. A locally integrable function f is said to belong to the Campanato spaces $\mathcal{E}^{\alpha,p}(\mathbb{R}^n)$ if

$$\|f\|_{\mathcal{E}^{\alpha,p}(\mathbb{R}^n)} := \sup_Q \frac{1}{|Q|^{\alpha/n}} \left(\frac{1}{|Q|} \int_Q |f(x) - f_Q|^p dx \right)^{1/p} < \infty.$$

Remark 2.3.

$$\mathcal{E}^{\alpha,p}(\mathbb{R}^n) = \begin{cases} Lip_\alpha(\mathbb{R}^n), & \text{for } 0 < \alpha < 1, \\ BMO(\mathbb{R}^n), & \text{for } \alpha = 0, \\ L^{p,n+p\alpha}(\mathbb{R}^n), & \text{for } -n/p \leq \alpha < 0. \end{cases}$$

Wiener in [Wiener (1930, 1932)] looked for a way to describe the behavior of a function at the infinity. The conditions he considered are related to appropriate weighted L^q spaces. Beurling [Benedek et al. (1962)] extended this idea and defined a pair of dual Banach spaces K_q and $B^{q'}$. To be precisely, K_q is a Banach algebra with respect to the convolution, expressed as a union of certain weighted L^q spaces. The space $B^{q'}$ is expressed as the intersection of the corresponding weighted $L^{q'}$ spaces.

Let us first introduce some notation. Let $B_k = \{x \in \mathbb{R}^n : |x| \leq 2^k\}$ and $C_k = B_k \backslash B_{k-1}$ for $k \in \mathbb{Z}$. Let $\chi_k = \chi_{C_k}$ for $k \in \mathbb{Z}$ and $\tilde{\chi}_k = \chi_k$ if $k \in \mathbb{N}$ and $\tilde{\chi}_0 = \chi_{B_0}$, where χ_{C_k} is the characteristic function of the set C_k.

Let $1 < q < \infty$. Feichtinger [Feichtinger (1987)] observed that the space B^q can be described by

$$\|f\|_{B^q(\mathbb{R}^n)} = \sup_{k \geq 0} \left(2^{-kn/q} \|f\chi_k\|_{L^q(\mathbb{R}^n)} \right) < \infty. \tag{2.1}$$

By duality, the Herz space $K_q(\mathbb{R}^n)$ in [García-Cuerva (1989)] and [Lu and Yang (1995b)] is expressed as the intersection of the corresponding weighted L^p spaces, called Beurling algebra now, can be described by

$$\|f\|_{K_q(\mathbb{R}^n)} = \sum_{k=0}^{\infty} \left(2^{kn/q'} \|f\chi_k\|_{L^q(\mathbb{R}^n)} \right) < \infty, \ 1 < q < \infty. \tag{2.2}$$

Let $\dot{B}^q(\mathbb{R}^n)$ and $\dot{K}_q(\mathbb{R}^n)$ be the homogeneous versions of $B^q(\mathbb{R}^n)$ and $K_q(\mathbb{R}^n)$ by taking $k \in \mathbb{Z}$ in (2.1) and (2.2) instead of $k \geq 0$ there.

In [Lu and Yang (1995c)] and [Lu and Yang (1992)], Lu and Yang introduced some new Hardy space $H\dot{K}_q(\mathbb{R}^n)$ (called Herz-Hardy space) related to the homogeneous Herz space $\dot{K}_q(\mathbb{R}^n)$, and obtained that the dual space of $H\dot{K}_q(\mathbb{R}^n)$ is the central bounded mean oscillation space $C\dot{M}O^q(\mathbb{R}^n)$ which satisfies the following condition: for $1 \leq p < \infty$,

$$\|f\|_{C\dot{M}O^q(\mathbb{R}^n)} = \sup_{R>0} \left(\frac{1}{|B(\mathbf{0}, R)|} \int_{B(\mathbf{0},R)} |f(x) - f_{B(\mathbf{0},R)}|^q dx \right)^{1/q} < \infty.$$

If $q = 1$, we denote $C\dot{M}O^1$ simply by $C\dot{M}O$. One can refer to [Lu and Yang (1997)] for the real-variable characterizations of the space $H\dot{K}_q(\mathbb{R}^n)$ and their applications.

The space $C\dot{M}O^q(\mathbb{R}^n)$ can be regarded as a local version of $BMO(\mathbb{R}^n)$ at the origin. But, they have quite different properties. By Remark 2.1, we have

$$\|b\|_* \sim \sup_{Q \subset \mathbb{R}^n} \left\{ \frac{1}{|Q|} \int_Q |b(x) - b_Q|^p dx \right\}^{1/p}.$$

However, the space $C\dot{M}O^q$ depends on q. If $q_1 < q_2$, then $C\dot{M}O^{q_2} \subsetneq C\dot{M}O^{q_1}$. Therefore, there is no analogy of the famous John-Nirenberg inequality of BMO for the space $C\dot{M}O^q$. One can imagine that the behavior of $C\dot{M}O^q$ may be quite different from that of BMO.

For the inhomogeneous version, Chen, Lau in [Chen and Lau (1989)] and García-Cuerva in [García-Cuerva (1989)] introduced an atomic space HA^q associated with the Beurling algebra A^q, and identified its dual as the space CMO^q defined by

$$\sup_{R \geq 1} \left(\frac{1}{|B(\mathbf{0}, R)|} \int_{B(\mathbf{0},R)} |f(x) - f_{B(\mathbf{0},R)}|^q dx \right)^{1/q} < \infty, \text{ for } 1 < q < \infty.$$

Alvarez, Guzmán-Partida and Lakey [Alvarez et al. (2000)] pointed out that B^q and BMO are roughly the bad part and the good part of CMO^q. They also gave an example to illuminate that BMO is strictly included in $\cap_{q>1}CMO^q$. In fact, if one considers the function

$$f(x) = \sum_{j=1}^{\infty} j\chi_{A_j}\operatorname{sgn}(x) + \chi_{\{|x|\geq 1\}}\operatorname{sgn}(x), \quad x \in \mathbb{R}^n,$$

where $A_j = \{4^{-j-1} < |x| \leq 4^j\}$ for $j \in \mathbb{N}$, then it is not difficult to show that $f \in \bigcap_{q>1} CMO^q$ but f is not in BMO.

Based on the above work, Alvarez, Guzmán-Partida and Lakey [Alvarez et al. (2000)] studied the relationship between central BMO spaces and Morrey spaces. Furthermore, they introduced λ-central bounded mean oscillation spaces and central Morrey spaces, respectively.

Definition 2.5. (The λ-central BMO space) Let $1 \leq q < \infty$ and $-1/q < \lambda < 1/n$. A function $f \in L_{\text{loc}}^q(\mathbb{R}^n)$ is said to belong to the λ-central bounded mean oscillation space $C\dot{M}O^{q,\lambda}(\mathbb{R}^n)$ if

$$\|f\|_{C\dot{M}O^{q,\lambda}(\mathbb{R}^n)}$$

$$= \sup_{R>0}\left(\frac{1}{|B(\mathbf{0},R)|^{1+\lambda q}}\int_{B(\mathbf{0},R)}|f(x) - f_{B(\mathbf{0},R)}|^q dx\right)^{1/q} < \infty. \quad (2.3)$$

Remark 2.4. If two functions which differ by a constant are regarded as a function in the space $C\dot{M}O^{q,\lambda}$, then $C\dot{M}O^{q,\lambda}$ becomes a Banach space. If $\lambda = 0$, then the space $C\dot{M}O^{q,\lambda}$ is just the space $C\dot{M}O^q$ defined above. Apparently, (2.3) is equivalent to the following condition

$$\sup_{R>0}\inf_{c\in\mathbb{C}}\left(\frac{1}{|B(\mathbf{0},R)|^{1+\lambda q}}\int_{B(\mathbf{0},R)}|f(x) - c|^q dx\right)^{1/q} < \infty.$$

Definition 2.6. (The central Morrey space) Let $1 \leq q < \infty$ and $-1/q \leq \lambda \leq 0$. The central Morrey space $\dot{B}^{q,\lambda}(\mathbb{R}^n)$ is defined by

$$\|f\|_{\dot{B}^{q,\lambda}(\mathbb{R}^n)} = \sup_{R>0}\left(\frac{1}{|B(\mathbf{0},R)|^{1+\lambda q}}\int_{B(\mathbf{0},R)}|f(x)|^q dx\right)^{1/q} < \infty. \quad (2.4)$$

Remark 2.5. It follows from (2.3) and (2.4) that $\dot{B}^{q,\lambda}$ is a Banach space continuously included in $C\dot{M}O^{q,\lambda}$. One can easily check that $\dot{B}^{q,0}$ is just the space $\dot{B}^q$ defined above.

We denote by $CMO^{q,\lambda}(\mathbb{R}^n)$ and $B^{q,\lambda}(\mathbb{R}^n)$ the inhomogeneous versions of the λ-central bounded mean oscillation space and the central Morrey space by taking the supremum over $R \geq 1$ in Definitions 2.5 and 2.6 instead of $R > 0$ there. Obviously, $C\dot{M}O^{q,\lambda}(\mathbb{R}^n) \subset CMO^{q,\lambda}(\mathbb{R}^n)$ for $\lambda < 1/n$ and $1 < q < \infty$, and $\dot{B}^{q,\lambda}(\mathbb{R}^n) \subset B^{q,\lambda}(\mathbb{R}^n)$ for $\lambda \geq -1/q$ and $1 < q < \infty$.

Remark 2.6. $\dot{B}^{q,\lambda}(\mathbb{R}^n)$ and $B^{q,\lambda}(\mathbb{R}^n)$ reduce to $\{0\}$ when $\lambda < -1/q$, and it is true that $\dot{B}^{q,-1/q} = B^{q,-1/q} = L^q$. $C\dot{M}O^{q,\lambda}(\mathbb{R}^n)$ and $CMO^{q,\lambda}(\mathbb{R}^n)$ reduce to the space of constant functions when $\lambda < -1/q$, and they coincide with L^q modulo constants when $\lambda = -1/q$.

Remark 2.7. When $\lambda_1 < \lambda_2$, it follows from the property of monotone functions that $B^{q,\lambda_1} \subset B^{q,\lambda_2}$ and $CMO^{q,\lambda_1} \subset CMO^{q,\lambda_2}$ for $1 < q < \infty$. If $1 < q_1 < q_2 < \infty$, then by Hölder's inequality, we know that $\dot{B}^{q_2,\lambda} \subset \dot{B}^{q_1,\lambda}$, $B^{q_2,\lambda} \subset B^{q_1,\lambda}$ for $\lambda \in \mathbb{R}$, and $C\dot{M}O^{q_2,\lambda} \subset C\dot{M}O^{q_1,\lambda}$, $CMO^{q_2,\lambda} \subset CMO^{q_1,\lambda}$ for $\lambda < 1/n$.

Since Herz spaces are natural generalization of the Lebesgue spaces with power weight, researchers are also interested in studying the boundedness of n-dimensional Hardy operators on Herz spaces. The definition of the Herz spaces is given below, see also [Lu and Yang (1995a)].

Definition 2.7. Let $\alpha \in \mathbb{R}$ and $0 < p, q \leq \infty$. Then
(1) The homogenous Herz spaces $\dot{K}_q^{\alpha,p}(\mathbb{R}^n)$ is defined by

$$\dot{K}_q^{\alpha,p}(\mathbb{R}^n) := \left\{ f \in L_{\mathrm{loc}}^q(\mathbb{R}^n \backslash \{\mathbf{0}\}) : \|f\|_{\dot{K}_q^{\alpha,p}(\mathbb{R}^n)} < \infty \right\},$$

where

$$\|f\|_{\dot{K}_q^{\alpha,p}(\mathbb{R}^n)} = \left\{ \sum_{k=-\infty}^{\infty} 2^{k\alpha p} \|f\chi_k\|_{L^q(\mathbb{R}^n)}^p \right\}^{1/p}.$$

with the usual modifications made when $p = \infty$ and/or $q = \infty$.
(2) The non-homogeneous Herz spaces $K_q^{\alpha,p}(\mathbb{R}^n)$ is defined by

$$K_q^{\alpha,p}(\mathbb{R}^n) := \left\{ f \in L_{\mathrm{loc}}^q(\mathbb{R}^n) : \|f\|_{K_q^{\alpha,p}(\mathbb{R}^n)} < \infty \right\},$$

where

$$\|f\|_{K_q^{\alpha,p}(\mathbb{R}^n)} = \left\{ \sum_{k=0}^{\infty} 2^{k\alpha p} \|f\tilde{\chi}_k\|_{L^q(\mathbb{R}^n)}^p \right\}^{1/p},$$

with the usual modifications made when $p = \infty$ and/or $q = \infty$.

Remark 2.8. $\dot{K}_p^{0,p}(\mathbb{R}^n) = K_p^{0,p}(\mathbb{R}^n) = L^p(\mathbb{R}^n)$, $\dot{K}_p^{\alpha/p,p}(\mathbb{R}^n) = K_p^{\alpha/p,p}(\mathbb{R}^n)$ $= L^p(|x|^\alpha dx)$ for all $0 < p \leq \infty$ and $\alpha \in \mathbb{R}$.

It is easy to see that $\dot{B}^q$ and $\dot{K}_q$ are just the classical homogeneous Herz spaces $\dot{K}_q^{-1/q,\infty}$ and $\dot{K}_q^{1/q',1}$, respectively.

In [Lu and Xu (2005)], Lu and Xu introduced the Morrey-Herz spaces.

Definition 2.8. Let $\alpha \in \mathbb{R}$, $0 < p \leq \infty$, $0 < q \leq \infty$ and $\lambda \geq 0$.

(1) The homogeneous Morrey-Herz space $M\dot{K}_{p,q}^{\alpha,\lambda}(\mathbb{R}^n)$ is defined by

$$M\dot{K}_{p,q}^{\alpha,\lambda}(\mathbb{R}^n) := \{f \in L_{\text{loc}}^q(\mathbb{R}^n\backslash\{\mathbf{0}\}) : \|f\|_{M\dot{K}_{p,q}^{\alpha,\lambda}(\mathbb{R}^n)} < \infty\},$$

where

$$\|f\|_{M\dot{K}_{p,q}^{\alpha,\lambda}(\mathbb{R}^n)} := \sup_{k_0 \in \mathbb{Z}} 2^{-k_0\lambda} \left\{\sum_{k=-\infty}^{k_0} 2^{k\alpha p}\|f\chi_k\|_{L^q(\mathbb{R}^n)}^p\right\}^{1/p}$$

with the usual modifications made when $p = \infty$ and/or $q = \infty$.

(2) The non-homogeneous Morrey-Herz space $MK_{p,q}^{\alpha,\lambda}(\mathbb{R}^n)$ is defined by

$$MK_{p,q}^{\alpha,\lambda}(\mathbb{R}^n) := \{f \in L_{\text{loc}}^q(\mathbb{R}^n) : \|f\|_{MK_{p,q}^{\alpha,\lambda}(\mathbb{R}^n)} < \infty\},$$

where

$$\|f\|_{MK_{p,q}^{\alpha,\lambda}(\mathbb{R}^n)} := \sup_{k_0 \in \mathbb{N}} 2^{-k_0\lambda} \left\{\sum_{k=0}^{k_0} 2^{k\alpha p}\|f\widetilde{\chi}_k\|_{L^q(\mathbb{R}^n)}^p\right\}^{1/p}$$

with the usual modifications made when $p = \infty$ and/or $q = \infty$.

Remark 2.9. It is easy to see that

$$M\dot{K}_{p,q}^{\alpha,0}(\mathbb{R}^n) = \dot{K}_q^{\alpha,p}(\mathbb{R}^n)$$

and

$$L^{q,\lambda}(\mathbb{R}^n) \subset M\dot{K}_{q,q}^{0,\lambda}(\mathbb{R}^n).$$

Theorem 2.1. *Suppose that* $1 < q < \infty$ *and* $-1/q \leq \lambda \leq 0$. *Then*

$$\|\mathcal{H}f\|_{\dot{B}^{q,\lambda}(\mathbb{R}^n)} \leq \frac{1}{1+\lambda}\|f\|_{\dot{B}^{q,\lambda}(\mathbb{R}^n)}$$

and for $-1/q \leq \lambda < 0$,

$$\|\mathcal{H}^*f\|_{\dot{B}^{q,\lambda}(\mathbb{R}^n)} \leq -\frac{1}{\lambda}\|f\|_{\dot{B}^{q,\lambda}(\mathbb{R}^n)}.$$

Moreover,

$$\|\mathcal{H}\|_{\dot{B}^{q,\lambda}(\mathbb{R}^n)\to\dot{B}^{q,\lambda}(\mathbb{R}^n)} = \frac{1}{1+\lambda}, \quad \|\mathcal{H}^*\|_{\dot{B}^{q,\lambda}(\mathbb{R}^n)\to\dot{B}^{q,\lambda}(\mathbb{R}^n)} = -\frac{1}{\lambda}.$$

Particularly, when $\lambda = 0$, the following result is obtained.

Corollary 2.1. *Let* $1 < q < \infty$. *Then*

$$\|\mathcal{H}f\|_{\dot{B}^q(\mathbb{R}^n)} \leq \|f\|_{\dot{B}^q(\mathbb{R}^n)}.$$

Moreover,

$$\|\mathcal{H}\|_{\dot{B}^q(\mathbb{R}^n) \to \dot{B}^q(\mathbb{R}^n)} = 1.$$

Theorem 2.2. *Let* $1 < q < \infty$ *and* $-1/q < \lambda < 1/n$. *Then*

$$\|\mathcal{H}f\|_{C\dot{M}O^{q,\lambda}(\mathbb{R}^n)} \leq \frac{1}{1+\lambda} \|f\|_{C\dot{M}O^{q,\lambda}(\mathbb{R}^n)} \tag{2.5}$$

and for $-1/q < \lambda < 0$,

$$\|\mathcal{H}^* f\|_{C\dot{M}O^{q,\lambda}(\mathbb{R}^n)} \leq -\frac{1}{\lambda} \|f\|_{C\dot{M}O^{q,\lambda}(\mathbb{R}^n)}.$$

Moreover,

$$\|\mathcal{H}\|_{C\dot{M}O^{q,\lambda}(\mathbb{R}^n) \to C\dot{M}O^{q,\lambda}(\mathbb{R}^n)} = \frac{1}{1+\lambda},$$

$$\|\mathcal{H}^*\|_{C\dot{M}O^{q,\lambda}(\mathbb{R}^n) \to C\dot{M}O^{q,\lambda}(\mathbb{R}^n)} = -\frac{1}{\lambda}.$$

Specially, when $\lambda = 0$, the following interesting result is obtained.

Corollary 2.2. *Let* $1 < q < \infty$. *Then*

$$\|\mathcal{H}f\|_{C\dot{M}O^q(\mathbb{R}^n)} \leq \|f\|_{C\dot{M}O^q(\mathbb{R}^n)}.$$

Moreover,

$$\|\mathcal{H}\|_{C\dot{M}O^q(\mathbb{R}^n) \to C\dot{M}O^q(\mathbb{R}^n)} = 1.$$

Remark 2.10. Comparing to the result of Xiao in [Xiao (2001)], we see that

$$\|\mathcal{H}\|_{C\dot{M}O^q(\mathbb{R}^n) \to C\dot{M}O^q(\mathbb{R}^n)} = 1 = \|H\|_{BMO(\mathbb{R}^n) \to BMO(\mathbb{R}^n)}.$$

Remark 2.11. The above results remain true for the inhomogeneous versions of λ-central bounded mean oscillation spaces and central Morrey spaces.

Similar to Theorems 2.1 and 2.2, we have

Theorem 2.3. *Let* f *be a radial function,* $1 < q < \infty$ *and* $-1/q \leq \lambda < 0$. *Then*

$$\|\mathcal{H}f\|_{L^{q,\lambda}(\mathbb{R}^n)} \leq \frac{1}{1+\lambda} \|f\|_{L^{q,\lambda}(\mathbb{R}^n)} \tag{2.6}$$

and

$$\|\mathcal{H}^* f\|_{L^{q,\lambda}(\mathbb{R}^n)} \leq -\frac{1}{\lambda}\|f\|_{L^{q,\lambda}(\mathbb{R}^n)}. \tag{2.7}$$

Moreover, the constants $\frac{1}{1+\lambda}$ in (2.6) and $-\frac{1}{\lambda}$ in (2.7) are the best possible, respectively.

Theorem 2.4. *Let f be a radial function, $1 < p < \infty$ and $-n/p < \lambda < 1$. Then*

$$\|\mathcal{H}f\|_{\mathcal{E}^{p,\lambda}(\mathbb{R}^n)} \leq \frac{1}{1+\lambda}\|f\|_{\mathcal{E}^{p,\lambda}(\mathbb{R}^n)} \tag{2.8}$$

and for $-n/p < \lambda < 0$,

$$\|\mathcal{H}^* f\|_{\mathcal{E}^{p,\lambda}(\mathbb{R}^n)} \leq -\frac{1}{\lambda}\|f\|_{\mathcal{E}^{p,\lambda}(\mathbb{R}^n)}. \tag{2.9}$$

Conjecture 2.1. *The constants $\frac{1}{1+\lambda}$ in (2.8) and $-\frac{1}{\lambda}$ in (2.9) are the best possible, respectively.*

Below we obtain the boundedness of n-dimensional Hardy operators on Herz spaces, and give the estimates for the corresponding operator norms.

Theorem 2.5. *Let $1 < p, q < \infty$ and $\alpha < n(1 - 1/q)$. Then the operator norm of $\mathcal{H}$ on $\dot{K}_q^{\alpha,p}(\mathbb{R}^n)$ satisfies*

$$\frac{qn}{qn - q\alpha - n} \leq \|\mathcal{H}\|_{\dot{K}_q^{\alpha,p}(\mathbb{R}^n) \to \dot{K}_q^{\alpha,p}(\mathbb{R}^n)} \leq \frac{(1 + 2^{|\alpha|})qn}{qn - q\alpha - n}.$$

By Remark 2.8, we have the following result.

Corollary 2.3. *Let $1 < p < \infty$ and $\alpha < n(p-1)$. Then the operator norm of $\mathcal{H}$ on $L^p(|x|^\alpha\, dx)$ satisfies*

$$\frac{pn}{n(p-1) - \alpha} \leq \|\mathcal{H}\|_{L^p(|x|^\alpha\, dx) \to L^p(|x|^\alpha\, dx)} \leq \frac{(1 + 2^{|\alpha|/p})pn}{n(p-1) - \alpha}.$$

Similarly, we can obtain the following results for the operator $\mathcal{H}^*$.

Theorem 2.6. *Let $1 < p, q < \infty$ and $\alpha > -n/q$. Then the operator norm of $\mathcal{H}^*$ on $\dot{K}_q^{\alpha,p}(\mathbb{R}^n)$ satisfies*

$$\frac{qn}{q\alpha + n} \leq \|\mathcal{H}^*\|_{\dot{K}_q^{\alpha,p}(\mathbb{R}^n) \to \dot{K}_q^{\alpha,p}(\mathbb{R}^n)} \leq \frac{(1 + 2^{|\alpha|})qn}{q\alpha + n}.$$

Corollary 2.4. *Let* $1 < p < \infty$ *and* $\alpha > -n$. *Then the operator norm of* $\mathcal{H}^*$ *on* $L^p(|x|^\alpha dx)$ *satisfies*

$$\frac{pn}{\alpha + n} \leq \|\mathcal{H}^*\|_{L^p(|x|^\alpha dx) \to L^p(|x|^\alpha dx)} \leq \frac{(1 + 2^{|\alpha|/p})pn}{\alpha + n}.$$

Remark 2.12. The above results remain true for the inhomogeneous versions of Herz spaces.

In what follows, we obtain the boundedness of Hardy operators on Morrey-Herz spaces, and give the estimates for the corresponding operator norms.

Theorem 2.7. *Let* $1 < p, q < \infty$, $\lambda > 0$ *and* $\alpha < n(1 - 1/q) - \lambda$. *Then the operator norm of* $\mathcal{H}$ *on* $M\dot{K}_{p,q}^{\alpha,\lambda}(\mathbb{R}^n)$ *satisfies*

$$\frac{qn}{qn - q\alpha - n - q\lambda} \leq \|\mathcal{H}\|_{M\dot{K}_{p,q}^{\alpha,\lambda}(\mathbb{R}^n) \to M\dot{K}_{p,q}^{\alpha,\lambda}(\mathbb{R}^n)}$$

$$\leq \frac{(1 + 2^{|\alpha - \lambda|})qn}{qn - q\alpha - n - q\lambda}.$$

Similarly, we have the following result.

Theorem 2.8. *Let* $1 < p, q < \infty$, $\lambda > 0$ *and* $\alpha > -n/q + \lambda$. *Then the operator norm of* $\mathcal{H}^*$ *on* $M\dot{K}_{p,q}^{\alpha,\lambda}(\mathbb{R}^n)$ *satisfies*

$$\frac{qn}{q\alpha + n - q\lambda} \leq \|\mathcal{H}^*\|_{M\dot{K}_{p,q}^{\alpha,\lambda}(\mathbb{R}^n) \to M\dot{K}_{p,q}^{\alpha,\lambda}(\mathbb{R}^n)} \leq \frac{(1 + 2^{|\alpha - \lambda|})qn}{q\alpha + n - q\lambda}.$$

Proof of Theorem 2.1. We only consider the case $\lambda > -1/q$ below since $\dot{B}^{q,-1/q} = L^q$.

Similar to the proof of Theorem 1.2, the operator $\mathcal{H}$ and its restriction to radial functions have the same operator norm in $\dot{B}^{q,\lambda}(\mathbb{R}^n)$.

By changing variables, we have

$$\mathcal{H}f(x) = \frac{1}{\nu_n} \int_{B(0,1)} f(t|x|)dt.$$

Using Minkowski's inequality, we have

$$\left(\frac{1}{|B(0,R)|^{1+\lambda q}} \int_{B(0,R)} \left| \frac{1}{\nu_n} \int_{B(0,1)} f(t|x|)dt \right|^q dx \right)^{1/q}$$

$$\leq \frac{1}{\nu_n} \int_{B(0,1)} \left(\frac{1}{|B(0,R)|^{1+\lambda q}} \int_{B(0,R)} |f(|t|x)|^q dx \right)^{1/q} dt$$

$$= \frac{1}{\nu_n} \int_{B(0,1)} \left(\frac{1}{|B(0,|t|R)|^{1+\lambda q}} \int_{B(0,|t|R)} |f(x)|^q dx \right)^{1/q} |t|^{n\lambda} dt$$

$$\leq \frac{\|f\|_{\dot{B}^{q,\lambda}(\mathbb{R}^n)}}{\nu_n} \int_{B(0,1)} |t|^{n\lambda} dt$$

$$\leq \frac{1}{1+\lambda} \|f\|_{\dot{B}^{q,\lambda}(\mathbb{R}^n)}.$$

On the other hand, take

$$f_0(x) = |x|^{n\lambda}, \quad x \in \mathbb{R}^n.$$

Then

$$f_0 \in \dot{B}^{q,\lambda}(\mathbb{R}^n),$$

and hence

$$\mathcal{H}(f_0)(x) = \frac{1}{1+\lambda} f_0(x).$$

The desired result can be obtained soon.

Similarly, we can obtain

$$\|\mathcal{H}^*\|_{\dot{B}^{q,\lambda} \to \dot{B}^{q,\lambda}} = -\frac{1}{\lambda}.$$

$\square$

Proof of Theorem 2.2. Similar to the proof of Theorem 2.1, we set a radial function g_f as follows:

$$g_f(x) = \frac{1}{\omega_n} \int_{|\xi|=1} f(|x|\xi) \, d\xi, \quad x \in \mathbb{R}^n.$$

By Minkowski's inequality, we obtain

$$\left(\frac{1}{|B(0,R)|^{1+\lambda q}} \int_{B(0,R)} |g(x) - g_{B(0,R)}|^q dx \right)^{1/q}$$

$$= \left(\frac{1}{|B(0,R)|^{1+\lambda q}} \int_{B(0,R)} |g(x) - f_{B(0,R)}|^q dx \right)^{1/q}$$

$$\leq \frac{1}{\omega_n} \int_{|\xi|=1} \left(\frac{1}{|B(0,R)|^{1+\lambda q}} \int_{B(0,R)} |f(|x|\xi) - f_{B(0,R)}|^q \, dx \right)^{1/q} d\xi$$

$$\leq \|f\|_{C\dot{M}O^{q,\lambda}}.$$

Thus
$$\frac{\|\mathcal{H}f\|_{C\dot{M}O^{q,\lambda}}}{\|f\|_{C\dot{M}O^{q,\lambda}}} \le \frac{\|\mathcal{H}g\|_{C\dot{M}O^{q,\lambda}}}{\|g\|_{C\dot{M}O^{q,\lambda}}},$$

which implies that the operator $\mathcal{H}$ and its restriction to radial functions have the same operator norm in $C\dot{M}O^{q,\lambda}$. So, in what follows, we may assume that f is a radial function.

For each real number $t > 0$ and $B(\mathbf{0}, R)$ in $\mathbb{R}^n$, we denote by $B(\mathbf{0}, tR) = tB(\mathbf{0}, R)$. For any ball $B(\mathbf{0}, R)$, we use Fubini's theorem to establish

$$
(\mathcal{H}f)_{B(\mathbf{0},R)} = \frac{1}{\nu_n} \int_{B(\mathbf{0},1)} \left(\frac{1}{|B(\mathbf{0},R)|} \int_{B(\mathbf{0},R)} f(t|x|)dx \right) dt
$$
$$
= \frac{1}{\nu_n} \int_{B(\mathbf{0},1)} f_{B(\mathbf{0},|t|R)} dt.
$$

Thus, using Minkowski's inequality, we have

$$
\left(\frac{1}{|B(\mathbf{0},R)|^{1+\lambda q}} \int_{B(\mathbf{0},R)} |\mathcal{H}f(x) - (\mathcal{H}f)_{B(\mathbf{0},R)}|^q dx \right)^{1/q}
$$
$$
= \frac{1}{\nu_n} \left(\frac{1}{|B(\mathbf{0},R)|^{1+\lambda q}} \int_{B(\mathbf{0},R)} \left| \int_{B(\mathbf{0},1)} \left(f(|t|x) - f_{B(\mathbf{0},|t|R)} \right) dt \right|^q dx \right)^{1/q}
$$
$$
\le \frac{1}{\nu_n} \int_{B(\mathbf{0},1)} \left(\frac{1}{|B(\mathbf{0},R)|^{1+\lambda q}} \int_{B(\mathbf{0},R)} |f(|t|x) - f_{B(\mathbf{0},|t|R)}|^q dx \right)^{1/q} dt
$$
$$
= \frac{1}{\nu_n} \int_{B(\mathbf{0},1)} \left(\frac{1}{|B(\mathbf{0},|t|R)|^{1+\lambda q}} \int_{B(\mathbf{0},|t|R)} |f(x) - f_{B(\mathbf{0},|t|R)}|^q dx \right)^{1/q} |t|^{n\lambda} dt
$$
$$
\le \frac{1}{\nu_n} \|f\|_{C\dot{M}O^{q,\lambda}(\mathbb{R}^n)} \int_{B(\mathbf{0},1)} |t|^{n\lambda} dt
$$
$$
\le \frac{1}{1+\lambda} \|f\|_{C\dot{M}O^{q,\lambda}(\mathbb{R}^n)},
$$

which implies (2.5) holds.

Take
$$
f_0(x) = \begin{cases} |x|^{n\lambda} & x \in \mathbb{R}_r^n, \\ (-1)^n |x|^{n\lambda} & x \in \mathbb{R}_l^n, \end{cases}
$$

where $\mathbb{R}_r^n$ and $\mathbb{R}_l^n$ denote the right and the left halves of $\mathbb{R}^n$, separated by the hyperplane $x_1 = 0$, where x_1 is the first coordinate of $x \in \mathbb{R}^n$.

Thus, by a standard calculation, $(f_0)_{B(\mathbf{0},R)} = 0$ and

$$
\|f_0\|_{C\dot{M}O^{q,\lambda}(\mathbb{R}^n)} = \left(\frac{\Gamma(1+n/2)}{\pi^{n/2}} \right)^{q\lambda} \cdot \frac{1}{q\lambda+1},
$$

and

$$\mathcal{H}(f_0)(x) = \frac{1}{1+\lambda} f_0(x).$$

The desired result is obtained.   □

Using the idea in the proof of Theorem 2.2, one can obtain the following result.

Theorem 2.9. *Let f be a radial function. Then*

$$\|\mathcal{H}f\|_{BMO(\mathbb{R}^n)} \leq 1 \cdot \|f\|_{BMO(\mathbb{R}^n)}.$$

Moreover, the constant 1 is the best possible in the above inequalities.

Proof of Theorem 2.3. Similar to the proof of Theorem 2.1, we take

$$f_0(x) = |x|^{n\lambda}, \quad x \in \mathbb{R}^n,$$

then

$$f_0 \in L^{q,\lambda}(\mathbb{R}^n).$$

In fact, we will consider the following two cases.

(i) if $|a| > 2R$, then $|x| > R$. For $-1/q < \lambda < 0$, we have

$$\frac{1}{|Q(a,R)|^{1+\lambda q}} \int_{Q(a,R)} |x|^{n\lambda q} dx \leq \frac{1}{|Q(a,R)|^{1+\lambda q}} \int_{Q(a,R)} R^{n\lambda q} dx = 1.$$

(ii) if $|a| < 2R$, then $Q(a,R) \subset Q(0,3R)$, we have

$$\frac{1}{|Q(a,R)|^{1+\lambda q}} \int_{Q(a,R)} |x|^{n\lambda q} dx \leq \frac{1}{|Q(a,R)|^{1+\lambda q}} \int_{Q(0,3R)} |x|^{n\lambda q} dx$$
$$\leq C 3^{n(1+q\lambda)}.$$

□

Proofs of Theorems 2.5 and 2.6. Similar to the proof of Theorem 1.2, the operator $\mathcal{H}$ and its restriction to radial functions have the same operator norm in $\dot{K}_q^{\alpha p}$. So, we can assume that f is a radial function.

By Minkowski's inequality, we have

$$\nu_n \|(\mathcal{H}f)\chi_k\|_{L^q} = \left(\int_{C_k} \left| \int_{B(0,1)} f(t|x|) dt \right|^q dx \right)^{1/q}$$
$$\leq \int_{B(0,1)} \left(\int_{C_k} |f(|t|x|)|^q dx \right)^{1/q} dt$$

$$= \int_{B(0,1)} \left(\int_{2^{k-1}|t|<|x|\leq 2^k|t|} |f(x)|^q \, dx \right)^{1/q} |t|^{-n/q} \, dt.$$

Since for any $|t| \in (0,1)$, there exists an $m \in \mathbb{Z}$ such that $2^{m-1} < |t| \leq 2^m$, by the Minkowski inequality, we get

$$\nu_n \|(\mathcal{H}f)\chi_k\|_{L^q} \leq \int_{B(0,1)} \left(\int_{2^{k+m-2}<|x|\leq 2^{k+m}} |f(x)|^q \, dx \right)^{1/q} |t|^{-n/q} \, dt$$

$$\leq \int_{B(0,1)} (\|f\chi_{k+m-1}\|_{L^q} + \|f\chi_{k+m}\|_{L^q}) \, |t|^{-n/q} \, dt.$$

Thus

$$\nu_n \|\mathcal{H}f\|_{\dot{K}_q^{\alpha,p}}$$

$$\leq \left(\sum_{k=-\infty}^{\infty} 2^{k\alpha p} \left(\int_{B(0,1)} (\|f\chi_{k+m-1}\|_{L^q} + \|f\chi_{k+m}\|_{L^q})|t|^{-n/q} \, dt \right)^p \right)^{1/p}.$$

Using the Minkowski inequality again, we obtain

$$\nu_n \|\mathcal{H}f\|_{\dot{K}_q^{\alpha,p}} \leq \int_{B(0,1)} \left(\sum_{k=-\infty}^{\infty} 2^{k\alpha p} \|f\chi_{k+m-1}\|_{L^q}^p |t|^{-np/q} \right)^{1/p} \, dt$$

$$+ \int_{B(0,1)} \left(\sum_{k=-\infty}^{\infty} 2^{k\alpha p} \|f\chi_{k+m}\|_{L^q}^p |t|^{-np/q} \right)^{1/p} \, dt$$

$$\leq \|f\|_{\dot{K}_q^{\alpha,p}} \int_{B(0,1)} (2^{(1-m)\alpha} + 2^{-m\alpha})|t|^{-n/q} \, dt$$

$$\leq (1 + 2^{|\alpha|}) \|f\|_{\dot{K}_q^{\alpha,p}} \int_{B(0,1)} |t|^{-(\alpha+n/q)} \, dt.$$

Therefore

$$\|\mathcal{H}f\|_{\dot{K}_q^{\alpha,p}} \leq \frac{(1 + 2^{|\alpha|})qn}{qn - q\alpha - n} \|f\|_{\dot{K}_q^{\alpha,p}}.$$

On the other hand, for any $0 < \varepsilon < 1$, set

$$f_\varepsilon(x) = \begin{cases} 0 & |x| \leq 1, \\ |x|^{-(\alpha+n/q+\varepsilon)} & |x| > 1. \end{cases}$$

Obviously, when $k = 0, -1, -2, \ldots$, $\|f_\varepsilon \chi_k\|_{L^q(\mathbb{R}^n)} = 0$ and

$$\|f_\varepsilon \chi_k\|_{L^q(\mathbb{R}^n)}^q = \int_{2^{k-1}<|x|\leq 2^k} |x|^{-(\alpha+n/q+\varepsilon)q} \, dx$$

$$= 2^{-k(\alpha+\varepsilon)q} \left| \frac{\omega_n(2^{(\alpha+\varepsilon)q} - 1)}{(\alpha+\varepsilon)q} \right|.$$

By a standard integral calculation, we have

$$\|f_\varepsilon\|_{\dot{K}_q^{\alpha,p}(\mathbb{R}^n)} = 2^{-\varepsilon} \left(\frac{1}{1 - 2^{-\varepsilon p}} \right)^{1/p} \left| \frac{c_n(2^{(\alpha+\varepsilon)q} - 1)}{(\alpha+\varepsilon)q} \right|^{1/q}.$$

When $|x| \leq 1$, for any $|t| \in [0,1]$, $|tx| \leq 1$, hence $(\mathcal{H}f_\varepsilon)(x) = 0$.
When $|x| > 1$, it is easy to check that

$$\nu_n(\mathcal{H}f_\varepsilon)(x) = \int_{1/|x|<|t|<1} f_\varepsilon(t|x|) \, dt$$

$$= |x|^{-(\alpha+n/q+\varepsilon)} \int_{1/|x|<|t|<1} |t|^{-(\alpha+n/q+\varepsilon)} \, dt.$$

Select $\delta = \varepsilon^{-1} > 1$ ($0 < \varepsilon < \min\{1, n - n/q - \alpha\}$), there exists an $l \in \mathbb{Z}_+$ such that $2^{l-1} \leq \delta < 2^l$. Hence, we have

$$\nu_n^p \|\mathcal{H}f_\varepsilon\|_{\dot{K}_q^{\alpha,p}}^p = \nu_n^p \sum_{k=1}^{\infty} 2^{k\alpha p} \|(\mathcal{H}f_\varepsilon)\chi_k\|_{L^q}^p$$

$$= \sum_{k=1}^{\infty} 2^{k\alpha p} \left(\int_{|x|>1} \left(|x|^{-(\alpha+\frac{n}{q}+\varepsilon)} \chi_k(x) \int_{1/|x|<|t|<1} |t|^{-(\alpha+\frac{n}{q}+\varepsilon)} \, dt \right)^q dx \right)^{p/q}$$

$$\geq \sum_{k=1}^{\infty} 2^{k\alpha p} \left(\int_{|x|>\delta} |x|^{-(\alpha+\frac{n}{q}+\varepsilon)q} \chi_k(x) \, dx \right)^{\frac{p}{q}} \left(\int_{1/\delta<|t|<1} |t|^{-(\alpha+\frac{n}{q}+\varepsilon)} dt \right)^p$$

$$\geq \omega_n^{p/q} \left(\int_{1/\delta<|t|<1} |t|^{-(\alpha+\frac{n}{q}+\varepsilon)} \, dt \right)^p \sum_{k=l+1}^{\infty} 2^{k\alpha p} \left(\int_{2^{k-1}}^{2^k} r^{-(\alpha+\varepsilon)q-1} \, dr \right)^{p/q}$$

$$= \omega_n^{p/q} \left| \frac{2^{(\alpha+\varepsilon)q} - 1}{(\alpha+\varepsilon)q} \right|^{p/q} \frac{2^{-\varepsilon(l+1)p}}{1 - 2^{-\varepsilon p}} \left(\int_{1/\delta<|t|<1} |t|^{-(\alpha+n/q+\varepsilon)} \psi(t) \, dt \right)^p$$

$$\geq \varepsilon^{\varepsilon p} \|f_\varepsilon\|_{\dot{K}_q^{\alpha,p}}^p \left(\int_{\varepsilon<|t|<1} |t|^{-(\alpha+n/q+\varepsilon)} \, dt \right)^p$$

$$= \nu_n^p \varepsilon^{\varepsilon p} \left(\frac{qn(1 - \varepsilon^{n-(\alpha+n/q+\varepsilon)})}{qn - q\alpha - n - q\varepsilon} \right)^p \|f_\varepsilon\|_{\dot{K}_q^{\alpha,p}}^p.$$

Thus, we have

$$\|\mathcal{H}f_\varepsilon\|_{\dot{K}_q^{\alpha,p}} \geq \varepsilon^{\varepsilon} \frac{qn(1 - \varepsilon^{n-(\alpha+n/q+\varepsilon)})}{qn - q\alpha - n - q\varepsilon} \|f_\varepsilon\|_{\dot{K}_q^{\alpha,p}}.$$

Let $\varepsilon \to 0^+$. We get

$$\|\mathcal{H}f\|_{\dot{K}_q^{\alpha,p}} \geq \frac{qn}{qn - q\alpha - n}\|f\|_{\dot{K}_q^{\alpha,p}}.$$

This finishes the proof of Theorem 2.5. Similarly, the proof of Theorem 2.6 can be obtained. $\square$

Proof of Theorem 2.7. By the proof of Theorem 2.5, we have

$$\nu_n\|(\mathcal{H}f)\chi_k\|_{L^q} \leq \int_{B(0,1)} \left(\int_{2^{k+m-2}<|x|\leq 2^{k+m}} |f(x)|^q\,dx\right)^{1/q} |t|^{-n/q}\,dt$$

$$\leq \int_{B(0,1)} (\|f\chi_{k+m-1}\|_{L^q} + \|f\chi_{k+m}\|_{L^q})\,|t|^{-n/q}\,dt.$$

Thus

$$\nu_n\|\mathcal{H}f\|_{M\dot{K}_{p,q}^{\alpha,\lambda}} \leq \sup_{k_0\in\mathbb{Z}} 2^{-k_0\lambda}$$

$$\times \left(\sum_{k=-\infty}^{k_0} 2^{k\alpha p}\left(\int_{B(0,1)} |t|^{-n/q}(\|f\chi_{k+m-1}\|_{L^q} + \|f\chi_{k+m}\|_{L^q})dt\right)^p\right)^{1/p}.$$

By the Minkowski inequality, we obtain

$$\nu_n\|\mathcal{H}f\|_{M\dot{K}_{p,q}^{\alpha,\lambda}}$$

$$\leq \sup_{k_0\in\mathbb{Z}} 2^{-k_0\lambda} \int_{B(0,1)} \left(\sum_{k=-\infty}^{k_0} 2^{k\alpha p}\|f\chi_{k+m-1}\|_{L^q}^p |t|^{-np/q}\right)^{1/p} dt$$

$$+ \sup_{k_0\in\mathbb{Z}} 2^{-k_0\lambda} \int_{B(0,1)} \left(\sum_{k=-\infty}^{k_0} 2^{k\alpha p}\|f\chi_{k+m}\|_{L^q}^p |t|^{-np/q}\right)^{1/p} dt$$

$$\leq \|f\|_{M\dot{K}_{p,q}^{\alpha,\lambda}} \int_{B(0,1)} (2^{(1-m)(\alpha-\lambda)} + 2^{-m(\alpha-\lambda)})|t|^{-n/q}\,dt$$

$$\leq (1 + 2^{|\alpha-\lambda|})\|f\|_{M\dot{K}_{p,q}^{\alpha,\lambda}} \int_{B(0,1)} |t|^{\lambda-(\alpha+n/q)}\,dt.$$

Therefore

$$\|\mathcal{H}f\|_{M\dot{K}_{p,q}^{\alpha,\lambda}} \leq \frac{(1 + 2^{|\alpha-\lambda|})qn}{qn - q\alpha - n + \lambda n}\|f\|_{M\dot{K}_{p,q}^{\alpha,\lambda}}.$$

On the other hand, set

$$f(x) = |x|^{-(\alpha+n/q-\lambda)}, \quad x \in \mathbb{R}^n.$$

Then

$$\nu_n \mathcal{H} f(x) = \int_{B(0,1)} f(t|x|)\, dt$$

$$= |x|^{-(\alpha+n/q-\lambda)} \int_{B(0,1)} |t|^{-(\alpha+n/q-\lambda)}\, dt$$

$$= \frac{\omega_n}{n - (\alpha + n/q - \lambda)} f(x).$$

We need to consider the following two cases. If $\alpha \neq \lambda$, then it can be obtained that

$$\|f\chi_k\|_{L^q}^q = \int_{2^{k-1} < |x| \leq 2^k} |x|^{-(\alpha+n/q-\lambda)q}\, dx = 2^{-k(\alpha-\lambda)q} \left| \frac{\omega_n(2^{(\alpha-\lambda)q} - 1)}{(\alpha-\lambda)q} \right|.$$

Thus we have

$$\|f\|_{M\dot{K}_{p,q}^{\alpha,\lambda}} = \frac{2^\lambda}{(2^{\lambda p} - 1)^{1/p}} \left| \frac{\omega_n(2^{(\alpha-\lambda)q} - 1)}{(\alpha-\lambda)q} \right|^{1/q}.$$

This implies that $f \in M\dot{K}_{p,q}^{\alpha,\lambda}$. Therefore, we have

$$\|\mathcal{H}f\|_{M\dot{K}_{p,q}^{\alpha,\lambda}} = \frac{qn}{qn - q\alpha - n + q\lambda} \|f\|_{M\dot{K}_{p,q}^{\alpha,\lambda}}.$$

If $\alpha = \lambda$, then we have

$$\|f\chi_k\|_{L^q}^q = \int_{2^{k-1} < |x| \leq 2^k} |x|^{-n}\, dx = \omega_n \ln 2$$

and also have

$$\|f\|_{M\dot{K}_{p,q}^{\alpha,\lambda}} = \frac{2^\lambda(\omega_n \ln 2)^{1/q}}{(2^{\lambda p} - 1)^{1/p}}.$$

Hence we have

$$\|\mathcal{H}f\|_{M\dot{K}_{p,q}^{\alpha,\lambda}} = \frac{qn}{qn - n} \|f\|_{M\dot{K}_{p,q}^{\alpha,\lambda}},$$

which finishes the proof of Theorem 2.7. $\qquad\square$

Similar to the proof of Theorem 2.7, we can get the proof of Theorem 2.8. We omit the details here.

2.2 m-linear Hardy's inequality on (central) Morrey spaces

Theorem 2.10. *Let $m \in \mathbb{N}$, f_i be in $\dot{B}^{p_i,\lambda_i}(\mathbb{R}^n)$, $1 < p_i < \infty$, $1 < p < \infty$, $1/p = 1/p_1 + \cdots + 1/p_m$, $-1/p_i \le \lambda_i < 0$, $i = 1, \ldots, m$, and $\lambda = \lambda_1 + \cdots + \lambda_m$. Then $\mathcal{H}^m$ is bounded from $\dot{B}^{p_1,\lambda_1} \times \cdots \times \dot{B}^{p_m,\lambda_m}$ to $\dot{B}^{p,\lambda}$. Moreover, if $\lambda p = \lambda_1 p_1 = \cdots = \lambda_m p_m$, then*

$$\|\mathcal{H}^m\|_{\dot{B}^{p_1,\lambda_1} \times \cdots \times \dot{B}^{p_m,\lambda_m} \to \dot{B}^{p,\lambda}} = \frac{\omega_n^m}{\omega_{mn}} \cdot \frac{m}{\lambda + m} \cdot \frac{1}{2^{m-1}} \cdot \frac{\prod_{i=1}^m \Gamma(\frac{n}{2}(\lambda_i + 1))}{\Gamma(\frac{n}{2}(m + \lambda))}.$$

Proof. We only prove the case $m = 2$. The proof for the case $m \ge 3$ is similar. We first note that the operator $\mathcal{H}^2$ and its restriction to radial functions have the same operator norm in $\dot{B}^{p,\lambda}$. Choosing radial functions f_1 and f_2, we then write

$$\mathcal{H}^2(f_1, f_2)(x) = \frac{1}{\nu_{2n}} \int_{|(z_1,z_2)|<1} f_1(|x|z_1) f_2(|x|z_2) \, dz_1 dz_2.$$

Using the Minkowski integral inequality and the Hölder inequality, we have

$$\left(\frac{1}{|B(0,R)|^{1+\lambda p}} \int_{B(0,R)} \left| \int_{|(y_1,y_2)|<1} f_1(y_1|x|) f_2(y_2|x|) \, dy_1 dy_2 \right|^p dx \right)^{1/p}$$

$$\le \int_{|(y_1,y_2)|<1} \left(\frac{1}{|B(0,R)|^{1+\lambda p}} \int_{B(0,R)} \left| f_1(|y_1|x) f_2(|y_2|x) \right|^p dx \right)^{1/p}$$

$$\le \int_{|(y_1,y_2)|<1} \left(\frac{1}{|B(0,R)|^{1+\lambda_1 p_1}} \int_{B(0,R)} \left| f_1(|y_1|x) \right|^{p_1} dx \right)^{1/p_1}$$

$$\times \left(\frac{1}{|B(0,R)|^{1+\lambda_2 p_2}} \int_{B(0,R)} \left| f_2(|y_2|x) \right|^{p_2} dx \right)^{1/p_2} dy_1 dy_2$$

$$= \int_{|(y_1,y_2)|<1} \left(\frac{1}{|B(0,|y_1|R)|^{1+\lambda_1 p_1}} \int_{B(0,|y_1|R)} \left| f_1(x) \right|^{p_1} dx \right)^{1/p_1}$$

$$\times \left(\frac{1}{|B(0,|y_2|R)|^{1+\lambda_2 p_2}} \int_{B(0,|y_2|R)} \left| f_2(x) \right|^{p_2} dx \right)^{1/p_2} \prod_{i=1}^2 |y_i|^{n\lambda_i} dy_1 dy_2$$

$$\le \int_{|(y_1,y_2)|<1} \prod_{i=1}^2 |y_i|^{n\lambda_i} dy_1 dy_2 \, \|f_1\|_{\dot{B}^{p_1,\lambda_1}(\mathbb{R}^n)} \|f_2\|_{\dot{B}^{p_2,\lambda_2}(\mathbb{R}^n)},$$

where $\lambda = \lambda_1 + \lambda_2$. A calculation yields that the value of the integral is

$$\int_{|(y_1,y_2)|<1} \prod_{i=1}^2 |y_i|^{n\lambda_i} dy_1 dy_2 = \frac{\omega_n^2}{2n} \cdot \frac{1}{2+\lambda} B\left(\frac{n}{2}(1+\lambda_1), \frac{n}{2}(1+\lambda_2) \right), \quad (2.10)$$

and this proves one direction in the claimed identity.

On the other hand, taking $\widetilde{f}_i(x_i) = |x_i|^{n\lambda_i}$ for $x_i \in \mathbb{R}^n$, $i = 1, 2$, we obtain

$$\frac{1}{|B(0,R)|^{1+\lambda_i p_i}} \int_{B(0,R)} \left|\widetilde{f}_i(x_i)\right|^p dx_i = \frac{1}{|B(0,R)|^{1+\lambda_i p_i}} \int_{B(0,R)} |x_i|^{n\lambda_i p_i} dx_i$$

$$= \frac{\omega_n}{R^{n(1+\lambda_i p_i)}} \int_0^R r^{n\lambda_i p_i + n - 1} dr = \frac{\omega_n}{n(1+\lambda_i p_i)}.$$

It is easy to verify that $\widetilde{f}_i \in \dot{B}^{p_i,\lambda_i}$, $i = 1, 2$. By a simple calculation, we obtain that

$$\mathcal{H}^2(\widetilde{f}_1, \widetilde{f}_2)(x) = \frac{1}{\nu_{2n}} \int_{|(y_1,y_2)|<1} |y_1|^{n\lambda_1} |y_2|^{n\lambda_2} dy_1 dy_2 \, \widetilde{f}_1(x) \widetilde{f}_2(x).$$

This observation together with (2.10) concludes the proof in the case $m = 2$. $\qquad\square$

Next we have the following result concerning best constants on the subspaces of $L^{p_i,\lambda_i}(\mathbb{R}^n)$ consisting of radial functions.

Proposition 2.1. *Let* $m \in \mathbb{N}$, f_i *be radial functions in* $L^{p_i,\lambda_i}(\mathbb{R}^n)$, $i = 1,\dots,m$, $1 < p_i < \infty$, $-1/p_i \le \lambda_i < 0$, $i = 1,\dots,m$, $1 \le p < \infty$, $1/p = 1/p_1 + \cdots + 1/p_m$, *and* $\lambda = \lambda_1 + \cdots + \lambda_m$. *Then*

$$\|\mathcal{H}^m(f_1,\dots,f_m)\|_{L^{p,\lambda}} \le C_{n,m,\lambda,\lambda_1,\dots,\lambda_m} \prod_{i=1}^m \|f_i\|_{L^{p_i,\lambda_i}}. \tag{2.11}$$

Moreover, the constant

$$C_{n,m,\lambda,\lambda_1,\dots,\lambda_m} = \frac{\omega_n^m}{\omega_{mn}} \cdot \frac{m}{\lambda + m} \cdot \frac{1}{2^{m-1}} \cdot \frac{\prod_{i=1}^m \Gamma(\frac{n}{2}(\lambda_i + 1))}{\Gamma(\frac{n}{2}(m + \lambda))}$$

is the same with that in Theorem 2.10. If $\lambda p = \lambda_1 p_1 = \cdots = \lambda_m p_m$, *then the best possible in inequality (2.11) for radial functions.*

Proof. We only consider the case $m = 2$. As in the proof of Theorem 2.10, we let $f_1 \in L^{p_1,\lambda_1}$ and $f_2 \in L^{p_2,\lambda_2}$ be radial functions. By Minkowski's integral inequality and Hölder's inequality, we have

$$\left(\frac{1}{|Q(a,R)|^{1+\lambda p}} \int_{Q(a,R)} \left|\int_{|(y_1,y_2)|<1} f_1(y_1|x|) f_2(y_2|x|) dy_1 dy_2\right|^p dx\right)^{1/p}$$

$$\le \int_{|(y_1,y_2)|<1} \left(\frac{1}{|Q(a,R)|^{1+\lambda p}} \int_{Q(a,R)} \left|f_1(|y_1|x) f_2(|y_2|x)\right|^p dx\right)^{1/p} dy_1 dy_2$$

$$\leq \int_{|(y_1,y_2)|<1} \left(\frac{1}{|Q(a,R)|^{1+\lambda_1 p_1}} \int_{Q(a,R)} \left| f_1(|y_1|x) \right|^{p_1} dx \right)^{1/p_1}$$

$$\times \left(\frac{1}{|Q(a,R)|^{1+\lambda_2 p_2}} \int_{Q(a,R)} \left| f_2(|y_2|x) \right|^{p_2} dx \right)^{1/p_2} dy_1 dy_2$$

$$= \int_{|(y_1,y_2)|<1} \left(\frac{1}{|Q(a|y_1|,|y_1|R)|^{1+\lambda_1 p_1}} \int_{Q(a|y_1|,|y_1|R)} \left| f_1(x) \right|^{p_1} dx \right)^{1/p_1}$$

$$\times \left(\frac{1}{|Q(a|y_2|,|y_2|R)|^{1+\lambda_2 p_2}} \int_{Q(a|y_2|,|y_2|R)} \left| f_2(x) \right|^{p_2} dx \right)^{1/p_2}$$

$$\times \prod_{i=1}^{2} |y_i|^{n\lambda_i} dy_1 dy_2$$

$$\leq \int_{|(y_1,y_2)|<1} |y_1|^{n\lambda_1} |y_2|^{n\lambda_2} dy_1 dy_2 \|f_1\|_{L^{p_1,\lambda_1}} \|f_2\|_{L^{p_2,\lambda_2}}$$

$$= C_{n,\lambda,\lambda_1,\lambda_2} \|f_1\|_{L^{p_1,\lambda_1}} \|f_2\|_{L^{p_2,\lambda_2}}.$$

On the other hand, for $i = 1,2$, taking $\widetilde{f}_i(x_i) = |x_i|^{n\lambda_i}, x_i \in \mathbb{R}^n$, we easily verify $\widetilde{f}_i \in L^{p_i,\lambda_i}$ by considering the cases $|a| > 2R$ and $|a| < 2R$. Then the desired conclusion follows via the method in the proof of Theorem 2.10. We omit the details. $\square$

2.3 $(H^1(\mathbb{R}^n), L^1(\mathbb{R}^n))$ bounds of Hardy operators

In Chapter 1, we mainly focus on the strong type estimations of the n-dimensional Hardy operators on Lebesgue spaces. We now investigate the endpoint estimates for n-dimensional Hardy operators on Hardy spaces.

In order to give the definition of Hardy space, we begin with the definition of atoms on $\mathbb{R}^n$.

Definition 2.9 (see [Lu (1995)]). A function $a \in L^\infty(\mathbb{R}^n)$ is called a $(1, \infty, 0)$-atom, if it satisfies the following conditions: (1) supp $a \subset B(x_0,r)$; (2) $\|a\|_{L^\infty} \leq |B(x_0,r)|^{-1}$; (3) $\int_{\mathbb{R}^n} a(x)dx = 0$.

As a proper subspace of $L^1(\mathbb{R}^n)$, the atomic Hardy space $H^1(\mathbb{R}^n)$ is defined by

$$H^1(\mathbb{R}^n) = \left\{ f : f(x) \overset{\mathcal{S}'}{=} \sum_k \lambda_k a_k(x), \text{ and } \sum_k |\lambda_k| < \infty \right\},$$

where a_k is a $(1,\infty,0)$-atom, and f is a tempered distribution. Then the H^1 norm of f is defined by

$$\|f\|_{H^1(\mathbb{R}^n)} := \inf \sum_{k}^{\infty} |\lambda_k|,$$

where the infimum is taken over all the decompositions of $f = \sum_k \lambda_k a_k$ as above.

Motivated by the counterexample in [Golubov (1997)], we choose

$$b(x) = \begin{cases} 1 - 2^n & \text{for } |x| \leq 1, \\ 1 & \text{for } 1 < |x| \leq 2, \\ 0 & \text{other}, \end{cases}$$

then

$$\mathcal{H}(b)(x) = \begin{cases} 1 - 2^n & \text{for } |x| \leq 1, \\ 1 - 2^n |x|^{-n} & \text{for } 1 < |x| \leq 2, \\ 0 & \text{other}. \end{cases}$$

Obviously, b is an atom of $H^1(\mathbb{R}^n)$, however,

$$\int_{\mathbb{R}^n} \mathcal{H}(b)(x)dx = -n2^n \ln 2 \neq 0,$$

therefore $\mathcal{H}(b) \notin H^1(\mathbb{R}^n)$.

Since $\mathcal{H}$ is not bounded on $L^1(\mathbb{R}^n)$ or $H^1(\mathbb{R}^n)$, a natural question is: does $\mathcal{H}$ map $H^1(\mathbb{R}^n)$ into $L^1(\mathbb{R}^n)$? The answer is confirmed. Our proof is based on the atomic decomposition and certain beautiful and elegant ideas.

Theorem 2.11. $\mathcal{H}$ *maps* $H^1(\mathbb{R}^n)$ *into* $L^1(\mathbb{R}^n)$.

Proof. Assume that a is an atom of $H^1(\mathbb{R}^n)$ and satisfies the following conditions: (i) supp $a \subset B(x_0, r)$, (ii) $\|a\|_{L^\infty} \leq |B(x_0,r)|^{-1}$ and (iii) $\int_{\mathbb{R}^n} a(x)dx = 0$. Let $\tilde{a}(x) = a(x + x_0)$. Then $\tilde{a}$ satisfies the following conditions: (i) supp $\tilde{a} \subset B(\mathbf{0}, r)$, (ii) $\|\tilde{a}\|_{L^\infty} \leq |B(\mathbf{0},r)|^{-1}$ and (iii) $\int_{\mathbb{R}^n} \tilde{a}(x)dx = 0$.

It is enough to prove that $\int_{\mathbb{R}^n} |\mathcal{H}\tilde{a}(x)|dx < C$, where C is independent of $\tilde{a}$. Suppose that supp $\tilde{a} \subset B(\mathbf{0}, r)$ for $r > 0$. Then

$$\int_{\mathbb{R}^n} \upsilon_n |\mathcal{H}(\tilde{a})(x)| dx$$

$$= \int_{B(\mathbf{0},2r)} \left| \frac{1}{|x|^n} \int_{|y|<|x|} \tilde{a}(y)dy \right| dx + \int_{\mathbb{R}^n \setminus B(\mathbf{0},2r)} \left| \frac{1}{|x|^n} \int_{|y|<|x|} \tilde{a}(y)dy \right| dx$$

$$= \int_{B(0,2r)} \left| \frac{1}{|x|^n} \int_{\{y:|y|<|x|\} \cap B(0,r)} \tilde{a}(y) dy \right| dx$$

$$+ \int_{\mathbb{R}^n \setminus B(0,2r)} \left| \frac{1}{|x|^n} \int_{\{y:|y|<|x|\} \cap B(0,r)} \tilde{a}(y) dy \right| dx$$

$$:= I_1 + I_2,$$

where we have used that the condition supp $\tilde{a} \subset B(0,r)$. For I_1, we have the following estimate

$$I_1 \leq \int_{B(0,2r)} \frac{1}{|x|^n} \int_{\{y:|y|<|x|\} \cap B(0,r)} |\tilde{a}(y)| \, dy dx$$

$$\leq \int_{B(0,2r)} \frac{1}{|x|^n} \left(\frac{1}{|B(0,r)|} \int_{|y|<|x|} dy \right) dx$$

$$= \frac{1}{|B(0,r)|} \int_{B(0,2r)} \frac{\nu_n |x|^n}{|x|^n} dx = 2^n \nu_n.$$

For I_2, since $x \in \mathbb{R}^n \setminus B(0,2r)$, we have $\{y \in \mathbb{R}^n : |y| < |x|\} \cap \{y \in \mathbb{R}^n : y \in B(0,r)\} = \{y \in \mathbb{R}^n : y \in B(0,r)\}$. Then

$$I_2 = \int_{\mathbb{R}^n \setminus B(0,2r)} \left| \frac{1}{|x|^n} \int_{\{y:|y|<|x|\} \cap B(0,r)} \tilde{a}(y) dy \right| dx$$

$$= \int_{\mathbb{R}^n \setminus B(0,2r)} \frac{1}{|x|^n} \left| \int_{B(0,r)} \tilde{a}(y) dy \right| dx = 0.$$

$\square$

Remark 2.13. If we divide $\mathbb{R}^n$ into two parts: $B(0, c_0 r)$ and $\mathbb{R}^n \setminus B(0, c_0 r)$. Here $c_0 > 0$. Repeating the previous argument, we obtain that for $c_0 \geq 1$,

$$I_1 \leq c_0^n \nu_n, \quad I_2 = 0.$$

But for $0 < c_0 < 1$, we have

$$I_1 \leq c_0^n \nu_n, \quad I_2 = \infty.$$

Therefore, we can obtain that the supremum whose the Hardy operator maps from H^1 to L^1 is 1 when $c_0 = 1$.

Definition 2.10. Let $1 < p < \infty$. A function a on $\mathbb{R}^n$ is said to be a central $(1,p)$-atom if a satisfies the following conditions: (i) supp $a \subset B(0,R)$; (ii) $\|a\|_{L^p} \leq |B(0,r)|^{1/p-1}$; (iii) $\int_{\mathbb{R}^n} a(x) dx = 0$.

We also need the following lemma [Lu and Yang (1995b)].

Lemma 2.1. *Let $f \in L^1(\mathbb{R}^n)$ and $1 < p < \infty$. Then f belongs to the Herz-Hardy space $H\dot{K}_p(\mathbb{R}^n)$ if and only if f can be represented as*

$$f(x) = \sum_j \lambda_j a_j(x),$$

each a_j is a central $(1,p)$-atom and $\sum_j |\lambda_j| < \infty$. Moreover,

$$\|f\|_{H\dot{K}_p(\mathbb{R}^n)} \sim \inf\left\{\sum_i |\lambda_i|\right\},$$

where the infimum is taken over all decompositions of f as above.

In fact, $H\dot{K}_p$ is the localization of H^1 at the origin. It is easy to see that the relation between $H\dot{K}_p$ and $\dot{K}_p$ is similar to that of H^1 and L^1. Thus we have the following result.

Theorem 2.12. *Let $1 < p < \infty$. Then $\mathcal{H}$ maps from $H\dot{K}_p(\mathbb{R}^n)$ to $\dot{K}_p(\mathbb{R}^n)$.*

To prove this theorem, we need the following proposition from [Lu and Yang (1995b)].

Proposition 2.2. *Let $r > 0$. Then*

$$\|f(r\cdot)\|_{\dot{K}_p(\mathbb{R}^n)} \sim r^{-n}\|f\|_{\dot{K}_p(\mathbb{R}^n)}.$$

Proof of Theorem 2.12. By Lemma 2.1, it is easy to see that the proof of Theorem 2.12 is reduced to show that for any central $(1,p)$-atom a,

$$\|\mathcal{H}(a)\|_{\dot{K}_p(\mathbb{R}^n)} \leq C,$$

where C is independent of a. Let supp $a \subset B(\mathbf{0}, r)$. If we write $\tilde{a}(x) = r^n a(rx)$, then $\tilde{a}(x)$ is a central $(1,p)$-atom supported on the unit ball $B(\mathbf{0}, 1)$. Substituting $y = rz$ and $dy = r^n dz$, we have

$$\mathcal{H}(a)(rx) = \frac{1}{\nu_n |rx|^n} \int_{|y|<|rx|} a(y)\,dy$$

$$= r^{-n} \frac{1}{\nu_n |x|^n} \int_{|z|<|x|} a(zr) r^n\,dz$$

$$= r^{-n}\mathcal{H}(\tilde{a})(x).$$

By Proposition 2.2, we obtain

$$\|\mathcal{H}(a)\|_{\dot{K}_p(\mathbb{R}^n)} \sim \|\mathcal{H}(\tilde{a})\|_{\dot{K}_p(\mathbb{R}^n)}.$$

Therefore, it suffices to show that

$$\|\mathcal{H}(\widetilde{a})\|_{\dot{K}_p(\mathbb{R}^n)} \leq C,$$

where C is independent of $\widetilde{a}$. By the definition of $\dot{K}_p$, we write

$$\|\mathcal{H}(\widetilde{a})\|_{\dot{K}_p(\mathbb{R}^n)} = \sum_{k \in \mathbb{Z}} 2^{kn/p'} \|\mathcal{H}(\widetilde{a})\chi_k\|_{L^p(\mathbb{R}^n)}$$

$$= \sum_{k \leq 1} 2^{kn/p'} \|\mathcal{H}(\widetilde{a})\chi_k\|_{L^p(\mathbb{R}^n)} + \sum_{k > 1} 2^{kn/p'} \|\mathcal{H}(\widetilde{a})\chi_k\|_{L^p(\mathbb{R}^n)}$$

$$:= J_1 + J_2.$$

The $L^p(\mathbb{R}^n)$ boundedness of the n-dimensional Hardy operator leads to

$$J_1 \leq \frac{p}{p-1} \sum_{k \leq 1} 2^{kn/p'} \|\widetilde{a}\|_{L^p(\mathbb{R}^n)}$$

$$\leq \nu_n^{1/p-1} \frac{p}{p-1} \sum_{k \leq 1} 2^{kn/p'}$$

$$= \nu_n^{1/p-1} 2^{n/p'} p' \frac{2^{n/p'}}{2^{n/p'} - 1}.$$

For $k > 1$ and $|x| > 2^{k-1} \geq 1$, it follows from $\{y : |y| < |x|\} \supseteq \{y : |y| < 1\}$ that

$$\|\mathcal{H}(\widetilde{a})\chi_k\|_{L^p(\mathbb{R}^n)}^p = \int_{2^{k-1} < |x| \leq 2^k} \left| \frac{1}{\nu_n |x|^n} \int_{|y| < 1} \widetilde{a}(y) dy \right|^p dx = 0,$$

where we have used the fact that

$$\int_{B(0,1)} \widetilde{a}(y) dy = 0.$$

$\square$

2.4 Notes

Muckenhoupt in [Muckenhoupt (1972)] obtained the weighted L^p boundedness of the one-dimensional Hardy operator H for $1 < p < \infty$. Andersen and Muckenhoupt in [Andersen and Muckenhoupt (1982)] obtained the weighted weak type $(1, 1)$ inequalities of H. As applications, they also got the weighted weak type $(1, 1)$ inequalities of Hilbert transform and Hardy-Littlewood maximal function. Sawyer in [Sawyer (1984, 1985)] proved the weighted Lebesgue and Lorentz norm inequalities for the Hardy operators and obtained the weighted inequalities for the two-dimensional Hardy operators. Golubov in [Golubov (1997)] studied the boundedness of the Hardy operators in the Hardy spaces. Martín-Reyes and Ortega [Martín-Reyes and Ortega (1998)] got the weighted weak type inequalities for modified Hardy operators. The fact that the Hardy operator H is bounded on BMO space was proved by many authors, see [Siskakis (1987, 1990)] and [Stempak (1994)]. In particular, Xiao in [Xiao (2001)] obtained that $\|H\|_{BMO(\mathbb{R}^1) \to BMO(\mathbb{R}^1)} = 1$.

The proofs of Theorems 2.11 and 2.12 were obtained by Zhao, Fu and Lu in [Zhao et al. (2012)]. Fu in [Fu (2008)] established the boundedness of n-dimensional Hardy operators on the other function spaces, such as central Morrey space, λ-central BMO space and Morrey-Herz space. The proofs of Theorem 2.10 and Proposition 2.1 came from [Fu et al. (2012)] by Fu, Grafakos and Lu et al. using the rotation method.

Chapter 3

Weighted inequalities for Hardy type operators

3.1 Weighted norm inequalities for Hardy operators

In this chapter, we focus on the n-dimensional generalization of H and its weighted theory. In order to characterize the generalized weights (here and in what follows, a weight w will be a nonnegative locally integrable function) w, v, for which the one-dimensional Hardy inequality

$$\int_0^\infty |Hf(x)|^p w(x)dx \le C \int_0^\infty |f(x)|^p v(x)dx, \quad 1 \le p \le \infty \tag{3.1}$$

holds, Muckenhoupt [Muckenhoupt (1972)] gave a nice and simple proof. See also the related work of Tomaselli [Tomaselli (1969)] and Talenti [Talenti (1969)].

The corresponding characterization of the weights w, v for the n-dimensional Hardy inequality holds just like (3.1) was obtained by Drábek et al. in [Drábek et al. (1995)],

$$\int_{\mathbb{R}^n} |\mathcal{H}f(x)|^p w(x)dx \le C \int_{\mathbb{R}^n} |f(x)|^p v(x)dx, \quad 1 < p < \infty. \tag{3.2}$$

The good weights (w, v) for $\mathcal{H}$ are now called M_p weights $(1 \le p < \infty)$. We now give the definition of M_1 weight.

Definition 3.1. A pair of nonnegative functions (w, v) is called an M_1 weight if for almost $x \in \mathbb{R}^n$,

$$\int_{|x|>r} |x|^{-n} w(x)dx \le C_1 \operatorname{essinf}_{|x|<r} v(x)$$

for some constant C_1.

Definition 3.2. [Drábek et al. (1995)] Let $1 < p < \infty$. A pair of nonnegative functions (w, v) is said to be of class M_p weight if

$$C_2 = \sup_{0 < r < \infty} \left(\int_{|x|>r} |x|^{-np} w(x) dx \right)^{1/p} \left(\int_{|x|<r} v(x)^{1-p'} dx \right)^{1/p'} < \infty.$$

If $w = v$, we say that $w \in M_p (1 \le p < \infty)$. Drábek, et al.'s result in [Drábek et al. (1995)] can be stated as follows.

Proposition 3.1. *Let w and v be nonnegative measurable functions on $\mathbb{R}^n$. For $1 \le p < \infty$, the inequality (3.2) holds for $f \ge 0$ if and only if $(w, v) \in M_p$. Moreover, if C is the smallest constant for which (3.2) holds, then*

$$C_2 \le C\nu_n \le C_2 (p')^{1/p'} p^{1/p}.$$

The first goal of this section is to study the new properties of the weights w and v in (3.2) related to the n-dimensional Hardy operator. As we know, $\mathcal{H}(|f|)(x) \le CMf(x)$, where M is the Hardy-Littlewood maximal operator. Recall that a pair of weights (w, v) belongs to the class A_p $(1 < p < \infty)$, if

$$\sup_Q \left(\frac{1}{|Q|} \int_Q w(y) dy \right) \left(\frac{1}{|Q|} \int_Q v(y)^{1-p'} dy \right)^{p-1} < \infty.$$

And a pair of weights (w, v) belongs to the class A_1 if there is a constant C such that

$$\frac{1}{|Q|} \int_Q w(y) dy \le C \inf_{y \in Q} v(y).$$

If $w = v$, we say that $w \in A_p (1 \le p < \infty)$. It is well known that (cf. [Muckenhoupt (1978)]) $M : L^p(v) \to L^{p,\infty}(w)$ if and only if (w, v) is A_p weights. And the theory of Muckenhoupt weights was studied by many mathematicians, such as Cerdà and Martín [Cerdà and Martín (2000)], Rosenblum [Rosenblum (1962)], Fefferman and Stein [Fefferman and Stein (1971)], Muckenhoupt [Muckenhoupt (1978)]. However, the properties of M_p weights seem not to be researched systematically as before especially for the higher-dimensional case.

It is also well known that ([Drábek et al. (1995)]) $\mathcal{H}^* : L^p(v) \to L^p(w)$ if and only if

$$C_3 = \sup_{0 < r < \infty} \left(\int_{|x|<r} w(x) dx \right)^{1/p} \left(\int_{|x|>r} v(x)^{1-p'} |x|^{-np'} dx \right)^{1/p'} < \infty,$$

$$\tag{3.3}$$

where $1 < p < \infty$ and $\mathcal{H}^*$ is the adjoint operator of $\mathcal{H}$ defined by

$$\mathcal{H}^* f(x) = \frac{1}{\nu_n} \int_{|y|>|x|} f(y)|y|^{-n} dy.$$

The pair of nonnegative measurable functions (w, v) is said to be of class M^p weight if the weight satisfies (3.3). We also give the definition of M^1 weight.

Definition 3.3. A pair of nonnegative measurable functions (w, v) is called an M^1 weight if for almost all $x \in \mathbb{R}^n$,

$$\int_{|x|<r} w(x) dx \leq C_4 \, \mathrm{essinf}_{|x|>r} \{|x|^n v(x)\}$$

for some constant C_4.

Recently, Lerner et al. in [Lerner et al. (2009)] gave a new maximal function $\mathcal{M}$ defined by

$$\mathcal{M}(f_1, f_2)(x) = \sup_{x \in Q} \prod_{i=1}^{2} \frac{1}{|Q|} \int_Q |f_i(y_i)| dy_i.$$

This operator is strictly smaller than the 2-fold product of the Hardy-Littlewood maximal function M. And it is used to obtain a precise control on bilinear singular integral operators of Calderón-Zygmund type and to build a theory of weights adapted to the bilinear setting. It was shown in [Lerner et al. (2009)] that if $\mathcal{M}$ is bounded from $L^{p_1}(w_1) \times L^{p_2}(w_2)$ into $L^{p,\infty}(w)$ if and only if

$$\sup_Q \left(\int_Q w(x) dx \right)^{1/p} \prod_{j=1}^{2} \left(\frac{1}{|Q|} \int_Q w_j(x)^{1-p_j'} dx \right)^{1/p_j'} < \infty,$$

where $1 \leq p_j < \infty$, $j = 1, 2$ and $1/p = 1/p_1 + 1/p_2$. From the definition of $\mathcal{H}^2$ and $\mathcal{M}$, it is easy to see that $\mathcal{H}^2(|f_1|, |f_2|)(x) \leq C\mathcal{M}(f_1, f_2)(x)$.

The second problem is whether we can characterize the operator $\mathcal{H}^2$ by means of a new class of weight functions. Highly inspired by [Drábek et al. (1995)] and [Lerner et al. (2009)], we first give an analogue of the M_p (resp. M^p) classes for multiple weights.

Definition 3.4. For $1 \leq p_1, p_2 < \infty$, $1 \leq p < \infty$ and $1/p = 1/p_1 + 1/p_2$. We say that the weights $(w(\cdot), w_1(\cdot), w_2(\cdot)) \in M_{(p_1, p_2)}$, that means the following conditions hold:

$$\widetilde{A}^n := \sup_{0<r<\infty} \left(\int_{|x|>r} |x|^{-2np} w(x) dx \right)^{\frac{1}{p}}$$

$$\times \left(\int_{|x|<r} w_1(x)^{1-p_1'} dx \right)^{\frac{1}{p_1'}} \left(\int_{|x|<r} w_2(x)^{1-p_2'} dx \right)^{\frac{1}{p_2'}} < \infty;$$

$$\tilde{B}^n := \sup_{0<r<\infty} \left(\int_{|x|<r} w_1(x)^{1-p_1'} dx \right)^{\frac{1}{p_1'}} \left(\int_{|y|>r} \left(\int_{|x|>|y|} |x|^{-2np} w(x) dx \right)^{\frac{p_1}{p}} \right.$$

$$\times \left. \left(\int_{|x|<|y|} w_2(x)^{1-p_2'} dx \right)^{\frac{p_1}{p'}} w_2(y)^{1-p_2'} dy \right)^{\frac{1}{p_1}} < \infty;$$

$$\tilde{C}^n := \sup_{0<r<\infty} \left(\int_{|x|<r} w_2(x)^{1-p_2'} dx \right)^{\frac{1}{p_2'}} \left(\int_{|y|>r} \left(\int_{|x|>|y|} |x|^{-2np} w(x) dx \right)^{\frac{p_2}{p}} \right.$$

$$\times \left. \left(\int_{|x|<|y|} w_1(x)^{1-p_1'} dx \right)^{\frac{p_2}{p'}} w_1(y)^{1-p_1'} dy \right)^{\frac{1}{p_2}} < \infty.$$

3.1.1   The properties of M_p weights

Corresponding to the A_p weight, we present some new properties for the M_p weight. We mainly focus on the case of $w = v$ for the M_p weight in this section. The following proposition can be found in [Lu et al. (2007), p. 35].

Proposition 3.2. *Let $x \in \mathbb{R}^n$. Then*
(i) *$|x|^\beta \in A_1$ if and only if $-n < \beta \le 0$.*
(ii) *For $1 < p < \infty$, $|x|^\beta \in A_p$ if and only if $-n < \beta < n(p-1)$.*

Similarly, we obtain the following result for M_p weights.

Proposition 3.3. *Let $x \in \mathbb{R}^n$. Then*
(i) *$|x|^\beta \in M_1$ if and only if $\beta < 0$.*
(ii) *For $1 < p < \infty$, $|x|^\beta \in M_p$ if and only if $\beta < n(p-1)$.*

Proof. (i) If $\beta < 0$, then

$$\int_{|x|>r} |x|^{\beta-n} dx = \frac{\omega_n}{-\beta} r^\beta \le \frac{\omega_n}{-\beta} \text{essinf}_{|x|<r} |x|^\beta,$$

which implies that $|x|^\beta \in M_1$.

On the other hand, if $|x|^\beta \in M_1$, then the function $|x|^\beta$ is integrable in the domain $\{x : |x| > r\}$. Therefore, the condition $\beta < 0$ is necessary.

(ii) If $\beta < n(p-1)$, then we have

$$\left(\int_{|x|>r} |x|^{-np} |x|^{\beta} dx \right)^{1/p} \left(\int_{|x|<r} |x|^{\beta(1-p')} dx \right)^{1/p'}$$

$$= \omega_n \left(\int_r^{\infty} t^{\beta - n(p-1)-1} dt \right)^{1/p} \left(\int_0^r t^{\beta(1-p')+n-1} dt \right)^{1/p'}$$

$$= \omega_n \left(\frac{r^{\beta - n(p-1)}}{n(p-1) - \beta} \right)^{1/p} \left(\frac{r^{n+\beta(1-p')}}{\beta(1-p') + n} \right)^{1/p'}$$

$$= \omega_n \left(\frac{1}{n(p-1) - \beta} \right)^{1/p} \left(\frac{1}{n + \beta(1-p')} \right)^{1/p'} < \infty.$$

Thus, the function $|x|^{\beta} \in M_p$ if $\beta < n(p-1)$.

Assume that $|x|^{\beta} \in M_p$. Then it is easy to verify the following conditions:

(a) The functions $|x|^{\beta-np}$ and $|x|^{\beta(1-p')}$ are integrable on their domains, respectively.

(b) $C_5 = \sup_{0 < r < \infty} \left(\int_{|x|>r} |x|^{-np+\beta} dx \right)^{1/p} \left(\int_{|x|<r} |x|^{\beta(1-p')} dx \right)^{1/p'} < \infty.$

A simple calculation yields that the condition (b) implies that $\beta < n(p-1)$. $\qquad\square$

We need point out that for any constant $C < \infty$, $C \in A_1$ and $C \notin M_1$. In fact, for any $r > 0$, $\int_{|x|>r} |x|^{-n} dx = \infty$. But the constant $C \in A_p$ and $C \in M_p$ for $1 < p < \infty$.

Proposition 3.4. *Let $x \in \mathbb{R}^n$. Then for $1 \le p < \infty$, $|x|^{\beta} \in M^p$ if and only if $\beta > -n$.*

We summarize some basic properties of M_p weights in the following.

Proposition 3.5. *Let $1 < p < \infty$.*
(i) *If $w_1 \in M_1$, $w_2 \in M^1$, then $w = w_1 w_2^{1-p} \in M_p$.*
(ii) *If $w_1 \in M^1$, $w_2 \in M_1$, then $w = w_1 w_2^{1-p} \in M^p$.*

Proof. We only give the proof of (i) for the similarity between (i) and (ii).

On the one hand

$$\left(\int_{|x|>r}|x|^{-np}w(x)dx\right)^{1/p} = \left(\int_{|x|>r}|x|^{-np}w_1(x)w_2(x)^{1-p}dx\right)^{1/p}$$

$$\leq \left(\int_{|x|>r}|x|^{-n}w_1(x)dx\right)^{1/p}\left(\text{essinf}_{|x|>r}|x|^n w_2(x)\right)^{-1/p'}.$$

On the other hand

$$\left(\int_{|x|<r}w(x)^{1-p'}dx\right)^{1/p'} = \left(\int_{|x|<r}(w_1(x)w_2(x)^{1-p})^{1-p'}dx\right)^{1/p'}$$

$$\leq \left(\int_{|x|<r}w_2(x)dx\right)^{1/p'}\left(\text{essinf}_{|x|<r}w_1(x)\right)^{-1/p}.$$

The assumption of $w_1 \in M_1$ and $w_2 \in M^1$ leads to

$$\left(\int_{|x|>r}|x|^{-np}w(x)dx\right)^{1/p}\left(\int_{|x|<r}w(x)^{1-p'}dx\right)^{1/p'} \leq C_1^{1/p}C_4^{1/p'}.$$

$\square$

Proposition 3.6. *Let $w \in M_p$ for some $1 \leq p < \infty$. Then*
(i) The class M_p are increasing as p increases; that is, for $1 \leq p < q < \infty$, we have $M_p \subsetneqq M_q$.
(ii) For $1 < p < \infty$, $w \in M_p$ if and only if $w^{1-p'} \in M^{p'}$.
(iii) For $1 \leq p < \infty$, $0 < \alpha, \varepsilon < 1$. If $w \in M_p$, then $w^\alpha \in M_{\alpha p + 1 - \varepsilon \alpha}$.
(iv) For $1 \leq p < \infty$, $0 \leq \alpha \leq 1$. If $w_1, w_2 \in M_p$, then $w_1^\alpha w_2^{1-\alpha} \in M_p$.
(v) If $w_1, w_2 \in M_p$, then $w_1 + w_2 \in M_p$.
(vi) For $\lambda \neq 0$, $\delta^\lambda(w) \in M_p$, where $\delta^\lambda(w)(x) = w(\lambda x_1, \dots, \lambda x_n)$.
(vii) For $\lambda > 0$, $\lambda w(x) \in M_p$.

Proof. (i) Let $w \in M_p(1 < p < \infty)$. Then

$$\left(\int_{|x|>r}|x|^{-np}w(x)dx\right)^{1/p}\left(\int_{|x|<r}w(x)^{1-p'}dx\right)^{1/p'} < \infty.$$

Applying Hölder's inequality with exponents $\frac{p'-1}{q'-1}$ and $\frac{p'-1}{p'-q'}$, we obtain

$$\int_{|x|<r}w(x)^{1-q'}dx \leq \left(\int_{|x|<r}w(x)^{1-p'}dx\right)^{\frac{q'-1}{p'-1}}(\nu_n r^n)^{\frac{p'-q'}{p'-1}}.$$

The above estimate yields that

$$\left(\int_{|x|>r}|x|^{-nq}w(x)dx\right)^{1/q}\left(\int_{|x|<r}w(x)^{1-q'}dx\right)^{1/q'}$$

$$=\left(\int_{|x|>r}|x|^{-np}|x|^{-n(q-p)}w(x)dx\right)^{1/q}\left(\int_{|x|<r}w(x)^{1-q'}dx\right)^{1/q'}$$

$$\leq (\nu_n r^n)^{\frac{p'-q'}{q'(p'-1)}}r^{\frac{n(p-q)}{q}}\left(\int_{|x|>r}|x|^{-np}w(x)dx\right)^{1/q}$$

$$\times\left(\int_{|x|<r}w(x)^{1-p'}dx\right)^{\frac{q'-1}{q'(p'-1)}}$$

$$=\nu_n^{\frac{p'-q'}{q'(p'-1)}}\left(\left(\int_{|x|>r}|x|^{-np}w(x)dx\right)^{1/p}\left(\int_{|x|<r}w(x)^{1-p'}dx\right)^{1/p'}\right)^{p/q}$$

$$\leq C_2^{p/q}\nu_n^{1-p/q}.$$

Thus, $w\in M_q$.

If $w\in M_1$, then for $1<q<\infty$, we have

$$\left(\int_{|x|>r}|x|^{-nq}w(x)dx\right)^{1/q}\left(\int_{|x|<r}w(x)^{1-q'}dx\right)^{1/q'}$$

$$=\left(\int_{|x|>r}|x|^{-n}|x|^{-n(q-1)}w(x)dx\right)^{1/q}\left(\int_{|x|<r}w(x)^{1-q'}dx\right)^{1/q'}$$

$$\leq r^{\frac{n(1-q)}{q}}\left(\int_{|x|>r}|x|^{-n}w(x)dx\right)^{1/q}\left(\int_{|x|<r}w(x)^{1-q'}dx\right)^{1/q'}$$

$$\leq C_1 r^{\frac{n(1-q)}{q}}\left(\mathrm{essinf}_{|x|<r}w(x)\right)^{1/q}\left(\int_{|x|<r}w(x)^{1-q'}dx\right)^{1/q'}$$

$$\leq C_1 r^{\frac{n(1-q)}{q}}\left(\int_{|x|<r}w(x)^{1-q'}w(x)^{q'/q}dx\right)^{1/q'}$$

$$\leq C_1\nu_n^{1/q'}.$$

By (ii) of Proposition 3.3, we know that $M_p\neq M_q$.

(ii) The $M^{p'}$ condition for $w^{1-p'}$ is

$$\left(\int_{|x|<r}w(x)^{1-p'}dx\right)^{1/p'}\left(\int_{|x|>r}w(x)^{(1-p')(1-p)}|x|^{-np}dx\right)^{1/p}<C_3$$

since $(1 - p')(1 - p) = 1$.

(iii) Let $q = \alpha p + 1 - \varepsilon\alpha$. Using Hölder's inequality with exponents $1/\alpha$ and $1/(1 - \alpha)$, we have

$$
\left(\int_{|x|>r} |x|^{-nq} w(x)^\alpha dx \right)^{1/q}
$$

$$
= \left(\int_{|x|>r} |x|^{-np\alpha} w(x)^\alpha |x|^{-n(1-\varepsilon\alpha)} dx \right)^{1/q}
$$

$$
\leq \left(\int_{|x|>r} |x|^{-np} w(x) dx \right)^{\alpha/q} \left(\int_{|x|>r} |x|^{-\frac{n(1-\varepsilon\alpha)}{1-\alpha}} dx \right)^{\frac{1-\alpha}{q}}
$$

$$
= \left(\int_{|x|>r} |x|^{-np} w(x) dx \right)^{\alpha/q} \left(\frac{\nu_n(1 - \alpha)}{\alpha(1 - \varepsilon)} r^{\frac{n\alpha(\varepsilon-1)}{1-\alpha}} \right)^{\frac{1-\alpha}{q}}.
$$

Hölder's inequality with exponents $\frac{1-p'}{\alpha(1-q')}$ and $\frac{1-p'}{1-p'-\alpha(1-q')}$ yields

$$
\left(\int_{|x|<r} w(x)^{\alpha(1-q')} dx \right)^{1/q'}
$$

$$
\leq \left(\int_{|x|<r} w(x)^{(1-p')} dx \right)^{\frac{\alpha(1-q')}{q'(1-p')}} \left(\int_{|x|<r} dx \right)^{\frac{1}{q'}(1- \frac{\alpha(1-q')}{1-p'})}
$$

$$
= \left(\int_{|x|<r} w(x)^{(1-p')} dx \right)^{\frac{\alpha(1-q')}{q'(1-p')}} (\nu_n r^n)^{\frac{1}{q'}(1- \frac{\alpha(1-q')}{1-p'})}.
$$

Since $\frac{1-q'}{q'(1-p')} = \frac{p}{p'q}$, we get

$$
\left(\int_{|x|>r} |x|^{-nq} w(x)^\alpha dx \right)^{1/q} \left(\int_{|x|<r} w(x)^{\alpha(1-q')} dx \right)^{1/q'}
$$

$$
\leq \left[\left(\int_{|x|>r} |x|^{-np} w(x) dx \right)^{1/p} \left(\int_{|x|<r} w(x)^{(1-p')} dx \right)^{1/p'} \right]^{\alpha/q}
$$

$$
\cdot \left(\frac{(1 - \alpha)}{\alpha(1 - \varepsilon)} \right)^{\frac{1-\alpha}{q}} \nu_n^{1- \frac{\alpha p}{q}}
$$

$$
\leq C_2^{\frac{1-\alpha}{q}} \nu_n^{1- \frac{\alpha p}{q}}.
$$

(iv) Using Hölder's inequality with exponents $1/\alpha$ and $1/(1-\alpha)$, we can prove it.

(v) We have

$$\left(\int_{|x|>r} |x|^{-np}(w_1(x)+w_2(x))dx\right)\left(\int_{|x|<r}(w_1(x)+w_2(x))^{1-p'}dx\right)^{p/p'}$$

$$\leq \sum_{j=1}^{2}\left(\int_{|x|>r}|x|^{-np}w_j(x)dx\right)\left(\int_{|x|<r}w_j(x)^{1-p'}dx\right)^{p/p'}.$$

(vi) Substituting $y = \lambda x$, we have

$$\left(\int_{|x|>r}|x|^{-np}w(\lambda x)dx\right)^{1/p}\left(\int_{|x|<r}w(\lambda x)^{1-p'}dx\right)^{1/p'}$$

$$=\left(\int_{|y|>|\lambda|r}|y|^{-np}w(y)dy\right)^{1/p}\left(\int_{|y|<|\lambda|r}w(y)^{1-p'}dy\right)^{1/p'}$$

$$\leq C_2.$$

(vii) It is obviously. $\qquad\square$

But for $w \in M_p$, the following property is different from the Muckenhoupt class A_p weights. The measure $w(x)dx$ is not doubling: precisely, for all $x_0 \in \mathbb{R}^n$ and all balls $B(x_0, 2r)(r > 0)$, the following estimate does not hold

$$w(B(x_0, 2r)) \leq Cw(B(x_0, r)).$$

In fact, if we take $w(x) = |x|^\alpha$, then $|x|^\alpha dx$ is a doubling measure when $\alpha > -n$ (cf. [Stein (1993)]). We now take $w_0(x) = |x|^{-2n}$. By (i) of Proposition 3.3, we have that $w_0(x) \in M_1$. But $w_0(x)dx$ is not a doubling measure.

Proposition 3.7. *Let $w \in M^p$ for some $1 \leq p < \infty$. Then*
(i) *For $1 < p < \infty$, $w \in M^p$ if and only if $w^{1-p'} \in M_{p'}$.*
(ii) *For $1 \leq p < \infty$, $0 < \alpha, \varepsilon < 1$. If $w \in M^p$, then $w^\alpha \in M^{\alpha p + 1 - \varepsilon\alpha}$.*
(iii) *For $1 \leq p < \infty$, $0 \leq \alpha \leq 1$. If $w_1, w_2 \in M^p$, then $w_1^\alpha w_2^{1-\alpha} \in M^p$.*
(iv) *If $w_1, w_2 \in M^p$, then $w_1 + w_2 \in M^p$.*
(v) *For $\lambda \neq 0$, $\delta^\lambda(w) \in M^p$, where $\delta^\lambda(w)(x) = w(\lambda x_1, \ldots, \lambda x_n)$.*
(vi) *For $\lambda > 0$, $\lambda w(x) \in M^p$.*

Similarly, for $w \in M^p$, the measure $w(x)dx$ is not double. In fact, if we take $w_0(x) = \exp(|x|)$. Then it is easy to verify that $w_0(x) \in M^1$, but $w_0(x)dx$ is not a doubling measure.

As applications, by (ii) of Proposition 3.6, we have

Corollary 3.1. *Let* $1 < p < \infty$. *The following three propositions are equivalent.*

(i) $w \in M_p$;

(ii) $\mathcal{H} : L^p(w) \to L^p(w)$;

(iii) $\mathcal{H}^* : L^{p'}(w^{1-p'}) \to L^{p'}(w^{1-p'})$.

In general, the results of (ii) and (iii) in Corollary 3.1 need to be proved independently. However, using the propositions of M^p and M_p, we know that they are equivalent.

Furthermore, making use of Proposition 3.6, we also have

Corollary 3.2. (i) *For* $1 \leq p < \infty$, $0 \leq \alpha \leq 1$. *If* $w_1, w_2 \in M_p$, *then*

$$\int_{\mathbb{R}^n} (\mathcal{H}f(x))^p w_1(x)^\alpha w_2(x)^{1-\alpha} dx \leq C \int_{\mathbb{R}^n} f^p(x) w_1(x)^\alpha w_2(x)^{1-\alpha} dx.$$

(ii) *For* $1 \leq p < \infty$, $0 < \alpha, \varepsilon < 1$. *If* $w \in M_p$, *then*

$$\int_{\mathbb{R}^n} (\mathcal{H}f(x))^q w(x)^\alpha dx \leq C \int_{\mathbb{R}^n} f^q(x) w(x)^\alpha dx,$$

where $q = \alpha p + 1 - \varepsilon \alpha$.

Similarly, in analogue with the n-dimensional Hardy operators, we can yield some weighted results for the adjoint n-dimensional Hardy operators $\mathcal{H}^*$, and we do not list them here.

3.1.2　*A weight characterization of bilinear Hardy inequality*

For convenience, we assume $(w(\sqrt{2}\cdot), w_1(\cdot), w_2(\cdot)) \in M_{(p_1,p_2)}$, that means the following conditions hold:

$$A^n := \sup_{0<r<\infty} \left(\int_{|x|>\sqrt{2}r} |x|^{-2np} w(x) dx \right)^{\frac{1}{p}}$$

$$\times \left(\int_{|x|<r} w_1(x)^{1-p_1'} dx \right)^{\frac{1}{p_1'}} \left(\int_{|x|<r} w_2(x)^{1-p_2'} dx \right)^{\frac{1}{p_2'}} < \infty; \qquad (3.4)$$

$$B^n := \sup_{0<r<\infty} \left(\int_{|x|<r} w_1(x)^{1-p_1'} dx \right)^{\frac{1}{p_1'}} \left(\int_{|y|>r} \left(\int_{|x|>\sqrt{2}|y|} \frac{w(x)}{|x|^{2np}} dx \right)^{\frac{p_1}{p}} \right.$$

$$\times \left. \left(\int_{|x|<|y|} w_2(x)^{1-p_2'} dx \right)^{\frac{p_1}{p_2'}} w_2(y)^{1-p_2'} dy \right)^{\frac{1}{p_1}} < \infty; \qquad (3.5)$$

$$C^n := \sup_{0<r<\infty} \left(\int_{|x|<r} w_2(x)^{1-p_2'} dx \right)^{\frac{1}{p_2'}} \left(\int_{|y|>r} \left(\int_{|x|>\sqrt{2}|y|} \frac{w(x)}{|x|^{2np}} dx \right)^{\frac{p_2}{p}} \right.$$

$$\times \left. \left(\int_{|x|<|y|} w_1(x)^{1-p_1'} dx \right)^{\frac{p_2}{p_1'}} w_1(y)^{1-p_1'} dy \right)^{\frac{1}{p_2}} < \infty. \qquad (3.6)$$

The results on bilinear Hardy operators $\mathcal{H}^2$ are as follows.

Theorem 3.1. *Let $1 \le p_1, p_2 < \infty$, $1 \le p < \infty$ and $1/p = 1/p_1 + 1/p_2$. Assume that w, w_1 and w_2 are nonnegative measurable functions on $\mathbb{R}^n$. If $w(\sqrt{2}x) \ge \lambda w(x)$ for $x \in \mathbb{R}^n$, $0 < \lambda < \infty$, then there exists a positive constant C such that the inequality*

$$\left(\int_{\mathbb{R}^n} (\mathcal{H}^2(f_1, f_2)(x))^p w(x) dx \right)^{\frac{1}{p}} \le C \prod_{j=1}^{2} \left(\int_{\mathbb{R}^n} f_j(x)^{p_j} w_j(x) \right)^{\frac{1}{p_j}} \qquad (3.7)$$

holds for all positive functions f_1 and f_2 if and only if $(w(\cdot), w_1(\cdot), w_2(\cdot)) \in M_{(p_1, p_2)}$.

Recently, Cañestroa, Salvadora and Torreblancab in [Cañestroa et al. (2012)] gave another version of n-dimensional bilinear Hardy operator $T(f_1, f_2)$ as follows, for $x \in \mathbb{R}^n \backslash \{0\}$,

$$T(f_1, f_2)(x) = \frac{1}{(\nu_n |x|^n)^2} \int_{|y_1|<|x|} \int_{|y_2|<|x|} f_1(y_1) f_2(y_2) dy_1 dy_2$$

$$= \mathcal{H}(f_1)(x) \mathcal{H}(f_2)(x),$$

where f_1 and f_2 are nonnegative measurable functions on $\mathbb{R}^n$. This form can be regarded as the extension of the following Hardy type operator H_2 (cf. [Bényi and Oh (2006)])

$$H_2(f_1, f_2)(x) = \frac{1}{x^2} \int_0^x \int_0^x f_1(s) f_2(t) ds dt,$$

where f_1 and f_2 are nonnegative functions and $x > 0$.

It is clear that

$$\frac{\nu_n^2}{2^n \nu_{2n}} T(f_1, f_2)(\frac{\sqrt{2}}{2}x) \leq \mathcal{H}^2(f_1, f_2)(x) \leq \frac{\nu_n^2}{\nu_{2n}} T(f_1, f_2)(x).$$

In [Cañestroa et al. (2012)], Cañestroa et al. obtained the following result.

Lemma 3.1 ([Cañestroa et al. (2012)]). *Let $1 \leq p_1, p_2 < \infty$, $1 \leq p < \infty$ and $1/p = 1/p_1 + 1/p_2$. Assume that w, w_1 and w_2 are nonnegative measurable functions on $\mathbb{R}^n$. Then there exists a positive constant C such that the inequality*

$$\left(\int_{\mathbb{R}^n} (T(f_1, f_2)(x))^p w(x) dx \right)^{1/p} \leq C \prod_{j=1}^{2} \left(\int_{\mathbb{R}^n} f_j(x)^{p_j} w_j(x) dx \right)^{1/p_j} \quad (3.8)$$

holds for all positive functions f_1 and f_2 if and only if $(w(\cdot), w_1(\cdot), w_2(\cdot)) \in M_{(p_1, p_2)}$.

The following result is essential to the proof of Theorem 3.1.

Lemma 3.2. *Let $1 \leq p_1, p_2 < \infty$, $1 \leq p < \infty$ and $1/p = 1/p_1 + 1/p_2$. Assume that w, w_1 and w_2 are nonnegative measurable functions on $\mathbb{R}^n$.*

(i) *If $(w(\sqrt{2}\cdot), w_1(\cdot), w_2(\cdot)) \in M_{(p_1, p_2)}$, then there exists a positive constant C such that the inequality*

$$\left(\int_{\mathbb{R}^n} (\mathcal{H}^2(f_1, f_2)(x))^p w(\sqrt{2}x) dx \right)^{1/p} \leq C \prod_{j=1}^{2} \left(\int_{\mathbb{R}^n} f_j(x)^{p_j} w_j(x) dx \right)^{1/p_j}$$

holds for all nonnegative functions f_1 and f_2.

(ii) *If there exists a positive constant C such that the inequality (3.7) holds for all nonnegative functions f_1 and f_2, then $(w(\sqrt{2}\cdot), w_1(\cdot), w_2(\cdot)) \in M_{(p_1, p_2)}$.*

Proof of Lemma 3.2. (i) For the sufficiency of the conditions, we only give the outline and omit the details. Substituting $x = \sqrt{2}z$ and letting $u(z) = w(\sqrt{2}z)$ for the first integral, (3.4) can be rewritten as

$$A^n := 2^{n(\frac{1}{2p} - 1)} \sup_{0 < r < \infty} \left(\int_{|z| > r} |z|^{-2np} u(z) dz \right)^{1/p}$$

$$\times \left(\int_{|x| < r} w_1(x)^{1 - p_1'} dx \right)^{1/p_1'} \left(\int_{|x| < r} w_2(x)^{1 - p_2'} dx \right)^{1/p_2'} < \infty.$$

Using the same method, we yield that

$$B^n := 2^{n(\frac{1}{2p}-1)} \sup_{0<r<\infty} \left(\int_{|x|<r} w_1(x)^{1-p_1'} dx \right)^{\frac{1}{p_1'}} \left(\int_{|y|>r} \right.$$

$$\left. \left(\int_{|z|>|y|} \frac{u(z)}{|z|^{2np}} dz \right)^{\frac{p_1}{p}} \left(\int_{|x|<|y|} w_2(x)^{1-p_2'} dx \right)^{\frac{p_1}{p'}} w_2(y)^{1-p_2'} dy \right)^{\frac{1}{p_1}} < \infty;$$

$$C^n := 2^{n(\frac{1}{2p}-1)} \sup_{0<r<\infty} \left(\int_{|x|<r} w_2(x)^{1-p_2'} dx \right)^{\frac{1}{p_2'}} \left(\int_{|y|>r} \right.$$

$$\left. \left(\int_{|z|>|y|} \frac{u(z)}{|z|^{2np}} dz \right)^{\frac{p_2}{p}} \left(\int_{|x|<|y|} w_1(x)^{1-p_1'} dx \right)^{\frac{p_2}{p'}} w_1(y)^{1-p_1'} dy \right)^{\frac{1}{p_2}} < \infty.$$

By Lemma 3.1, we obtain that (3.8) holds, that is

$$\left(\int_{\mathbb{R}^n} (T(f_1, f_2)(x))^p u(x) dx \right)^{1/p}$$

$$\leq C \prod_{j=1}^{2} \left(\int_{\mathbb{R}^n} f_j(x)^{p_j} w_j(x) dx \right)^{1/p_j}. \qquad (3.9)$$

We rewrite $u(x) = w(\sqrt{2}x)$ and obtain that the left-hand side of (3.9) is equal to

$$\left(\int_{\mathbb{R}^n} (T(f_1, f_2)(x))^p w(\sqrt{2}x) dx \right)^{1/p}.$$

Together with the observation that $\mathcal{H}^2(f_1, f_2)(x) \leq \frac{\nu_n^2}{\nu_{2n}} T(f_1, f_2)(x)$, we prove that

$$\left(\int_{\mathbb{R}^n} (\mathcal{H}^2(f_1, f_2)(x))^p w(\sqrt{2}x) dx \right)^{1/p} \leq C \prod_{j=1}^{2} \left(\int_{\mathbb{R}^n} f_j(x)^{p_j} w_j(x) dx \right)^{1/p_j}.$$

(ii) Proof of the necessity.

(3.7) $\Rightarrow$ (3.4) Assume that (3.7) holds. Take $f_j(x) = w_j(x)^{1-p_j'} \chi_{(0,r)}(|x|)$ for $0 < r < \infty$, and denote by $E = \{y_2 : |y_1|^2 + |y_2|^2 < |x|^2, |y_1| < r, |y_2| < r, |x| > \sqrt{2}r\}$. The conditions $|x| > \sqrt{2}r$ and $|y_1| < r$ imply that $|x|^2 - |y_1|^2 > r^2$. So we have $E = \{y_2 : |y_2| < r\}$. Then by $(1-p_j')p_j + 1 = 1 - p_j'$,

$$C \prod_{j=1}^{2} \left(\int_{\mathbb{R}^n} |f_j(x)|^{p_j} w_j(x) dx \right)^{1/p_j} = C \prod_{j=1}^{2} \left(\int_{|x|<r} w_j(x)^{1-p_j'} dx \right)^{1/p_j}$$

$$\geq \frac{1}{\nu_{2n}} \left(\int_{\mathbb{R}^n} \left| \frac{1}{|x|^{2n}} \int_{\substack{|y_1|^2+|y_2|^2<|x|^2 \\ |y_1|<r \\ |y_2|<r}} \prod_{j=1}^2 w_j(y_j)^{1-p_j'} dy_1 dy_2 \right|^p w(x) dx \right)^{1/p}$$

$$\geq \frac{1}{\nu_{2n}} \left(\int_{|x|\geq\sqrt{2}r} \left| \int_{\substack{|y_1|^2+|y_2|^2<|x|^2 \\ |y_1|<r \\ |y_2|<r}} \prod_{j=1}^2 w_j(y_j)^{1-p_j'} dy_1 dy_2 \right|^p |x|^{-2np} w(x) dx \right)^{1/p}$$

$$= \frac{1}{\nu_{2n}} \left(\int_{|x|\geq\sqrt{2}r} |x|^{-2np} w(x) dx \right)^{1/p} \prod_{j=1}^2 \int_{|x|<r} w_j(x)^{1-p_j'} dx.$$

Thus, (3.4) holds.

(3.7) $\Rightarrow$ (3.5) It follows from the condition $\frac{\nu_n^2}{2^n \nu_{2n}} T(f_1, f_2)(\frac{\sqrt{2}}{2} x) \leq \mathcal{H}^2(f_1, f_2)(x)$ that

$$\left(\int_{\mathbb{R}^n} \left(T(f_1, f_2) \left(\sqrt{2}x/2 \right) \right)^p w(x) dx \right)^{1/p}$$

$$\leq C \prod_{j=1}^2 \left(\int_{\mathbb{R}^n} f_j(x)^{p_j} w_j(x) dx \right)^{1/p_j}. \tag{3.10}$$

Letting $y = \sqrt{2}x/2$ with y, we obtain that the left-hand side of (3.10) is equal to

$$2^{\frac{n}{2p}} \left(\int_{\mathbb{R}^n} (T(f_1, f_2)(y))^p w(\sqrt{2}y) dy \right)^{1/p}.$$

Let $u(x) = w(\sqrt{2}x)$. Making use of Lemma 3.1, we have

$$\sup_{0<r<\infty} \left(\int_{|x|<r} w_1(x)^{1-p_1'} dx \right)^{1/p_1'} \left(\int_{|y|>r} \left(\int_{|x|>|y|} \frac{u(x) dx}{|x|^{-2np}} \right)^{p_1/p} \right.$$

$$\times \left(\int_{|x|<|y|} w_2(x)^{1-p_2'} dx \right)^{p_1/p'} w_2(y)^{1-p_2'} dy \right)^{1/p_1}$$

$$= \sup_{0<r<\infty} \left(\int_{|x|<r} w_1(x)^{1-p_1'} dx \right)^{1/p_1'} \left(\int_{|y|>r} \left(\int_{|x|>|y|} \frac{w(\sqrt{2}x) dx}{|x|^{2np}} \right)^{p_1/p} \right.$$

$$\times \left(\int_{|x|<|y|} w_2(x)^{1-p_2'} dx \right)^{p_1/p'} w_2(y)^{1-p_2'} dy \right)^{1/p_1}$$

$$= 2^{\frac{n(2p-1)}{2p}} \sup_{0<r<\infty} \left(\int_{|x|<r} w_1(x)^{1-p_1'} dx \right)^{1/p_1'}$$

$$\times \left(\int_{|y|>r} \left(\int_{|x|>\sqrt{2}|y|} \frac{w(x)dx}{|x|^{2np}} \right)^{p_1/p} \right.$$

$$\times \left. \left(\int_{|x|<|y|} w_2(x)^{1-p_2'} dx \right)^{p_1/p'} w_2(y)^{1-p_2'} dy \right)^{1/p_1}$$

$$= 2^{n(1-\frac{1}{2p})} B^n < \infty.$$

Using the same method, we can get that (3.6) holds by means of the hypothesis (3.7). $\qquad\square$

Proof of Theorem 3.1. Suppose that the inequality (3.7) holds. It follows from (ii) of Lemma 3.2 that $(w(\sqrt{2}\cdot), w_1(\cdot), w_2(\cdot)) \in M_{(p_1,p_2)}$. By means of the condition $w(\sqrt{2}x) \geq \lambda w(x)$, it is easy to check that $(w(\cdot), w_1(\cdot), w_2(\cdot)) \in M_{(p_1,p_2)}$. For the proof of sufficiency, assume that $(w(\cdot), w_1(\cdot), w_2(\cdot)) \in M_{(p_1,p_2)}$, then $(w(\sqrt{2}\cdot), w_1(\cdot), w_2(\cdot)) \in M_{(p_1,p_2)}$. It follows from (i) of Lemma 3.2 that

$$\left(\int_{\mathbb{R}^n} (\mathcal{H}^2(f_1, f_2)(x))^p w(\sqrt{2}x) dx \right)^{1/p} \leq C \prod_{j=1}^{2} \left(\int_{\mathbb{R}^n} f_j(x)^{p_j} w_j(x) dx \right)^{1/p_j}.$$

Thus the estimate for (3.7) can be easily obtained provided $w(\sqrt{2}x) \geq \lambda w(x)$. $\qquad\square$

We now give some further comments. From the proof of Theorem 3.1, it is easy to see that $(w(\cdot), w_1(\cdot), w_2(\cdot)) \in M_{(p_1,p_2)}$ is equivalent to $(w(\sqrt{2}\cdot), w_1(\cdot), w_2(\cdot)) \in M_{(p_1,p_2)}$ if w satisfies $w(\sqrt{2}x) \geq \lambda w(x)$. Next, we shall give two examples to show our result essentially cover power functions and exponential functions, respectively. The condition $w(\sqrt{2}x) = \lambda w(x)$ contains a class of functions, such as the power functions $|x|^\alpha$ with $\alpha < n(2p-1)$ and $x \in \mathbb{R}^n$. And the condition $w(\sqrt{2}x) > \lambda w(x)$ contains a class of functions, such as the exponential functions $e^{|x|}$ with $x \in \mathbb{R}^n$. It is still an interesting question whether the condition $w(\sqrt{2}x) \geq \lambda w(x)$ can be removed.

3.2 Weighted Knopp inequalities

The classical one-dimensional Knopp inequality [Hardy et al. (1952), Theorem 335, p. 250]

$$\int_0^\infty \exp\left(\frac{1}{x}\int_0^x \ln f(t)dt\right)dx \le e\int_0^\infty f(x)dx, \text{ for } f > 0 \qquad (3.11)$$

can be regarded as a continuous analogue of Carleman's inequality, where the discrete Carleman inequality [Hardy et al. (1952), Theorem 334, p. 249] is of the form

$$\sum_{k=1}^\infty (a_1 a_2 \cdots a_k)^{1/k} \le e\sum_{k=1}^\infty a_k, \text{ for all } a_k > 0, \qquad (3.12)$$

and the constant e in (3.11) and (3.12) is the best possible.

The Knopp inequality is closely connected with the classical Hardy inequality

$$\int_0^{+\infty}\left(\frac{1}{x}\int_0^x f(t)dt\right)^p dx \le \left(\frac{p}{p-1}\right)^p\int_0^\infty f^p(x)dx \qquad (3.13)$$

provided that $f > 0$ and $1 < p < \infty$. In fact, the one-dimensional Knopp inequality can be obtained when we replace f by the function $f^{1/p}$ in (3.13) and let p tend to infinity. From this observation, the Knopp inequality may be regarded as a limit case (as $p \to \infty$) of the Hardy inequality.

As we mentioned in Section 1.5, we let $x = (x_1, x_2) \in \mathbb{R}_+^2 := (0, \infty) \times (0, \infty)$ and $dx = dx_1 dx_2$. Pachpatte [Pachpatte (1992)] considered the two-dimensional Hardy operator given by

$$P^2(f)(x) = \frac{1}{x_1 x_2}\int_0^{x_1}\int_0^{x_2} f(y)dy, \quad f > 0$$

and proved that

$$\int_{\mathbb{R}_+^2}(P^2(f)(x))^p dx \le \left(\frac{p}{p-1}\right)^{2p}\int_{\mathbb{R}_+^2} f^p(x)dx \qquad (3.14)$$

for $1 < p < \infty$. Taking $p \to \infty$ and replacing f by $f^{1/p}$ in (3.14), Heinig, Kerman, and Krbec [Heinig et al. (2001)] characterized two positive weights u and v for the following inequality

$$\int_{\mathbb{R}_+^2} u(x)\exp\left(P^2(\log f)\right)(x)dx \le C\int_{\mathbb{R}_+^2} v(x)f(x)dx \qquad (3.15)$$

holds if and only if for any $\alpha_1, \alpha_2 > 0$,

$$\sup_{y_1>0,y_2>0} y_1^{\alpha_1}y_2^{\alpha_2}\int_{y_1}^\infty\int_{y_2}^\infty x_1^{-(1+\alpha_1)}x_2^{-(1+\alpha_2)}w(x)dx < \infty, \qquad (3.16)$$

where $w(x) = u(x) \exp(P^2(\log(1/v)))(x)$. The authors in [Heinig et al. (2001)] also pointed out that the higher-dimensional result carries over in the same way. Precisely, the results similar to (3.15) and (3.16) are also valid for the n-dimensional Hardy operator P^n defined as in (1.33).

Drábek, Heinig and Kufner [Drábek et al. (1995)] obtained the characterization for the n-dimensional Knopp inequality

$$\int_{\mathbb{R}^n} u(x) \exp \left(\mathcal{H}(\log f)\right)(x)dx \leq C \int_{\mathbb{R}^n} v(x)f(x)dx,$$

with two positive weights u and v on $\mathbb{R}^n$, which holds if and only if

$$\sup_{t>0} t^n \int_{|x|\geq t} \frac{w(x)}{|x|^{2n}} < \infty,$$

where $w(x) = u(x) \exp(\mathcal{H}(\log(1/v)))(x)$.

We shall use the following notation: $m \in \mathbb{Z}_+$, $n_k \in \mathbb{Z}_+$ for $1 \leq k \leq m$, $x = (x_1, \ldots, x_m) \in \mathbb{R}^{n_1} \times \cdots \times \mathbb{R}^{n_m}$, $y = (y_1, \ldots, y_m) \in \mathbb{R}^{n_1} \times \cdots \times \mathbb{R}^{n_m}$ and $dy = dy_m \cdots dy_1$. Lu, Yan and Zhao in [Lu et al. (2013)] gave the definition of the Hardy operator $\mathcal{H}_m$ on higher-dimensional product spaces, where

$$\mathcal{H}_m f(x) = \left(\prod_{k=1}^m \frac{1}{|B(\mathbf{0}, |x_k|)|} \right) \int_{B(\mathbf{0}, |x_1|)} \cdots \int_{B(\mathbf{0}, |x_m|)} f(y)dy, \ f > 0,$$

and $\prod_{i=1}^m |x_i| \neq 0$, see also Definition 1.2 in Section 1.5. It is easy to check that

$$\lim_{p \to \infty} \left(\mathcal{H}_m(f^{1/p})(x) \right)^p = \exp(\mathcal{H}_m(\log f))(x).$$

Inspired by the above work, it is natural to ask whether we can give a characterization for the weighted multi-dimensional Knopp type inequality on higher-dimensional product spaces. We formulate the results as follows.

Theorem 3.2. *Let f, U, V be positive measurable functions defined on $\mathbb{R}^{n_1} \times \cdots \times \mathbb{R}^{n_m}$. Define W by*

$$W(x) = U(x) \exp\left(\mathcal{H}_m(\log \frac{1}{V})\right)(x).$$

Then the inequality

$$\int_{\mathbb{R}^{n_1}} \cdots \int_{\mathbb{R}^{n_m}} U(x) \exp \left(\mathcal{H}_m (\log f)\right)(x)dx \leq C_1 \int_{\mathbb{R}^{n_1}} \cdots \int_{\mathbb{R}^{n_m}} V(x)f(x)dx \tag{3.17}$$

holds if and only if for any $\alpha_1, \ldots, \alpha_m > 0$,

$$A_1 := \sup_{t_k > 0, 1 \leq k \leq m} \prod_{k=1}^m t_k^{n_k \alpha_k} \int_{|x_1| \geq t_1} \cdots \int_{|x_m| \geq t_m} \frac{W(x)}{\prod_{j=1}^m |x_j|^{n_j(\alpha_j+1)}} dx < \infty. \tag{3.18}$$

Let $\alpha_k > 0$ for $1 \le k \le m$. We define the analogous adjoint Hardy type operator $\mathcal{H}_m^*$ as

$$\prod_{k=1}^{m} \left(\frac{\alpha_k}{n_k} |B(\mathbf{0}, |x_k|)|^{\frac{\alpha_k}{n_k}} \right)$$

$$\times \int_{\mathbb{R}^{n_1} \backslash B(\mathbf{0}, |x_1|)} \cdots \int_{\mathbb{R}^{n_m} \backslash B(\mathbf{0}, |x_m|)} \frac{f(y)}{\prod_{j=1}^{m} |B(\mathbf{0}, |y_j|)|^{(1+\frac{\alpha_j}{n_j})}} dy.$$

A similar characterization is given for the higher-dimensional Knopp inequality with the operator $\mathcal{H}_m^*$.

Theorem 3.3. *Let f, U, V be positive measurable functions defined on $\mathbb{R}^{n_1} \times \cdots \times \mathbb{R}^{n_m}$. Define W^* by*

$$W^*(x) = U(x) \exp \left(\mathcal{H}_m^* \left(\log \frac{1}{V} \right) \right) (x).$$

Then the inequality

$$\int_{\mathbb{R}^{n_1}} \cdots \int_{\mathbb{R}^{n_m}} U(x) \exp \left(\mathcal{H}_m^* \left(\log f \right) \right) (x) dx$$

$$\le C_2 \int_{\mathbb{R}^{n_1}} \cdots \int_{\mathbb{R}^{n_m}} V(x) f(x) dx \qquad (3.19)$$

holds if and only if for any $\alpha_1, \ldots, \alpha_m > 0$,

$$A_2 := \sup_{t_k > 0, 1 \le k \le m} \prod_{k=1}^{m} t_k^{-\alpha_k} \int_{|x_1| \le t_1} \cdots \int_{|x_m| \le t_m} \frac{W^*(x)}{\prod_{j=1}^{m} |x_j|^{n_j - \alpha_j}} dx < \infty. \qquad (3.20)$$

Especially, if $m = 1$ and $n_1 = n$, then $\mathcal{H}_1^*$ is the analogous adjoint Hardy type operator $\mathcal{H}^*$ given by

$$\mathcal{H}^* f(x) = \frac{\alpha}{n} |B(\mathbf{0}, |x|)|^{\frac{\alpha}{n}} \int_{\mathbb{R}^n \backslash B(\mathbf{0}, |x|)} \frac{f(y)}{|B(\mathbf{0}, |y|)|^{(1+\frac{\alpha}{n})}} dy, \quad \alpha > 0.$$

From Theorem 3.3, we have

Corollary 3.3. *Let f, U, V be positive measurable functions defined on $\mathbb{R}^n$. The inequality*

$$\int_{\mathbb{R}^n} U(x) \exp \left(\mathcal{H}^* (\log f) \right) (x) dx \le C_3 \int_{\mathbb{R}^n} V(x) f(x) dx$$

holds if and only if for any $\alpha > 0$,

$$A_3 := \sup_{t > 0} t^{-\alpha} \int_{|x| \le t} \frac{W^*(x)}{|x|^{n-\alpha}} dx < \infty,$$

where

$$W^*(x) = U(x) \exp \left(\mathcal{H}^* \left(\log \frac{1}{V} \right) \right) (x).$$

Proof of Theorem 3.2. For simplicity, we only provide the proof in the case $m = 2$. The case $m \geq 3$ presents only notational differences and does not require any new ideas.

We firstly assume that (3.18) holds and we shall prove (3.17). Writing $Vf = g$, we find that (3.17) is equivalent to

$$\int_{\mathbb{R}^{n_1+n_2}} W(x) \exp\left(\mathcal{H}_2(\log g)\right)(x) dx \leq c_1 \int_{\mathbb{R}^{n_1+n_2}} g(x) dx, \qquad (3.21)$$

where $x = (x_1, x_2) \in \mathbb{R}^{n_1} \times \mathbb{R}^{n_2}$, $dx = dx_2 dx_1$, and $\mathbb{R}^{n_1+n_2} = \mathbb{R}^{n_1} \times \mathbb{R}^{n_2}$. To shorten notation, write $B(r) = B(\mathbf{0}, r)$ for $r > 0$. By changing variables, we can write the left-hand side of (3.21) as

$$\int_{\mathbb{R}^{n_1+n_2}} W(x) \exp\left(\frac{1}{\nu_{n_1}\nu_{n_2}} \int_{B(1)} \int_{B(1)} \log g(|x_1|z_1|, |x_2|z_2|) dz_2 dz_1\right) dx.$$

Here and in what follows, $\nu_N = \frac{\pi^{N/2}}{\Gamma(1+N/2)}$ is the volume of the unit ball in $\mathbb{R}^N$ and $\omega_N = N\nu_N$ denotes the area of the unit sphere $\mathbb{S}^{N-1}$ for any positive integer $N \geq 2$.

An easy calculation shows that for any $\alpha_1, \alpha_2 > 0$,

$$\frac{1}{\nu_{n_1}\nu_{n_2}} \int_{B(1)} \int_{B(1)} \log(|z_1|^{n_1\alpha_1} |z_2|^{n_2\alpha_2}) dz_2 dz_1 = -(\alpha_1 + \alpha_2).$$

Hence the left-hand side of (3.21) becomes

$$\frac{e^{(\alpha_1+\alpha_2)}}{\nu_{n_1}\nu_{n_2}} \int_{\mathbb{R}^{n_1+n_2}} W(x)$$

$$\times \exp\left(\int_{B(1)} \int_{B(1)} \log(\prod_{i=1}^{2} |z_i|^{n_i\alpha_i} g(|x_1|z_1|, |x_2|z_2|)) dz_2 dz_1\right) dx$$

$$\leq \frac{e^{(\alpha_1+\alpha_2)}}{\nu_{n_1}\nu_{n_2}} \int_{\mathbb{R}^{n_1+n_2}} W(x)$$

$$\times \left(\int_{B(1)} \int_{B(1)} |z_1|^{n_1\alpha_1} |z_2|^{n_2\alpha_2} g(|x_1|z_1|, |x_2|z_2|) dz_2 dz_1\right) dx,$$

where the last estimate follows by Jensen's inequality. Furthermore, the last integral above can be written in the polar coordinates form:

$$\int_{\mathbb{S}^{n_1-1}} \int_{\mathbb{S}^{n_2-1}} \int_{\mathbb{S}^{n_1-1}} \int_{\mathbb{S}^{n_2-1}} \int_0^\infty \int_0^\infty t_1^{n_1-1} t_2^{n_2-1} W(t_1 x_1', t_2 x_2')$$

$$\times \int_0^1 \int_0^1 \prod_{i=1}^{2} s_i^{n_i\alpha_i+n_i-1} g(t_1 s_1 z_1', t_2 s_2 z_2') ds_2 ds_1 dt_2 dt_1$$

$$\times d\sigma(z_2') d\sigma(z_1') d\sigma(x_2') d\sigma(x_1'). \qquad (3.22)$$

By Fubini's theorem and changing variables $t_i s_i = r_i$, we see that

$$\int_0^\infty \int_0^\infty \int_0^1 \int_0^1 \prod_{i=1}^2 t_i^{n_i-1} s_i^{n_i\alpha_i+n_i-1} W(t_1 x_1', t_2 x_2') g(t_1 s_1 z_1', t_2 s_2 z_2')$$

$$\times\, ds_2 ds_1 dt_2 dt_1$$

$$= \int_0^\infty \int_0^\infty \int_0^1 \int_0^1 \prod_{i=1}^2 r_i^{n_i-1} s_i^{n_i\alpha_i-1} g(r_1 z_1', r_2 z_2') W\left(\frac{r_1}{s_1} x_1', \frac{r_2}{s_2} x_2'\right)$$

$$\times\, ds_2 ds_1 dr_2 dr_1$$

$$\left(\text{substituting } \frac{r_i}{s_i} = \tau_i,\ ds_i = -\frac{r_i}{\tau_i^2} d\tau_i\right)$$

$$= \int_0^\infty \int_0^\infty \int_{r_1}^\infty \int_{r_2}^\infty \prod_{i=1}^2 r_i^{n_i\alpha_i+n_i-1} \tau_i^{-n_i\alpha_i-1} g(r_1 z_1', r_2 z_2') W(\tau_1 x_1', \tau_2 x_2')$$

$$\times\, d\tau_2 d\tau_1 dr_2 dr_1.$$

Substituting into (3.22) shows that

$$\int_{\mathbb{R}^{n_1+n_2}} W(x) \exp\left(\frac{1}{\prod_{k=1}^2 |B(|x_k|)|} \int_{|y_1|\le|x_1|} \int_{|y_2|\le|x_2|} \log g(y_1, y_2) dy_2 dy_1\right) dx$$

multiplying $\frac{\nu_{n_1}\nu_{n_2}}{e^{(\alpha_1+\alpha_2)}}$ is not larger than

$$\int_{\mathbb{S}^{n_1-1}} \int_{\mathbb{S}^{n_2-1}} \int_{\mathbb{S}^{n_1-1}} \int_{\mathbb{S}^{n_2-1}} \int_0^\infty \int_0^\infty r_1^{n_1-1} r_2^{n_2-1} g(r_1 z_1', r_2 z_2') r_1^{n_1\alpha_1} r_2^{n_2\alpha_2}$$

$$\times \int_{r_1}^\infty \int_{r_2}^\infty \frac{\tau_1^{n_1-1}\tau_2^{n_2-1} W(\tau_1 x_1', \tau_2 x_2')}{\tau_1^{n_1(\alpha_1+1)}\tau_2^{n_2(\alpha_2+1)}} d\tau_2 d\tau_1 dr_2 dr_1$$

$$\times\, d\sigma(x_2')d\sigma(x_1')d\sigma(z_2')d\sigma(z_1')$$

$$= \int_{\mathbb{R}^{n_1+n_2}} g(x)$$

$$\times \left(|x_1|^{n_1\alpha_1}|x_2|^{n_2\alpha_2} \int_{|y_1|\ge|x_1|} \int_{|y_2|\ge|x_2|} \frac{W(y_1, y_2)}{|y_1|^{n_1(\alpha_1+1)}|y_2|^{n_2(\alpha_2+1)}} dy_2 dy_1\right) dx$$

$$\le A_1 \int_{\mathbb{R}^{n_1+n_2}} g(x) dx,$$

where in the last inequality we have used the condition (3.18).

Conversely, suppose that (3.21) holds, we shall show that (3.18) holds for $m = 2$. Fixing $t_1, t_2 > 0$, for $x = (x_1, x_2) \in \mathbb{R}^{n_1+n_2}$, we take the piecewise function

$$g(x) = \begin{cases} \dfrac{1}{t_1^{n_1} t_2^{n_2}} & |x_1| \le t_1, |x_2| \le t_2, \\[2mm] \dfrac{t_2^{n_2\alpha_2}}{t_1^{n_1} e^{n_2(1+\alpha_2)}|x_2|^{n_2(1+\alpha_2)}} & |x_1| \le t_1, |x_2| > t_2, \\[2mm] \dfrac{t_1^{n_1\alpha_1}}{t_2^{n_2} e^{n_1(1+\alpha_1)}|x_1|^{n_1(1+\alpha_1)}} & |x_1| > t_1, |x_2| \le t_2, \\[2mm] \dfrac{t_1^{n_1\alpha_1} t_2^{n_2\alpha_2}}{e^{n_1(1+\alpha_1)+n_2(1+\alpha_2)}|x_1|^{n_1(1+\alpha_1)}|x_2|^{n_2(1+\alpha_2)}} & |x_1| > t_1, |x_2| > t_2. \end{cases}$$

Obviously, g is radial and the right-hand side of (3.21) is

$$c_1 \int_{\mathbb{S}^{n_1-1}} \int_{\mathbb{S}^{n_2-1}} \int_0^\infty \int_0^\infty s_1^{n_1-1} s_2^{n_2-1} g(s_1, s_2) ds_2 ds_1 d\sigma(x_2') d\sigma(x_1')$$

$$= c_1 \omega_{n_1} \omega_{n_2} \left(\int_0^{t_1} \int_0^{t_2} + \int_0^{t_1} \int_{t_2}^\infty + \int_{t_1}^\infty \int_0^{t_2} + \int_{t_1}^\infty \int_{t_2}^\infty \right)$$

$$\times \prod_{i=1}^2 s_i^{n_i-1} g(s_1, s_2) ds_2 ds_1$$

$$= c_1 \nu_{n_1} \nu_{n_2} \left(1 + \frac{e^{-n_2(1+\alpha_2)}}{\alpha_2} + \frac{e^{-n_1(1+\alpha_1)}}{\alpha_1} + \frac{e^{-n_1(1+\alpha_1)-n_2(1+\alpha_2)}}{\alpha_1 \alpha_2} \right)$$

$$=: C_1(n_1, n_2, \alpha_1, \alpha_2).$$

Hence, the left-hand side of (3.21) satisfies

$$C_1(n_1, n_2, \alpha_1, \alpha_2)$$

$$\geq \int_{\mathbb{R}^{n_1+n_2}} W(x) \exp \left(\frac{1}{\prod_{k=1}^2 |B(|x_k|)|} \prod_{k=1}^2 \int_{|y_k| \leq |x_k|} \frac{dy_2 dy_1}{(\log g(y_1, y_2))^{-1}} \right) dx$$

$$= \int_{\mathbb{S}^{n_1-1}} \int_{\mathbb{S}^{n_2-1}} \int_0^\infty \int_0^\infty s_1^{n_1-1} s_2^{n_2-1} W(s_1 x_1', s_2 x_2')$$

$$\times \exp \left(\frac{n_1 n_2}{s_1^{n_1} s_2^{n_2}} \int_0^{s_1} \int_0^{s_2} r_1^{n_1-1} r_2^{n_2-1} \frac{dr_2 dr_1}{(\log g(r_1, r_2))^{-1}} \right)$$

$$\times ds_2 ds_1 d\sigma(x_2') d\sigma(x_1')$$

$$= \int_{\mathbb{S}^{n_1-1}} \int_{\mathbb{S}^{n_2-1}} \sum_{j=1}^4 I_j(x_1', x_2') d\sigma(x_2') d\sigma(x_1'),$$

where

$$I_1(x_1', x_2') = \int_0^{t_1} \int_0^{t_2} \prod_{j=1}^2 s_j^{n_j-1} W(s_1 x_1', s_2 x_2') E(s_1, s_2) ds_2 ds_1,$$

$$I_2(x_1', x_2') = \int_0^{t_1} \int_{t_2}^\infty \prod_{j=1}^2 s_j^{n_j-1} W(s_1 x_1', s_2 x_2') E(s_1, s_2) ds_2 ds_1,$$

$$I_3(x_1', x_2') = \int_{t_1}^\infty \int_0^{t_2} \prod_{j=1}^2 s_j^{n_j-1} W(s_1 x_1', s_2 x_2') E(s_1, s_2) ds_2 ds_1,$$

$$I_4(x_1', x_2') = \int_{t_1}^\infty \int_{t_2}^\infty \prod_{j=1}^2 s_j^{n_j-1} W(s_1 x_1', s_2 x_2') E(s_1, s_2) ds_2 ds_1$$

and

$$E(s_1, s_2) = \exp\left(\frac{n_1 n_2}{s_1^{n_1} s_2^{n_2}} \int_0^{s_1} \int_0^{s_2} \prod_{i=1}^{2} r_i^{n_i - 1} \log g(r_1, r_2) dr_2 dr_1\right).$$

Since each integral $\int_{\mathbb{S}^{n_1-1}} \int_{\mathbb{S}^{n_2-1}} I_j(x_1', x_2') d\sigma(x_2') d\sigma(x_1')$ is positive, we have

$$C_1(n_1, n_2, \alpha_1, \alpha_2) \geq \int_{\mathbb{S}^{n_1-1}} \int_{\mathbb{S}^{n_2-1}} I_4(x_1', x_2') d\sigma(x_2') d\sigma(x_1').$$

It is enough to estimate the term

$$\int_{\mathbb{S}^{n_1-1}} \int_{\mathbb{S}^{n_2-1}} I_4(x_1', x_2') d\sigma(x_2') d\sigma(x_1').$$

Split the integral $\int_0^{s_1} \int_0^{s_2}$ into four parts,

$$\int_0^{s_1} \int_0^{s_2} = \int_0^{t_1} \int_0^{t_2} + \int_0^{t_1} \int_{t_2}^{s_2} + \int_{t_1}^{s_1} \int_0^{t_2} + \int_{t_1}^{s_1} \int_{t_2}^{s_2}$$

and write

$$E_1(s_1, s_2) = \exp\left(\prod_{k=1}^{2} \frac{n_k}{s_k^{n_k}} \int_0^{t_1} \int_0^{t_2} \prod_{i=1}^{2} r_i^{n_i - 1} \log(t_1^{-n_1} t_2^{-n_2}) dr_2 dr_1\right),$$

$$E_2(s_1, s_2) =$$
$$\exp\left(\prod_{k=1}^{2} \frac{n_k}{s_k^{n_k}} \int_0^{t_1} \int_{t_2}^{s_2} \prod_{i=1}^{2} \frac{\log(t_1^{-n_1} e^{-n_2(1+\alpha_2)} r_2^{-n_2(1+\alpha_2)} t_2^{n_2\alpha_2}) dr_2 dr_1}{r_i^{1-n_i}}\right),$$

$$E_3(s_1, s_2) =$$
$$\exp\left(\prod_{k=1}^{2} \frac{n_k}{s_k^{n_k}} \int_{t_1}^{s_1} \int_0^{t_2} \prod_{i=1}^{2} \frac{\log(t_2^{-n_2} e^{-n_1(1+\alpha_1)} r_1^{-n_1(1+\alpha_1)} t_1^{n_1\alpha_1}) dr_2 dr_1}{r_i^{1-n_i}}\right),$$

$$E_4(s_1, s_2) =$$
$$\exp\left(\prod_{k=1}^{2} \frac{n_k}{s_k^{n_k}} \int_{t_1}^{s_1} \int_{t_2}^{s_2} \prod_{i=1}^{2} \frac{\log\left(\prod_{i=1}^{2} e^{-n_i(1+\alpha_i)} r_i^{-n_i(1+\alpha_i)} t_i^{n_i\alpha_i}\right) dr_2 dr_1}{r_i^{1-n_i}}\right).$$

We have

$$I_4(x_1', x_2') = \int_{t_1}^{\infty} \int_{t_2}^{\infty} s_1^{n_1 - 1} s_2^{n_2 - 1} W(s_1 x_1', s_2 x_2') \prod_{k=1}^{4} E_k(s_1, s_2) ds_2 ds_1.$$

It is easy to obtain

$$E_1(s_1, s_2) = \exp\left(\frac{t_1^{n_1} t_2^{n_2}}{s_1^{n_1} s_2^{n_2}} (-n_1 \log t_1 - n_2 \log t_2)\right).$$

An elementary calculation shows that

$$E_2(s_1, s_2) = \exp\left(\frac{t_1^{n_1} t_2^{n_2}}{s_1^{n_1} s_2^{n_2}}\big(n_1 \log t_1 + n_2 \log t_2 + (n_2 - 1)(1 + \alpha_2)\big)\right.$$

$$+ \left(\frac{t_1}{s_1}\right)^{n_1}\left(-n_1 \log t_1 + n_2\alpha_2 \log t_2 - n_2(\alpha_2 + 1)\log s_2 + \frac{(1 - n_2)}{(1 + \alpha_2)^{-1}}\right)\bigg),$$

$$E_3(s_1, s_2) = \exp\left(\frac{t_1^{n_1} t_2^{n_2}}{s_1^{n_1} s_2^{n_2}}\big(n_2 \log t_2 + n_1 \log t_1 + \frac{(n_1 - 1)}{(1 + \alpha_1)^{-1}}\big)\right.$$

$$+ \left(\frac{t_2}{s_2}\right)^{n_2}\left(-n_2 \log t_2 + n_1\alpha_1 \log t_1 - n_1(\alpha_1 + 1)\log s_1 + \frac{(1 - n_1)}{(1 + \alpha_1)^{-1}}\right)\bigg),$$

$$E_4(s_1, s_2) = \exp\left(\sum_{i=1}^{2}\big(n_i\alpha_i \log t_i - n_i(\alpha_i + 1)\log s_i + (1 - n_i)(1 + \alpha_i)\big)\right.$$

$$+ \frac{t_1^{n_1} t_2^{n_2}}{s_1^{n_1} s_2^{n_2}}\Big(\sum_{i=1}^{2}(1 - n_i)(1 + \alpha_i) - n_i \log t_i\Big)$$

$$+ \left(\frac{t_1}{s_1}\right)^{n_1}\Big(\sum_{i=1}^{2}\frac{(n_i - 1)}{(1 + \alpha_i)^{-1}} + n_1 \log t_1 + n_2(\alpha_2 + 1)\log s_2 - n_2\alpha_2 \log t_2\Big)$$

$$+ \left(\frac{t_2}{s_2}\right)^{n_2}\Big(\sum_{i=1}^{2}\frac{(n_i - 1)}{(1 + \alpha_i)^{-1}} + n_2 \log t_2 + n_1(\alpha_1 + 1)\log s_1 - n_1\alpha_1 \log t_1\Big)\bigg).$$

From these estimates we obtain that

$$\prod_{k=1}^{4} E_k(s_1, s_2) = \frac{t_1^{n_1\alpha_1} t_2^{n_2\alpha_2}}{s_1^{n_1(\alpha_1+1)} s_2^{n_2(\alpha_2+1)}} e^{\sum_{i=1}^{2}(1-n_i)(1+\alpha_i)}$$

$$\times \exp\left(\left(\frac{t_1}{s_1}\right)^{n_1}(n_1 - 1)(1 + \alpha_1) + \left(\frac{t_2}{s_2}\right)^{n_2}(n_2 - 1)(1 + \alpha_2)\right).$$

Clearly, we have

$$\int_{\mathbb{S}^{n_1-1}} \int_{\mathbb{S}^{n_2-1}} I_4(x_1', x_2')d\sigma(x_2')d\sigma(x_1') \geq e^{\sum_{i=1}^{2}(1-n_i)(1+\alpha_i)} t_1^{n_1\alpha_1} t_2^{n_2\alpha_2}$$

$$\times \int_{|x_1|\geq t_1} \int_{|x_2|\geq t_2} \frac{W(x_1, x_2)}{|x_1|^{n_1(\alpha_1+1)}|x_2|^{n_2(\alpha_2+1)}} dx_2 dx_1$$

and this implies (3.18) by taking the supremum over all $t_1, t_2 > 0$. $\qquad\square$

Proof of Theorem 3.3. Theorem 3.3 can be proved in a similar way as that of Theorem 3.2. Hence, we only point out some differences of the corresponding relations. It is easy to conclude that (3.19) is equivalent to

$$\int_{\mathbb{R}^{n_1+n_2}} W^*(x_1, x_2)\exp\big(\mathcal{H}_2^*(\log g)\big)(x_1, x_2)dx \leq c_2 \int_{\mathbb{R}^{n_1+n_2}} g(x_1, x_2)dx. \tag{3.23}$$

Assume that (3.20) holds. For any $\alpha_1, \alpha_2 > 0$,

$$\frac{\alpha_1 \alpha_2}{\omega_{n_1} \omega_{n_2}} \int_{|z_1| \geq 1} \int_{|z_2| \geq 1} \frac{\log(|z_1|^{n_1}|z_2|^{n_2})}{|z_1|^{n_1+\alpha_1}|z_2|^{n_2+\alpha_2}} dz_2 dz_1 = \frac{n_1}{\alpha_1} + \frac{n_2}{\alpha_2},$$

then by using Jensen's inequality, the integral becomes

$$\int_{\mathbb{R}^{n_1+n_2}} W^*(x_1, x_2) \exp\left(\mathcal{H}_2^*(\log g)\right)(x_1, x_2) dx$$

$$= e^{-\frac{n_1}{\alpha_1} - \frac{n_2}{\alpha_2}} \int_{\mathbb{R}^{n_1+n_2}} W^*(x_1, x_2)$$

$$\times \exp\left(\frac{\int_{|z_1| \geq 1} \int_{|z_2| \geq 1} \frac{\log(|z_1|^{n_1}|z_2|^{n_2} g(|x_1|z_1|,|x_2|z_2|))}{|z_1|^{n_1+\alpha_1}|z_2|^{n_2+\alpha_2}} dz_2 dz_1}{\frac{\omega_{n_1-1}\omega_{n_2-1}}{\alpha_1 \alpha_2}}\right) dx$$

$$\leq e^{-\frac{n_1}{\alpha_1} - \frac{n_2}{\alpha_2}} \frac{\alpha_1 \alpha_2}{\omega_{n_1-1}\omega_{n_2-1}} \int_{\mathbb{R}^{n_1+n_2}} W^*(x_1, x_2)$$

$$\times \int_{|z_1| \geq 1} \int_{|z_2| \geq 1} \frac{g(|x_1|z_1|, |x_2|z_2|)}{|z_1|^{\alpha_1}|z_2|^{\alpha_2}} dz_2 dz_1 dx.$$

By a similar argument as that of Theorem 3.2, the last integral above is expressed by the polar coordinates form:

$$\int_{\mathbb{S}^{n_1-1}} \int_{\mathbb{S}^{n_2-1}} \int_{\mathbb{S}^{n_1-1}} \int_{\mathbb{S}^{n_2-1}} \int_0^\infty \int_0^\infty t_1^{n_1-1} t_2^{n_2-1} W^*(t_1 x_1', t_2 x_2')$$

$$\times \int_1^\infty \int_1^\infty \prod_{i=1}^2 s_i^{n_i-1-\alpha_i}$$

$$\times g(t_1 s_1 z_1', t_2 s_2 z_2') ds_2 ds_1 dt_2 dt_1 d\sigma(z_2') d\sigma(z_1') d\sigma(x_2') d\sigma(x_1'). \qquad (3.24)$$

Interchanging the order of integration and changing variables $t_i s_i = r_i$, one has

$$\int_0^\infty \int_0^\infty \int_1^\infty \int_1^\infty \prod_{i=1}^2 t_i^{n_i-1} s_i^{n_i-1-\alpha_i}$$

$$\times W^*(t_1 x_1', t_2 x_2') g(t_1 s_1 z_1', t_2 s_2 z_2') ds_2 ds_1 dt_2 dt_1$$

$$= \int_0^\infty \int_0^\infty \int_1^\infty \int_1^\infty \prod_{i=1}^2 r_i^{n_i-1} s_i^{-1-\alpha_i} g(r_1 z_1', r_2 z_2')$$

$$\times W^*\left(\frac{r_1}{s_1} x_1', \frac{r_2}{s_2} x_2'\right) ds_2 ds_1 dr_2 dr_1$$

$$\left(\text{substituting } \frac{r_i}{s_i} = \tau_i, \ ds_i = -\frac{r_i}{\tau_i^2} d\tau_i\right)$$

$$= \int_0^\infty \int_0^\infty \int_0^{r_1} \int_0^{r_2} \prod_{i=1}^2 r_i^{n_i-1-\alpha_i} \tau_i^{\alpha_i-1} g(r_1 z_1', r_2 z_2')$$

$$\times W^*(\tau_1 x_1', \tau_2 x_2') d\tau_2 d\tau_1 dr_2 dr_1.$$

Substituting into (3.24) implies that

$$\int_{\mathbb{R}^{n_1+n_2}} W^*(x_1, x_2) \exp\big(\mathcal{H}_2^*(\log g)\big)(x_1, x_2) dx$$

$$\leq e^{-\frac{n_1}{\alpha_1} - \frac{n_2}{\alpha_2}} \frac{\alpha_1 \alpha_2}{\omega_{n_1} \omega_{n_2}} \int_{\mathbb{S}^{n_1-1}} \int_{\mathbb{S}^{n_2-1}} \int_{\mathbb{S}^{n_1-1}} \int_{\mathbb{S}^{n_2-1}} \int_0^\infty \int_0^\infty r_1^{n_1-1} r_2^{n_2-1}$$

$$\times g(r_1 z_1', r_2 z_2') \prod_{k=1}^2 r_k^{-\alpha_k} \int_0^{r_1} \int_0^{r_2} \frac{\prod_{i=1}^2 \tau_i^{n_i-1} W^*(\tau_1 x_1', \tau_2 x_2')}{\tau_1^{n_1-\alpha_1} \tau_2^{n_2-\alpha_2}}$$

$$d\tau_2 d\tau_1 dr_2 dr_1 d\sigma(z_2') d\sigma(z_1') d\sigma(x_2') d\sigma(x_1')$$

$$= e^{-\frac{n_1}{\alpha_1} - \frac{n_2}{\alpha_2}} \frac{\alpha_1 \alpha_2}{\omega_{n_1} \omega_{n_2}}$$

$$\times \int_{\mathbb{R}^{n_1+n_2}} g(x) \left(\frac{1}{|x_1|^{\alpha_1} |x_2|^{\alpha_2}} \int_{|y_1| \leq |x_1|} \int_{|y_2| \leq |x_2|} \frac{W^*(y_1, y_2) dy_2 dy_1}{|y_1|^{n_1-\alpha_1} |y_2|^{n_2-\alpha_2}} \right) dx$$

$$\leq e^{-\frac{n_1}{\alpha_1} - \frac{n_2}{\alpha_2}} \frac{\alpha_1 \alpha_2}{\omega_{n_1} \omega_{n_2}} A_2 \int_{\mathbb{R}^{n_1+n_2}} g(x) dx,$$

which conclude that (3.23) holds under the assumption of $A_2 < \infty$.

On the other hand, we take the radial piecewise function

$$g(x) = \begin{cases} e^{-2} t_1^{-\alpha_1} t_2^{-\alpha_2} |x_1|^{\alpha_1-n_1} |x_2|^{\alpha_2-n_2} & |x_1| \leq t_1, |x_2| \leq t_2, \\ e^{-1} t_1^{\alpha_1} t_2^{-\alpha_2} |x_1|^{-\alpha_1-n_1} |x_2|^{\alpha_2-n_2} & |x_1| > t_1, |x_2| \leq t_2, \\ e^{-1} t_1^{-\alpha_1} t_2^{\alpha_2} |x_1|^{\alpha_1-n_1} |x_2|^{-\alpha_2-n_2} & |x_1| \leq t_1, |x_2| > t_2, \\ t_1^{\alpha_1} t_2^{\alpha_2} |x_1|^{-\alpha_1-n_1} |x_2|^{-\alpha_2-n_2} & |x_1| > t_1, |x_2| > t_2 \end{cases}$$

with $t_1, t_2 > 0$ and $x = (x_1, x_2) \in \mathbb{R}^{n_1+n_2}$.

Obviously, the right-hand side of (3.23) is

$$c_2 \int_{\mathbb{S}^{n_1-1}} \int_{\mathbb{S}^{n_2-1}} \int_0^\infty \int_0^\infty s_1^{n_1-1} s_2^{n_2-1} g(s_1, s_2) ds_2 ds_1 d\sigma(x_2') d\sigma(x_1')$$

$$= \frac{c_2 \omega_{n_1} \omega_{n_2}}{\alpha_1 \alpha_2} (e^{-2} + 2e^{-1} + 1)$$

$$=: C_2(n_1, n_2, \alpha_1, \alpha_2).$$

Similar arguments as that of Theorem 3.2, we have

$$C_2(n_1, n_2, \alpha_1, \alpha_2) \geq \int_{\mathbb{S}^{n_1-1}} \int_{\mathbb{S}^{n_2-1}} \mathbf{J}_1(s_1, s_2) d\sigma(x_2') d\sigma(x_1'),$$

where $\mathbf{J}_1(s_1, s_2)$ is

$$\int_0^{t_1} \int_0^{t_2} s_1^{n_1-1} s_2^{n_2-1} W^*(s_1 x_1', s_2 x_2') \exp\big(\alpha_1 \alpha_2 s_1^{\alpha_1} s_2^{\alpha_2}$$

$$\left(\int_{s_1}^{t_1}\int_{s_2}^{t_2} + \int_{t_1}^{\infty}\int_{s_2}^{t_2} + \int_{s_1}^{t_1}\int_{t_2}^{\infty} + \int_{t_1}^{\infty}\int_{t_2}^{\infty}\right)\frac{\log g(r_1,r_2)dr_2dr_1}{r_1^{1+\alpha_1}r_2^{1+\alpha_2}}\right)ds_2ds_1$$

$$=\int_0^{t_1}\int_0^{t_2} s_1^{n_1-1}s_2^{n_2-1}W^*(s_1x_1',s_2x_2')\prod_{k=1}^4 F_k(s_1,s_2)ds_2ds_1.$$

By the tedious calculations without any technique, we obtain that

$$\prod_{k=1}^4 F_k(s_1,s_2) = e^{-2-\frac{n_1}{\alpha_1}-\frac{n_2}{\alpha_2}}\frac{t_1^{-\alpha_1}t_2^{-\alpha_2}}{s_1^{n_1-\alpha_1}s_2^{n_2-\alpha_2}}\exp\left(2 - \left(\frac{t_1}{s_1}\right)^{\alpha_1} - \left(\frac{t_2}{s_2}\right)^{\alpha_2}\right)$$

holds and then we have

$$C_2(n_1,n_2,\alpha_1,\alpha_2)$$

$$\geq \int_{\mathbb{S}^{n_1-1}}\int_{\mathbb{S}^{n_2-1}} \mathbf{J}_1(s_1,s_2)d\sigma(x_2')d\sigma(x_1')$$

$$\geq e^{-2-\frac{n_1}{\alpha_1}-\frac{n_2}{\alpha_2}}t_1^{-\alpha_1}t_2^{-\alpha_2}\int_{|x_1|\leq t_1}\int_{|x_2|\leq t_2}\frac{W^*(x_1,x_2)}{|x_1|^{n_1-\alpha_1}|x_2|^{n_2-\alpha_2}}dx_2dx_1.$$

Thus, we finish the proof by taking the supremum over all $t_1,t_2 > 0$.   □

3.3   Notes

The characterization of the weights w, v for the n-dimensional Hardy inequality holds was firstly obtained by Drábek et al. in [Drábek et al. (1995)]. The properties of M_p (resp. M^p) weights was systematically studied by Fu, Lu and Zhao [Zhao et al. (2014)]. The weight characterization of bilinear Hardy inequality were obtained by Cañestroa, Salvadora and Torreblancab in [Cañestroa et al. (2012)] and by Fu, Lu and Zhao [Zhao et al. (2014)] in different forms. Theorem 3.1 was due to Fu, Lu and Zhao [Zhao et al. (2014)]. The characterization for the n-dimensional Knopp inequality was obtained by Drábek, Heinig and Kufner [Drábek et al. (1995)]. Theorems 3.2 and 3.3 were obtained by Zhao and Ma [Zhao and Ma (2019)].

Chapter 4

Commutators of Hardy operators

In recent years, several authors have extended and studied Campanato spaces and Morrey spaces, and have considered the action of various operators on them. The excellent structures of the spaces $\mathcal{E}^{\alpha,p}(\mathbb{R}^n)$ and $L^{p,\lambda}(\mathbb{R}^n)$ render them useful in the study of the regularity theory of PDEs and the Sobolev embedding theorems, see for example [Adams and Xiao (2012a,b); Deng et al. (2005); Duong et al. (2007); Fan et al. (1998); Lu G. (1995); Yang et al. (2010); Yuan et al. (2010)]. In this chapter, we pay our attention mainly on the central version of these spaces.

Assume that $1 \le p < \infty$ and $-1/p \le \lambda < 1/n$. The central Campanato space can be defined by

$$\dot{\mathcal{E}}^{p,\lambda}(\mathbb{R}^n) = \left\{ f : \|f\|_{\dot{\mathcal{E}}^{p,\lambda}(\mathbb{R}^n)} < \infty \right\}$$

with

$$\|f\|_{\dot{\mathcal{E}}^{p,\lambda}(\mathbb{R}^n)} =: \sup_{r>0} \frac{1}{|B(\mathbf{0},r)|^\lambda} \left(\frac{1}{|B(\mathbf{0},r)|} \int_{B(\mathbf{0},r)} |f(x) - f_{B(\mathbf{0},r)}|^p dx \right)^{1/p}.$$

It is easy to check that $\mathcal{E}^{n\lambda,p}(\mathbb{R}^n) \subseteq \dot{\mathcal{E}}^{p,\lambda}(\mathbb{R}^n)$. If $\lambda = 0$, then $\dot{\mathcal{E}}^{p,0}(\mathbb{R}^n) = C\dot{M}O^p(\mathbb{R}^n)$. More information about the space $C\dot{M}O(\mathbb{R}^n)$ can be seen in [Chen and Lau (1989)]. When $0 < \lambda < 1/p$, $\dot{\mathcal{E}}^{p,\lambda}(\mathbb{R}^n)$ is the λ-central bounded mean oscillation space $C\dot{M}O^{p,\lambda}(\mathbb{R}^n)$. For the case $-1/p \le \lambda < 0$, $\dot{\mathcal{E}}^{p,\lambda}(\mathbb{R}^n) \supset \dot{B}^{p,\lambda}(\mathbb{R}^n)$. Let $\Omega \subset \mathbb{R}^n$ be an open domain. Denote by $diam\,\Omega$ the diameter of Ω and $\Omega_r = B(\mathbf{0},r) \cap \Omega$. If $diam\,\Omega < \infty$ and there is a positive constant A such that for $\mathbf{0} \in \Omega$ and $r \in (0, diam\,\Omega)$,

$$|\Omega_r| \ge Ar^n, \tag{4.1}$$

then a further use of the argument in [Campanato (1963)] shows that

$$\dot{B}^{p,\lambda}(\Omega) = \dot{\mathcal{E}}^{p,\lambda}(\Omega) \text{ for } -1/p \le \lambda < 0. \tag{4.2}$$

The domain Ω is said to satisfy A-condition if (4.1) holds.

For a locally integrable function b and an operator T, the commutator formed by T and b can be defined by

$$[b, T]f := b(Tf) - T(bf).$$

In the literature, b is also called the symbol function of $[b, T]$. Commutator theory, especially the boundedness of $[b, T]$ on different function spaces is now being applied to a variety of subjects, such as the regularity of solutions to elliptic equations ([Bramanti and Cerutti (1993); Chiarenza et al. (1993)]), and the characterization of function spaces ([Janson (1978); Paluszyński (1995); Shi and Lu (2013, 2014, 2015)]). There are many rich and significant results about the characterizations of $\mathcal{E}^{\alpha,p}(\mathbb{R}^n)$ via the boundedness of $[b, T]$ when T is the Calderón-Zygmund singular operator or fractional integral, see for example [Chanillo (1982); Ding (1997); Janson (1978); Paluszyński (1995); Shi and Lu (2013, 2014)].

In this chapter, we write the fractional Hardy operator in the form: let f be a locally integrable function f on $\mathbb{R}^n$, for $0 \le \beta < n$

$$\mathcal{H}_\beta f(x) = \frac{1}{|x|^{1-\frac{\beta}{n}}} \int_{|y|<|x|} f(y)dy, \quad x \in \mathbb{R}^n\backslash\{\mathbf{0}\} \tag{4.3}$$

and skip the ratio $\nu_n^{1-\beta/n}$ between (4.3) and (1.4). We also write $\mathcal{H}_0 = \mathcal{H}$.

The topic of this chapter is to study the spaces $\dot{\mathcal{E}}^{p,\lambda}(\mathbb{R}^n)$ and give these spaces some different characterizations via the boundedness and compactness of commutators formed by Hardy type operators. Let f and b be locally integrable functions on $\mathbb{R}^n$. The commutators of $\mathcal{H}$ and $\mathcal{H}^*$ are defined by

$$[b, \mathcal{H}]f = b(\mathcal{H}f) - \mathcal{H}(fb) \quad \text{and} \quad [b, \mathcal{H}^*]f = b(\mathcal{H}^*f) - \mathcal{H}^*(fb),$$

respectively. Some boundedness of $[b, \mathcal{H}]$ and $[b, \mathcal{H}^*]$ can be founded in [Komori (2003); Long and Wang (2002)]. As we shall see, the function spaces which are characterized by the boundedness of Hardy type operators are different from that of the Calderón-Zygmund singular integral operators even under the same conditions. Here and subsequently, for simplicity, we write $B_k = B(\mathbf{0}, 2^k) = \{x \in \mathbb{R}^n : |x| \le 2^k\}$, $C_k = B_k \setminus B_{k-1}$ for $k \in \mathbb{Z}$.

4.1 The boundedness characterizations

Since the Hardy operators are centrosymmetric, the function spaces characterized by the boundedness of $[b, \mathcal{H}]$ and $[b, \mathcal{H}^*]$ are central ones.

4.1.1 The characterization for $\lambda = 0$

Let $\varphi \in \mathcal{S}$ such that supp $\varphi \subset B(0,1)$, $\int_{B(0,1)} \varphi(x)dx \neq 0$ and $\varphi_t(x) = t^{-n}\varphi(x/t)$. Let f^+ be defined by

$$f^+(x) := \sup_{t>0} |f * \varphi_t(x)|.$$

The Hardy space $H\dot{A}_{q,1}(\mathbb{R}^n)$ is defined by

$$H\dot{A}_{q,1}(\mathbb{R}^n) := \{f \in L^1(\mathbb{R}^n) : f^+ \in \dot{K}_q^{n(1-1/q),1}(\mathbb{R}^n)\},$$

and $\|f\|_{H\dot{A}_{q,1}(\mathbb{R}^n)} = \|f^+\|_{\dot{K}_q^{n(1-1/q),1}(\mathbb{R}^n)}$.

Following Lu and Yang [Lu and Yang (1995c)], García-Cuerva [García-Cuerva (1989)], Chen and Lau [Chen and Lau (1989)], a function a is a central $H\dot{A}_{q,1}(\mathbb{R}^n)$-atom if there exists a ball $B(0,r)$ such that a satisfies the following conditions:

$$\text{supp } a \subset B(0,r), \quad \|a\|_{L^q(\mathbb{R}^n)} \leq |B(0,r)|^{1/q-1}, \quad \int_{\mathbb{R}^n} a(x)dx = 0.$$

To state the main result, some auxiliaries about the $H\dot{A}_{q,1}(\mathbb{R}^n)$-atom are needed. Firstly, the following three properties were obtained by García-Cuerva in [García-Cuerva (1989)] and Li, Yang in [Li and Yang (1996)].

Lemma 4.1. *If a function a is a central $H\dot{A}_{q,1}(\mathbb{R}^n)$-atom, then*

$$\|a\|_{H\dot{A}_{q,1}(\mathbb{R}^n)} \leq C_{n,q},$$

where $C_{n,q}$ is a positive constant depending only on n and q.

Lemma 4.2. *If $f \in H\dot{A}_{q,1}(\mathbb{R}^n)$, then f can be represented as*

$$f = \sum_{k=-\infty}^{\infty} \lambda_k a_k,$$

where a_k is a central $H\dot{A}_{q,1}(\mathbb{R}^n)$-atom supported on $B(0,2^k)$,

$$\sum_{k=-\infty}^{\infty} |\lambda_k| < \infty$$

and

$$\sum_{k=-\infty}^{\infty} |\lambda_k| \approx \|f\|_{H\dot{A}_{q,1}(\mathbb{R}^n)}.$$

Lemma 4.3. *The dual space of $H\dot{A}_{q,1}(\mathbb{R}^n)$ is $C\dot{M}O^{q'}(\mathbb{R}^n)$, where $1/q + 1/q' = 1$ for $1 < q < \infty$.*

Lemma 4.4. *The dual space of* $\dot{K}_q^{\alpha,p}(\mathbb{R}^n)$ *is* $\dot{K}_{q'}^{-\alpha,p'}(\mathbb{R}^n)$, *where* $1/p + 1/p' = 1$ *and* $1/q + 1/q' = 1$ *for* $1 < p, q < \infty$.

Lemma 4.5. *Let* b *be a central* $H\dot{A}_{q,1}(\mathbb{R}^n)$-*atom supported on* $B(0, 2^k)$, *and set* $h(x) = |x|^{-\beta}\chi_{B(0,3\cdot2^k)\setminus B(0,2^{k+1})}(x)$. *Then*

$$\|h\|_{\dot{K}_{q_2'}^{-\alpha,p_2'}(\mathbb{R}^n)}\|b\|_{\dot{K}_{q_1}^{\alpha,p_1}(\mathbb{R}^n)} \leq C,$$

where $\alpha, \beta \geq 0$, $0 < p_1 \leq p_2 < \infty$ *and* $1/q_1 - 1/q_2 = \beta/n$.

Proof. By a standard computation, we have

$$\|h\|_{\dot{K}_{q_2'}^{-\alpha,p_2'}(\mathbb{R}^n)} \leq C2^{-k\alpha}2^{kn(1/q_2'-\beta/n)}, \quad \|a\|_{\dot{K}_{q_1}^{\alpha,p_1}(\mathbb{R}^n)} \leq C2^{k\alpha}2^{kn(1/q_1-1)}.$$

Noting that $1/q_1 - 1/q_2 = \beta/n$, we get the desired result. $\qquad\square$

Lemma 4.6. *Let* b *be a* $C\dot{M}O(\mathbb{R}^n)$ *function and* $i, k \in \mathbb{Z}$. *Then*

$$|b(t) - b_{B_k}| \leq |b(t) - b_{B_i}| + C|i - k|\|b\|_{C\dot{M}O(\mathbb{R}^n)}.$$

Proof. For all $j \in \mathbb{Z}$, set

$$b_{B_j} = \frac{1}{|B_j|}\int_{B_j} b(x)dx.$$

We have

$$|b_{B_j} - b_{B_{j+1}}| \leq \frac{1}{|B_j|}\int_{B_j} |b(t) - b_{B_{j+1}}|dt$$

$$\leq \frac{C}{|B_{j+1}|}\int_{B_{j+1}} |b(t) - b_{B_{j+1}}|dt$$

$$= C\|b\|_{C\dot{M}O(\mathbb{R}^n)}.$$

If $k < i$, then

$$|b(t) - b_{B_k}| \leq |b(t) - b_{B_i}| + \sum_{j=k}^{i-1}|b_{B_j} - b_{B_{j+1}}|$$

$$\leq |b(t) - b_{B_i}| + C(i - k)\|b\|_{C\dot{M}O(\mathbb{R}^n)}.$$

If $i < k$, then

$$|b(t) - b_{B_k}| \leq |b(t) - b_{B_i}| + \sum_{j=i}^{k-1}|b_{B_j} - b_{B_{j+1}}|$$

$$\leq |b(t) - b_{B_i}| + C(k - i)\|b\|_{C\dot{M}O(\mathbb{R}^n)}.$$

$\qquad\square$

Theorem 4.1. *Let* $b \in C\dot{M}O^{\max\{p,p'\}}(\mathbb{R}^n)$. *Then both* $[b, \mathcal{H}]$ *and* $[b, \mathcal{H}^*]$ *are bounded on* $L^p(\mathbb{R}^n)$. *Conversely,*

(i) *If* $[b, \mathcal{H}]$ *is bounded on* $L^p(\mathbb{R}^n)$, *then* $b \in C\dot{M}O^{p'}(\mathbb{R}^n)$;

(ii) *If* $[b, \mathcal{H}^*]$ *is bounded on* $L^p(\mathbb{R}^n)$, *then* $b \in C\dot{M}O^p(\mathbb{R}^n)$.

Theorem 4.1 can be contained in the next theorem by choosing $\alpha = \beta = 0$ and $p_1 = p_2 = q_1 = q_2$.

Theorem 4.2. *Let* $0 \leq \beta < n, 0 < p_1 \leq p_2 < \infty, 1/q_1 - 1/q_2 = \beta/n, 1 < q_1 < \infty$ *and* $b \in C\dot{M}O^{\max(q_2, q_1')}(\mathbb{R}^n)$.

(i) *If* $\alpha < n/q_1'$, *then* $[b, \mathcal{H}_\beta]$ *is bounded from* $\dot{K}_{q_1}^{\alpha, p_1}(\mathbb{R}^n)$ *to* $\dot{K}_{q_2}^{\alpha, p_2}(\mathbb{R}^n)$.

(ii) *If* $\alpha > -n/q_2$, *then* $[b, \mathcal{H}_\beta^*]$ *is bounded from* $\dot{K}_{q_1}^{\alpha, p_1}(\mathbb{R}^n)$ *to* $\dot{K}_{q_2}^{\alpha, p_2}(\mathbb{R}^n)$.

Conversely,

(iii) *If* $[b, \mathcal{H}_\beta]$ *is bounded from* $\dot{K}_{q_1}^{\alpha, p_1}(\mathbb{R}^n)$ *to* $\dot{K}_{q_2}^{\alpha, p_2}(\mathbb{R}^n)$, *then* $b \in C\dot{M}O^{q_1'}(\mathbb{R}^n)$, *where* $\alpha \geq 0, 1 < p_2 < \infty$;

(iv) *If* $[b, \mathcal{H}_\beta^*]$ *is bounded from* $\dot{K}_{q_1}^{\alpha, p_1}(\mathbb{R}^n)$ *to* $\dot{K}_{q_2}^{\alpha, p_2}(\mathbb{R}^n)$, *then* $b \in C\dot{M}O^{q_2}(\mathbb{R}^n)$, *where* $\alpha \leq 0, 1 < p_2 < \infty$.

Proof. For simplicity, we write

$$\sum_{i=-\infty}^{\infty} f(x)\chi_i(x) = \sum_{i=-\infty}^{\infty} f_i(x).$$

When $\alpha < n/q_1'$, we get

$$\|([b, \mathcal{H}_\beta]f)\chi_k\|_{L^{q_2}(\mathbb{R}^n)}^{q_2} = \int_{C_k} \left| \int_{B(0,|x|)} f(t)(b(x) - b(t))dt \right|^{q_2} |x|^{q_2(\beta-n)} dx$$

$$\leq C \int_{C_k} \left(\int_{B_k} |f(t)(b(x) - b(t))| \, dt \right)^{q_2} 2^{-kq_2(n-\beta)} dx$$

$$= C2^{-kq_2(n-\beta)} \int_{C_k} \left(\sum_{i=-\infty}^{k} \int_{C_i} |f(t)(b(x) - b(t))| \, dt \right)^{q_2} dx$$

$$\leq C2^{-kq_2(n-\beta)} \int_{C_k} \left(\sum_{i=-\infty}^{k} \int_{C_i} |f(t)(b(x) - b_{B_k})| \, dt \right)^{q_2} dx$$

$$+ C \left(2^{-kn/q_1'} \sum_{i=-\infty}^{k} \int_{C_i} |f(t)(b(t) - b_{B_k})| \, dt \right)^{q_2}$$

$$:= I + J.$$

By Hölder's inequality, $1/q_1 + 1/q_1' = 1$, we get

$$I = C\left(2^{-kq_2(n-\beta)}\int_{C_k}|b(x) - b_{B_k}|^{q_2}\,dx\right)\left(\sum_{i=-\infty}^{k}\int_{C_i}|f(t)|\,dt\right)^{q_2}$$

$$\leq C 2^{-kq_2(n/q_2'-\beta)}\left(\frac{1}{|B_k|}\int_{B_k}|b(x) - b_{B_k}|^{q_2}\,dx\right)$$

$$\times\left(\sum_{i=-\infty}^{k}\left(\int_{C_i}|f(t)|^{q_1}\,dt\right)^{1/q_1}\left(\int_{C_i}dt\right)^{1/q_1'}\right)^{q_2}$$

$$\leq C\|b\|_{C\dot{M}O^{q_2}(\mathbb{R}^n)}^{q_2}2^{-kq_2(n/q_2'-\beta)}\left(\sum_{i=-\infty}^{k}2^{in/q_1'}\left(\int_{C_i}|f(t)|^{q_1}\,dt\right)^{1/q_1}\right)^{q_2}$$

$$= C\|b\|_{C\dot{M}O^{q_2}(\mathbb{R}^n)}^{q_2}\left(\sum_{i=-\infty}^{k}2^{(i-k)n/q_1'}\|f_i\|_{L^{q_1}(\mathbb{R}^n)}\right)^{q_2}.$$

By Lemma 4.6, it is true that

$$J = C\left(2^{-kn/q_1'}\sum_{i=-\infty}^{k}\int_{C_i}|(b(t) - b_{B_k})f(t)|\,dt\right)^{q_2}$$

$$\leq C\left(2^{-kn/q_1'}\sum_{i=-\infty}^{k}\int_{C_i}|(b(t) - b_{B_i})f(t)|\,dt\right)^{q_2}$$

$$+ C\|b\|_{C\dot{M}O(\mathbb{R}^n)}^{q_2}\left(2^{-kn/q_1'}\sum_{i=-\infty}^{k}(k-i)\int_{C_i}|f(t)|\,dt\right)^{q_2}$$

$$:= J_1 + J_2.$$

A further use of Hölder's inequality, we have

$$J_1 \leq C\left(2^{-kn/q_1'}\sum_{i=-\infty}^{k}\left(\int_{C_i}|(b(t) - b_{B_i})|^{q_1'}\,dt\right)^{1/q_1'}\left(\int_{C_i}|f(t)|^{q_1}\,dt\right)^{1/q_1}\right)^{q_2}$$

$$\leq C\sum_{i=-\infty}^{k}\left(2^{(i-k)n/q_1'}\left(\frac{1}{|B_i|}\int_{B_i}|(b(t) - b_{B_i})|^{q_1'}\,dt\right)^{1/q_1'}\|f_i\|_{L^{q_1}(\mathbb{R}^n)}\right)^{q_2}$$

$$= C\|b\|_{C\dot{M}O^{q_1'}(\mathbb{R}^n)}^{q_2}\left(\sum_{i=-\infty}^{k}2^{(i-k)n/q_1'}\|f_i\|_{L^{q_1}(\mathbb{R}^n)}\right)^{q_2}$$

and

$$J_2 = C\|b\|_{\dot{C}MO(\mathbb{R}^n)}^{q_2} \left(2^{-kn/q_1'} \sum_{i=-\infty}^{k} (k-i) \int_{C_i} |f(t)|dt \right)^{q_2}$$

$$\leq C\|b\|_{\dot{C}MO(\mathbb{R}^n)}^{q_2} \left(\sum_{i=-\infty}^{k} (k-i) 2^{(i-k)n/q_1'} \|f_i\|_{L^{q_1}(\mathbb{R}^n)} \right)^{q_2}.$$

Following the estimates of I, J_1 and J_2, it is true that

$$\|[b, \mathcal{H}_\beta]f\|_{\dot{K}_{q_2}^{\alpha,p_2}(\mathbb{R}^n)} = \left(\sum_{k=-\infty}^{\infty} 2^{k\alpha p_2} \|([b,\mathcal{H}_\beta]f)\chi_k\|_{L^{q_2}(\mathbb{R}^n)}^{p_2} \right)^{\frac{1}{p_2}}$$

$$\leq \left(\sum_{k=-\infty}^{\infty} 2^{k\alpha p_1} \|([b,\mathcal{H}_\beta]f)\chi_k\|_{L^{q_2}(\mathbb{R}^n)}^{p_1} \right)^{\frac{1}{p_1}}$$

$$\leq C\|b\|_{\dot{C}MO^{q_2}(\mathbb{R}^n)} \left(\sum_{k=-\infty}^{\infty} 2^{k\alpha p_1} \left(\sum_{i=-\infty}^{k} 2^{(i-k)n/q_1'} \|f_i\|_{L^{q_1}(\mathbb{R}^n)} \right)^{p_1} \right)^{\frac{1}{p_1}}$$

$$+ C\|b\|_{\dot{C}MO^{q_1'}(\mathbb{R}^n)} \left(\sum_{k=-\infty}^{\infty} 2^{k\alpha p_1} \left(\sum_{i=-\infty}^{k} 2^{(i-k)n/q_1'} \|f_i\|_{L^{q_1}(\mathbb{R}^n)} \right)^{p_1} \right)^{\frac{1}{p_1}}$$

$$+ C\|b\|_{\dot{C}MO(\mathbb{R}^n)} \left(\sum_{k=-\infty}^{\infty} 2^{k\alpha p_1} \left(\sum_{i=-\infty}^{k} (k-i)2^{(i-k)n/q_1'} \|f_i\|_{L^{q_1}(\mathbb{R}^n)} \right)^{p_1} \right)^{\frac{1}{p_1}}$$

$$:= S.$$

Thus, it follows that S is not bigger than

$$\left(\sum_{k=-\infty}^{\infty} 2^{k\alpha p_1} \left(\sum_{i=-\infty}^{k} (k-i)2^{\frac{(i-k)n}{q_1'}} \|f_i\|_{L^{q_1}(\mathbb{R}^n)} \right)^{p_1} \right)^{\frac{1}{p_1}}$$

multiplying the constant $C\|b\|_{\dot{C}MO^{\max(q_2,q_1')}(\mathbb{R}^n)}$. When $0 < p_1 \leq 1$, we obtain that S^{p_1} is equal to

$$C\|b\|_{\dot{C}MO^{\max(q_2,q_1')}(\mathbb{R}^n)}^{p_1} \sum_{k=-\infty}^{\infty} 2^{k\alpha p_1} \left(\sum_{i=-\infty}^{k} (k-i)2^{(i-k)n/q_1'} \|f_i\|_{L^{q_1}(\mathbb{R}^n)} \right)^{p_1}$$

$$= C\|b\|_{\dot{C}MO^{\max(q_2,q_1')}(\mathbb{R}^n)}^{p_1}$$

$$\times \sum_{k=-\infty}^{\infty} \left(\sum_{i=-\infty}^{k} 2^{i\alpha} \|f_i\|_{L^{q_1}(\mathbb{R}^n)} (k-i)2^{(i-k)(n/q_1'-\alpha)} \right)^{p_1}$$

$$\leq C\|b\|_{C\dot{M}O^{\max(q_2,q_1')}(\mathbb{R}^n)}^{p_1}$$

$$\times \sum_{k=-\infty}^{\infty} \sum_{i=-\infty}^{k} 2^{i\alpha p_1} \|f_i\|_{L^{q_1}(\mathbb{R}^n)}^{p_1} (k-i)^{p_1} 2^{(i-k)(n/q_1'-\alpha)p_1}$$

$$= C\|b\|_{C\dot{M}O^{\max(q_2,q_1')}(\mathbb{R}^n)}^{p_1} \sum_{i=-\infty}^{\infty} 2^{i\alpha p_1} \|f_i\|_{L^{q_1}(\mathbb{R}^n)}^{p_1} \sum_{k=i}^{\infty} (k-i)^{p_1} 2^{(i-k)(n/q_1'-\alpha)p_1}$$

$$= C\|b\|_{C\dot{M}O^{\max(q_2,q_1')}(\mathbb{R}^n)}^{p_1} \|f\|_{\dot{K}_{q_1}^{\alpha,p_1}(\mathbb{R}^n)}^{p_1}.$$

For the case $p_1 > 1$, it follows from Hölder's inequality that

$$S^{p_1} = C\|b\|_{C\dot{M}O^{\max(q_2,q_1')}(\mathbb{R}^n)}^{p_1}$$

$$\times \sum_{k=-\infty}^{\infty} \left(\sum_{i=-\infty}^{k} 2^{i\alpha} \|f_i\|_{L^{q_1}(\mathbb{R}^n)} (k-i) 2^{(i-k)(n/q_1'-\alpha)} \right)^{p_1}$$

$$\leq C\|b\|_{C\dot{M}O^{\max(q_2,q_1')}(\mathbb{R}^n)}^{p_1} \sum_{k=-\infty}^{\infty} \sum_{i=-\infty}^{k} 2^{i\alpha p_1} \|f_i\|_{L^{q_1}(\mathbb{R}^n)}^{p_1} 2^{(i-k)(n/q_1'-\alpha)p_1/2}$$

$$\times \left(\sum_{i=-\infty}^{k} (k-i)^{p_1'} 2^{(i-k)(n/q_1'-\alpha)p_1'/2} \right)^{p_1/p_1'}$$

$$= C\|b\|_{C\dot{M}O^{\max(q_2,q_1')}(\mathbb{R}^n)}^{p_1} \|f\|_{\dot{K}_{q_1}^{\alpha,p_1}(\mathbb{R}^n)}^{p_1}.$$

The proof of (i) is completed.

(ii) can be obtained with the same method as that of (i).

To prove (iii), let a be a central $H\dot{A}_{q_1,1}(\mathbb{R}^n)$-atom supported on B_k and set

$$g(x) = \left(\omega_n \log \frac{3}{2}\right)^{-1} a(x), \quad h(x) = |x|^{-\beta} \chi_{B(0,3\cdot 2^k)\setminus B(0,2^{k+1})}(x),$$

where $\omega_n = 2\pi^{n/2}/\Gamma(n/2)$. If $x \in B_k^c$, then $\mathcal{H}_\beta g(x) = 0$, and when $x \in B_k$, $h(x) = 0$, therefore $h(x)\mathcal{H}_\beta g(x) = 0$ on $\mathbb{R}^n$. By a simple computation, for $x \in B_k$, we have

$$\mathcal{H}_\beta^* h(x) = \omega_n \int_{2^{k+1}}^{3\cdot 2^k} \frac{dr}{r} = \omega_n \log \frac{3}{2} \quad \& \quad g(x)\mathcal{H}_\beta^* h(x) = a(x).$$

Therefore $a = g\mathcal{H}_\beta^* h - h\mathcal{H}_\beta g$.

We assume that the commutator $[b, \mathcal{H}_\beta]$ is a bounded operator from $\dot{K}_{q_1}^{\alpha,p_1}(\mathbb{R}^n)$ to $\dot{K}_{q_2}^{\alpha,p_2}(\mathbb{R}^n)$ and $f \in H\dot{A}_{q_1,1}(\mathbb{R}^n)$. Then by Lemmas 4.2, 4.4

and 4.5, we get

$$|\langle b, f \rangle| \leq \sum_k |\lambda_k| \left| \int_{\mathbb{R}^n} b(x)(g_k(x)\mathcal{H}_\beta^* h_k(x) - h_k(x)\mathcal{H}_\beta g_k(x))dx \right|$$

$$= \sum_k |\lambda_k| \left| \int_{\mathbb{R}^n} h_k(x)(b(x)\mathcal{H}_\beta g_k(x) - \mathcal{H}_\beta(bg_k)(x))dx \right|$$

$$\leq C \sum_k |\lambda_k| \|h_k\|_{\dot{K}_{q_2'}^{-\alpha, p_2'}(\mathbb{R}^n)} \|[b, \mathcal{H}_\beta]g_k\|_{\dot{K}_{q_2}^{\alpha, p_2}(\mathbb{R}^n)}$$

$$\leq C \sum_k |\lambda_k| \|h_k\|_{\dot{K}_{q_2'}^{-\alpha, p_2'}(\mathbb{R}^n)} \|g_k\|_{\dot{K}_{q_1}^{\alpha, p_1}(\mathbb{R}^n)} \|[b, \mathcal{H}_\beta]\|_{\dot{K}_{q_1}^{\alpha, p_1}(\mathbb{R}^n) \to \dot{K}_{q_2}^{\alpha, p_2}(\mathbb{R}^n)}$$

$$\leq C \|f\|_{H\dot{A}_{q_1, 1}(\mathbb{R}^n)} \|[b, \mathcal{H}_\beta]\|_{\dot{K}_{q_1}^{\alpha, p_1}(\mathbb{R}^n) \to \dot{K}_{q_2}^{\alpha, p_2}(\mathbb{R}^n)}.$$

It follows from the duality theorem between $H\dot{A}_{q_1, 1}(\mathbb{R}^n)$ and $C\dot{M}O^{q_1'}(\mathbb{R}^n)$ that the function b belongs to the space $C\dot{M}O^{q_1'}(\mathbb{R}^n)$ and $\|b\|_{C\dot{M}O^{q_1'}(\mathbb{R}^n)}$ is bounded by $\|\mathcal{H}_{\beta, b}\|_{\dot{K}_{q_1}^{\alpha, p_1}(\mathbb{R}^n) \to \dot{K}_{q_2}^{\alpha, p_2}(\mathbb{R}^n)}$, which is (iii). Similar arguments can be applied to (iv). $\qquad\square$

We end this section by a characterization of $C\dot{M}O(\mathbb{R}^n)$ on central Morrey spaces, which proof based on the boundedness of commutators on Lebesgue spaces.

Theorem 4.3. *Let $1 < p < \infty$, $-1/p < \lambda < 0$ and let $1/p + 1/p' = 1$. Then the following statements are equivalent:*
(a) $b \in C\dot{M}O^{\max(p, p')}(\mathbb{R}^n)$;
(b) $[b, \mathcal{H}]$ and $[b, \mathcal{H}^*]$ are bounded operators on both $\dot{B}^{p, \lambda}(\mathbb{R}^n)$ and $\dot{B}^{p', \lambda}(\mathbb{R}^n)$.

Proof. $(a) \Rightarrow (b)$ In this case, the task is to show that for a fixed ball B_{k_0} with $k_0 \in \mathbb{Z}$, there exist constants $C > 0$ such that

$$\frac{1}{|B_{k_0}|^{1+p\lambda}} \int_{B_{k_0}} |[b, \mathcal{H}]f(x)|^p dx \leq C \|f\|_{\dot{B}^{p, \lambda}(\mathbb{R}^n)}^p; \tag{4.4}$$

$$\frac{1}{|B_{k_0}|^{1+p\lambda}} \int_{B_{k_0}} |[b, \mathcal{H}^*]f(x)|^p dx \leq C \|f\|_{\dot{B}^{p, \lambda}(\mathbb{R}^n)}^p; \tag{4.5}$$

$$\frac{1}{|B_{k_0}|^{1+p'\lambda}} \int_{B_{k_0}} |[b, \mathcal{H}]f(x)|^{p'} dx \leq C \|f\|_{\dot{B}^{p', \lambda}(\mathbb{R}^n)}^{p'}; \tag{4.6}$$

$$\frac{1}{|B_{k_0}|^{1+p'\lambda}} \int_{B_{k_0}} |[b, \mathcal{H}^*]f(x)|^{p'} dx \leq C \|f\|_{\dot{B}^{p', \lambda}(\mathbb{R}^n)}^{p'}. \tag{4.7}$$

Decomposing $f = f\chi_{2B_{k_0}} + f\chi_{(2B_{k_0})^c} =: f_1 + f_2$ derives that

$$\frac{1}{|B_{k_0}|^{1+p\lambda}} \int_{B_{k_0}} |[b, \mathcal{H}]f(x)|^p dx \leq \frac{1}{|B_{k_0}|^{1+p\lambda}} \int_{B_{k_0}} |[b, \mathcal{H}]f_1(x)|^p dx$$

$$+ \frac{1}{|B_{k_0}|^{1+p\lambda}} \int_{B_{k_0}} |[b, \mathcal{H}]f_2(x)|^p dx$$

$$=: LL_1 + LL_2.$$

We conclude from Theorem 4.1 that

$$LL_1 \leq \frac{1}{|B_{k_0}|^{1+p\lambda}} \|[b, \mathcal{H}]f\chi_{B_{k_0}}\|^p_{L^p(\mathbb{R}^n)}$$

$$\leq \frac{C}{|B_{k_0}|^{1+p\lambda}} \|f\chi_{B_{k_0}}\|^p_{L^p(\mathbb{R}^n)} \leq C\|f\|^p_{\dot{B}^{p,\lambda}(\mathbb{R}^n)}.$$

For the term LL_2, by the definitions of $[b, \mathcal{H}]$ and $C\dot{M}O^p(\mathbb{R}^n)$, we have

$$LL_2 \leq \frac{1}{|B_{k_0}|^{1+p\lambda}} \int_{B_{k_0}} \left| \frac{1}{|x|^n} \int_{|y|<|x|} (b(x) - b(y)) f_2(y) dy \right|^p dx$$

$$\leq \frac{1}{|B_{k_0}|^{1+p\lambda}} \int_{B_{k_0}} \left| \sum_{k=2k_0}^{\infty} \frac{1}{|B_k|} \int_{B_k} (b(x) - c) f(y) dy \right|^p dx$$

$$+ \frac{1}{|B_{k_0}|^{1+p\lambda}} \int_{B_{k_0}} \left| \sum_{k=2k_0}^{\infty} \frac{1}{|B_k|} \int_{B_k} (b(y) - c) f(y) dy \right|^p dx$$

$$=: LL_{21} + LL_{22}.$$

Applying Hölder's inequality to p and p' can produce the estimate for LL_{21} as

$$LL_{21} \leq \frac{1}{|B_{k_0}|^{1+p\lambda}} \int_{B_{k_0}} \left| \sum_{k=2k_0}^{\infty} (b(x) - c)|B_k|^{-1/p} \left(\int_{B_k} |f(y)|^p dy \right)^{1/p} \right|^p dx$$

$$\leq C\|f\|^p_{\dot{B}^{p,\lambda}(\mathbb{R}^n)} \frac{1}{|B_{k_0}|^{1+p\lambda}} \int_{B_{k_0}} |b(x) - c|^p \left(\sum_{k=2k_0}^{\infty} |B_k|^\lambda \right)^p dx$$

$$\leq C\|b\|^p_{C\dot{M}O^p(\mathbb{R}^n)} \|f\|^p_{\dot{B}^{p,\lambda}(\mathbb{R}^n)}.$$

The same conclusion can be drawn for the term LL_{22} as

$$LL_{22} \leq \frac{1}{|B_{k_0}|^{1+p\lambda}}$$

$$\times \int_{B_{k_0}} \left(\sum_{k=2k_0}^{\infty} \frac{1}{|B_k|} \left(\int_{B_k} |b(y) - c|^{p'} dy \right)^{1/p'} \left(\int_{B_k} |f(y)|^p dy \right)^{1/p} \right)^p dx$$

$$\leq C\|b\|_{C\dot{M}O^{p'}(\mathbb{R}^n)}^p \|f\|_{\dot{B}^{p,\lambda}(\mathbb{R}^n)}^p \frac{1}{|B_{k_0}|^{1+p\lambda}} \int_{B_{k_0}} \left(\sum_{k=2k_0}^{\infty} |B_k|^{\lambda} \right)^p dx$$

$$\leq C\|b\|_{C\dot{M}O^{p'}(\mathbb{R}^n)}^p \|f\|_{\dot{B}^{p,\lambda}(\mathbb{R}^n)}^p \frac{1}{|B_{k_0}|^{p\lambda}} \left(\sum_{k=2k_0}^{\infty} |B_k|^{\lambda} \right)^p$$

$$\leq C\|b\|_{C\dot{M}O^{p'}(\mathbb{R}^n)}^p \|f\|_{\dot{B}^{p,\lambda}(\mathbb{R}^n)}^p.$$

Combining the above estimates for LL_1 and LL_2, (4.4) is proved. In the same manner, we can show (4.5). Then (4.6) and (4.7) can be handled in much the same way as that of (4.4) and (4.5). This produces the desired results.

$(b) \Rightarrow (a)$ This step will be divided into two cases.

CASE 1. $p > p'$. In this case, we want to show that there is a constant $C > 0$ such that for a fixed ball $B = B(\mathbf{0}, r)$,

$$\frac{1}{|B|} \int_B |b(y) - b_B|^p dy \leq C. \tag{4.8}$$

To reach this inequality, we note that

$$\frac{1}{|B|} \int_B |b(y) - b_B|^p dy \leq \frac{1}{|B|^{1+p}} \int_B \left| \int_{|z|<|y|} (b(y) - b(z)) \chi_B(z) dz \right|^p dy$$

$$+ \frac{1}{|B|^{1+p}} \int_B \left| \int_{|z| \geq |y|} (b(y) - b(z)) \chi_B(z) dz \right|^p dy$$

$$=: J_1 + J_2.$$

By the $(\dot{B}^{p,\lambda}(\mathbb{R}^n), \dot{B}^{p,\lambda}(\mathbb{R}^n))$ boundedness of $[b, \mathcal{H}]$, we have

$$J_1 \leq \frac{C}{|B|^{1+p}} \int_B |y|^{np} |[b, \mathcal{H}] \chi_B(\cdot)|^p dy$$

$$\leq C|B|^{p\lambda} \|[b, \mathcal{H}] \chi_B(\cdot)\|_{\dot{B}^{p,\lambda}(\mathbb{R}^n)}^p$$

$$\leq C|B|^{p\lambda} \|\chi_B(\cdot)\|_{\dot{B}^{p,\lambda}(\mathbb{R}^n)}^p$$

$$\leq C.$$

Applying the $(\dot{B}^{p,\lambda}(\mathbb{R}^n), \dot{B}^{p,\lambda}(\mathbb{R}^n))$ boundedness of $[b, \mathcal{H}^*]$ to the term J_2, the following estimate can be confirmed easily

$$J_2 \leq \frac{C}{|B|^{1+p}} \int_B \left(\int_{|z| \geq |y|} |z|^n \frac{|b(y) - b(z)|}{|z|^n} \chi_B(z) dz \right)^p dy$$

$$\leq \frac{C}{|B|} \int_B |[b, \mathcal{H}^*] \chi_B(\cdot)|^p \, dy$$

$$\leq C|B|^{p\lambda} \|[b, \mathcal{H}^*] \chi_B(\cdot)\|^p_{\dot{B}^{p,\lambda}(\mathbb{R}^n)}$$

$$\leq C|B|^{p\lambda} \|\chi_B(\cdot)\|^p_{\dot{B}^{p,\lambda}(\mathbb{R}^n)}$$

$$\leq C.$$

Thus (4.8) is a by-product of the estimates for J_1 and J_2.

CASE 2. $p < p'$. With the $(\dot{B}^{p,\lambda}(\mathbb{R}^n), \dot{B}^{p,\lambda}(\mathbb{R}^n))$ boundedness of $[b, \mathcal{H}]$ and $[b, \mathcal{H}^*]$ be replaced by the $(\dot{B}^{p',\lambda}(\mathbb{R}^n), \dot{B}^{p',\lambda}(\mathbb{R}^n))$ boundedness of $[b, \mathcal{H}]$ and $[b, \mathcal{H}^*]$, the similar arguments of Case 1 can be applied to this case and show that

$$\frac{1}{|B|} \int_B |b(y) - b_B|^{p'} dy \leq C,$$

which completes the proof of Theorem 4.3. $\qquad\square$

The corresponding characterization by $[b, \mathcal{H}_\beta]$ on $\dot{B}^{p',\lambda}(\mathbb{R}^n)$ is also true.

4.1.2 *The characterization for $\lambda > 0$*

Under some more conditions on λ, the space $C\dot{M}O^{p,\lambda}(\mathbb{R}^n)$ was characterized via the boundedness of $[b, \mathcal{H}]$ and $[b, \mathcal{H}^*]$ on Lebesgue spaces. We start with a technical lemma.

Lemma 4.7. *Let $1 < q < \infty$, $\lambda < 1/n$ and $i, k \in \mathbb{Z}$. If $b \in C\dot{M}O^{q,\lambda}(\mathbb{R}^n)$, then*

$$|b(y) - b_{B_k}| \leq |b(y) - b_{B_i}| + C \frac{2^{n\lambda}}{|1 - 2^{n\lambda}|} |2^{kn\lambda} - 2^{in\lambda}| \|b\|_{C\dot{M}O^{q,\lambda}(\mathbb{R}^n)}.$$

Proof. For all $j \in \mathbb{Z}$, set

$$b_{B_j} = \frac{1}{|B_j|} \int_{B_j} b(x) dx.$$

By the Hölder inequality, we have

$$|b_{B_j} - b_{B_{j+1}}| \leq \frac{1}{|B_j|} \int_{B_j} |b(y) - b_{B_{j+1}}| dy \leq \frac{1}{|B_j|} \int_{B_{j+1}} |b(y) - b_{B_{j+1}}| dy$$

$$\leq \frac{|B_{j+1}|^{1+\lambda}}{|B_j|} \left(\frac{1}{|B_{j+1}|^{1+\lambda q}} \int_{B_{j+1}} |b(y) - b_{B_{j+1}}|^q dy \right)^{1/q}$$

$$\leq C 2^{(j+1)n\lambda} \|b\|_{C\dot{M}O^{q,\lambda}}.$$

If $k < i$, then

$$|b(y) - b_{B_k}| \leq |b(y) - b_{B_i}| + \sum_{j=k}^{i-1} |b_{B_j} - b_{B_{j+1}}|$$

$$\leq |b(y) - b_{B_i}| + C 2^{n\lambda}(1 - 2^{n\lambda})^{-1}(2^{kn\lambda} - 2^{in\lambda})\|b\|_{C\dot{M}O^{q,\lambda}}.$$

If $i < k$, then

$$|b(y) - b_{B_k}| \leq |b(y) - b_{B_i}| + \sum_{j=i}^{k-1} |b_{B_j} - b_{B_{j+1}}|$$

$$\leq |b(y) - b_{B_i}| + C 2^{n\lambda}(1 - 2^{n\lambda})^{-1}(2^{in\lambda} - 2^{kn\lambda})\|b\|_{C\dot{M}O^{q,\lambda}},$$

which is the desired result. $\qquad \square$

Theorem 4.4. *Let* $1 < p, q < \infty$, $0 < \lambda = 1/q - 1/p < 1/n$. *Then* $b \in C\dot{M}O^{\max\{p,q'\},\lambda}(\mathbb{R}^n)$ *if and only if both* $[b, \mathcal{H}]$ *and* $[b, \mathcal{H}^*]$ *are bounded from* $L^q(\mathbb{R}^n)$ *to* $L^p(\mathbb{R}^n)$.

Proof. For simplicity, we write

$$\sum_{i=-\infty}^{\infty} f(x)\chi_i(x) = \sum_{i=-\infty}^{\infty} f_i(x).$$

Then we have

$$\int_{\mathbb{R}^n} |[b, \mathcal{H}]f(x)|^p dx = \int_{\mathbb{R}^n} \left| \frac{1}{|x|^n} \int_{|y|<|x|} (b(x) - b(y))f(y) dy \right|^p dx$$

$$\leq \sum_{k=-\infty}^{\infty} \int_{C_k} \left(\frac{1}{|x|^n} \int_{B_k} |(b(x) - b(y))f(y)| dy \right)^p dx$$

$$\leq C \sum_{k=-\infty}^{\infty} \int_{C_k} 2^{-kpn} \left(\sum_{i=-\infty}^{k} \int_{C_i} |b(x) - b_{B_k}| |f(y)| dy \right)^p dx$$

$$+ C \sum_{k=-\infty}^{\infty} \int_{C_k} 2^{-kpn} \left(\sum_{i=-\infty}^{k} \int_{C_i} |b(y) - b_{B_k}| |f(y)| dy \right)^p dx$$

$$=: I_1 + I_2.$$

By the definition of the space $C\dot{M}O^{p,\lambda}(\mathbb{R}^n)$ and Hölder's inequality, we get

$$I_1 = C \sum_{k=-\infty}^{\infty} 2^{-kpn} \int_{C_k} |b(x) - b_{B_k}|^p \left(\sum_{i=-\infty}^{k} \int_{C_i} |f(y)| \, dy \right)^p dx$$

$$\leq C\|b\|^p_{C\dot{M}O^{p,\lambda}(\mathbb{R}^n)} \sum_{k=-\infty}^{\infty} 2^{kpn(1/p+\lambda-1)} \left(\sum_{i=-\infty}^{k} 2^{in/q'} \|f_i\|_{L^q(\mathbb{R}^n)} \right)^p.$$

It follows from Lemma 4.7 that

$$I_2 \leq C \sum_{k=-\infty}^{\infty} \int_{C_k} 2^{-kpn} \left(\sum_{i=-\infty}^{k} \int_{C_i} |b(y) - b_{B_i}| \, |f(y)| \, dy \right)^p dx$$

$$+ C\|b\|^p_{C\dot{M}O^{p,\lambda}(\mathbb{R}^n)} \sum_{k=-\infty}^{\infty} \int_{C_k} 2^{-kpn} \left(\sum_{i=-\infty}^{k} \int_{C_i} \frac{2^{in\lambda} - 2^{kn\lambda}}{1 - 2^{n\lambda}} |f(y)| \, dy \right)^p dx$$

$$=: I_{21} + I_{22}.$$

Then the Hölder inequality yields that

$$I_{21} \leq C \sum_{k=-\infty}^{\infty} \int_{C_k} 2^{-kpn}$$

$$\times \left(\sum_{i=-\infty}^{k} \left(\int_{B_i} |b(y) - b_{B_i}|^{q'} dy \right)^{\frac{1}{q'}} \left(\int_{C_i} |f(y)|^q \, dy \right)^{\frac{1}{q}} \right)^p dx$$

$$\leq C\|b\|^p_{C\dot{M}O^{q',\lambda}(\mathbb{R}^n)} \sum_{k=-\infty}^{\infty} \int_{C_k} 2^{-kpn} \left(\sum_{i=-\infty}^{k} 2^{in(\lambda+1/q')} \|f_i\|_{L^q(\mathbb{R}^n)} \right)^p dx$$

$$= C\|b\|^p_{C\dot{M}O^{q',\lambda}(\mathbb{R}^n)} \sum_{k=-\infty}^{\infty} 2^{kn-kpn} \left(\sum_{i=-\infty}^{k} 2^{in(\lambda+1/q')} \|f_i\|_{L^q(\mathbb{R}^n)} \right)^p$$

and

$$I_{22} \leq C\|b\|^p_{C\dot{M}O^{p,\lambda}(\mathbb{R}^n)} \sum_{k=-\infty}^{\infty} \int_{C_k} 2^{-kpn}$$

$$\times \left(\sum_{i=-\infty}^{k} (2^{in\lambda} - 2^{kn\lambda})(1 - 2^{n\lambda})^{-1} \left(\int_{C_i} |f(y)|^q \, dy \right)^{\frac{1}{q}} 2^{\frac{in}{q'}} \right)^p dx$$

$$\leq C\|b\|^p_{C\dot{M}O^{p,\lambda}(\mathbb{R}^n)} \sum_{k=-\infty}^{\infty} 2^{kn-kpn} \left(\sum_{i=-\infty}^{k} \frac{2^{in\lambda} - 2^{kn\lambda}}{1 - 2^{n\lambda}} 2^{in/q'} \|f_i\|_{L^q(\mathbb{R}^n)} \right)^p.$$

The estimates of I_1, I_{21} and I_{22} lead to

$$\int_{\mathbb{R}^n} |[b, \mathcal{H}]f(x)|^p \, dx$$

$$\leq C\|b\|_{C\dot{M}O^{p,\lambda}(\mathbb{R}^n)}^p \sum_{k=-\infty}^{\infty} 2^{knq(1/p+\lambda-1)} \left(\sum_{i=-\infty}^{k} 2^{in/q'} \|f_i\|_{L^q(\mathbb{R}^n)} \right)^p$$

$$+ C\|b\|_{C\dot{M}O^{q',\lambda}(\mathbb{R}^n)}^p \sum_{k=-\infty}^{\infty} 2^{knq(1/p-1)} \left(\sum_{i=-\infty}^{k} 2^{in(\lambda+1/q')} \|f_i\|_{L^q(\mathbb{R}^n)} \right)^p$$

$$+ C\|b\|_{C\dot{M}O^{p,\lambda}(\mathbb{R}^n)}^p \sum_{k=-\infty}^{\infty} 2^{knp(1/p-1)} \left(\sum_{i=-\infty}^{k} \frac{2^{in\lambda} - 2^{kn\lambda}}{1 - 2^{n\lambda}} 2^{in/q'} \|f_i\|_{L^p(\mathbb{R}^n)} \right)^p$$

$$\leq C\|b\|_{C\dot{M}O^{\max(q',p),\lambda}(\mathbb{R}^n)}^p \sum_{k=-\infty}^{\infty} \|f_k\|_{L^q(\mathbb{R}^n)}^p$$

$$+ C\|b\|_{C\dot{M}O^{p,\lambda}(\mathbb{R}^n)}^p \sum_{k=-\infty}^{\infty} 2^{knp(1/p-1)} \left(\sum_{i=-\infty}^{k} \frac{2^{in\lambda} - 2^{kn\lambda}}{1 - 2^{n\lambda}} 2^{in/q'} \|f_i\|_{L^q(\mathbb{R}^n)} \right)^p$$

$$=: E_1 + E_2.$$

For E_1, it follows from $1/q - 1/p = \lambda > 0$ that $q < p$. This in turn implies

$$E_1 = C\|b\|_{C\dot{M}O^{\max(q',p),\lambda}(\mathbb{R}^n)}^p \left(\left(\sum_{k=-\infty}^{\infty} \left(\int_{C_k} |f(y)|^q dy \right)^{p/q} \right)^{q/p} \right)^{p/q}$$

$$\leq C\|b\|_{C\dot{M}O^{\max(q',p),\lambda}(\mathbb{R}^n)}^p \left(\sum_{k=-\infty}^{\infty} \int_{C_k} |f(y)|^q dy \right)^{p/q}$$

$$= C\|b\|_{C\dot{M}O^{\max(q',p),\lambda}(\mathbb{R}^n)}^p \|f\|_{L^q(\mathbb{R}^n)}^p.$$

The term E_2 can be controlled by

$$C\|b\|_{C\dot{M}O^{\max(q',p),\lambda}(\mathbb{R}^n)}^p$$

$$\times \sum_{k=-\infty}^{\infty} 2^{knp(1/p-1)} \left(\sum_{i=-\infty}^{k} \frac{2^{in\lambda} - 2^{kn\lambda}}{1 - 2^{n\lambda}} 2^{in/q'} \|f_i\|_{L^q(\mathbb{R}^n)} \right)^p$$

$$= C\|b\|_{C\dot{M}O^{\max(q',p),\lambda}(\mathbb{R}^n)}^p$$

$$\times \sum_{k=-\infty}^{\infty} \left(\sum_{i=-\infty}^{k} \frac{2^{in\lambda} - 2^{kn\lambda}}{1 - 2^{n\lambda}} 2^{kn(1/p-1)} 2^{in/q'} \|f_i\|_{L^q(\mathbb{R}^n)} \right)^p$$

$$\leq C\|b\|_{\dot{C}MO^{\max(q',p),\lambda}(\mathbb{R}^n)}^p \sum_{k=-\infty}^{\infty} \sum_{i=-\infty}^{k} \left(\frac{2^{in\lambda}-2^{kn\lambda}}{1-2^{n\lambda}}2^{kn(1/p-1)}2^{in/q'}\right)^{p/2}$$

$$\times \left(\sum_{i=-\infty}^{k} \left(\frac{2^{in\lambda}-2^{kn\lambda}}{1-2^{n\lambda}}2^{kn(1/p-1)}2^{in/q'}\right)^{p'/2}\right)^{p/p'}\|f_i\|_{L^q(\mathbb{R}^n)}^p$$

$$\leq C\|b\|_{\dot{C}MO^{\max(q',p),\lambda}(\mathbb{R}^n)}^p \sum_{i=-\infty}^{\infty}\|f_i\|_{L^q(\mathbb{R}^n)}^p$$

$$\leq C\|b\|_{\dot{C}MO^{\max(q',p),\lambda}(\mathbb{R}^n)}^p\|f\|_{L^q(\mathbb{R}^n)}^p,$$

where we have used the fact that $1/q - 1/p = \lambda > 0$ and $\lambda + 1/q' > 0$. This clearly forces

$$\|[b,\mathcal{H}]f\|_{L^p(\mathbb{R}^n)} \leq C\|b\|_{\dot{C}MO^{\max(q',p),\lambda}(\mathbb{R}^n)}\|f\|_{L^q(\mathbb{R}^n)}.$$

In the same manner we can see that $[b,\mathcal{H}^*]$ is bounded from $L^q(\mathbb{R}^n)$ to $L^p(\mathbb{R}^n)$.

Conversely, suppose that $[b,\mathcal{H}]$ and $[b,\mathcal{H}^*]$ are bounded from $L^q(\mathbb{R}^n)$ to $L^p(\mathbb{R}^n)$. We shall show that $b \in \dot{C}MO^{\max(p,q'),\lambda}(\mathbb{R}^n)$. Denote by $s = \max(p,q')$. Then we have

$$\frac{1}{|B(0,r)|^{1+\lambda s}} \int_{B(0,r)} |b(y) - b_{B(0,r)}|^s dy$$

$$= \frac{1}{|B(0,r)|^{1+(1+\lambda)s}} \int_{B(0,r)} \left|\int_{B(0,r)} (b(y)-b(z))dz\right|^s dy$$

$$\leq \frac{C}{|B(0,r)|^{1+(1+\lambda)s}} \int_{|y|<r} \left|\int_{|z|<|y|} (b(y)-b(z))\chi_{B(0,r)}(z)dz\right|^s dy$$

$$+ \frac{C}{|B(0,r)|^{1+(1+\lambda)s}} \int_{|y|<r} \left|\int_{|z|\geq|y|} (b(y)-b(z))\chi_{B(0,r)}(z)dz\right|^s dy$$

$$=: J_1 + J_2.$$

We have that J_1 is

$$\frac{C}{|B(0,r)|^{1+(1+\lambda)s}} \int_{|y|<r} |y|^{sn}\left|\frac{1}{|y|^n}\int_{|z|<|y|} (b(y)-b(z))\chi_{B(0,r)}(z)dz\right|^s dy$$

$$= \frac{Cr^{sn}}{|B(0,r)|^{1+(1+\lambda)s}} \int_{|y|<r} |([b,\mathcal{H}]\chi_{B(0,r)})(y)|^s dy.$$

For J_2, we have

$$J_2 = \frac{C}{|B(0,r)|^{1+(1+\lambda)s}} \int_{|y|<r} \left|\int_{|z|\geq|y|} \frac{(b(y)-b(z))\chi_{B(0,r)}(z)}{|z|^n}|z|^n dz\right|^s dy$$

$$\leq \frac{Cr^{sn}}{|B(\mathbf{0},r)|^{1+(1+\lambda)s}} \int_{|y|<r} \left| ([b,\mathcal{H}^*]\chi_{B(\mathbf{0},r)})(y) \right|^s dy.$$

The estimates for J_1 and J_2 can be divided into two cases.

CASE 1: $s = p$. Since both $[b,\mathcal{H}]$ and $[b,\mathcal{H}^*]$ are bounded from $L^q(\mathbb{R}^n)$ to $L^p(\mathbb{R}^n)$, we have

$$J_1 \leq \frac{Cr^{pn}}{|B(\mathbf{0},r)|^{1+(1+\lambda)p}} \int_{|y|<r} \left| ([b,\mathcal{H}]\chi_{B(\mathbf{0},r)})(y) \right|^p dy$$

$$\leq \frac{Cr^{pn}}{|B(\mathbf{0},r)|^{1+(1+\lambda)p}} \left(\int_{|y|<r} dy \right)^{p/q} \leq \frac{Cr^{pn}r^{np/q}}{r^{n(1+(1+\lambda)p)}} = C,$$

where we have used the condition $1/q - 1/p = \lambda$.

Similarly, we can deduce that $J_2 \leq C$.

CASE 2: $s = q'$. By duality, we know that both $[b,\mathcal{H}]$ and $[b,\mathcal{H}^*]$ map $L^{p'}(\mathbb{R}^n)$ into $L^{q'}(\mathbb{R}^n)$. Thus

$$J_1 \leq \frac{Cr^{nq'}}{|B(\mathbf{0},r)|^{1+(1+\lambda)q'}} \int_{|y|<r} \left| ([b,\mathcal{H}]\chi_{B(\mathbf{0},r)})(y) \right|^{q'} dy$$

$$\leq \frac{Cr^{nq'}}{|B(\mathbf{0},r)|^{1+(1+\lambda)q'}} \left(\int_{|y|<r} dy \right)^{q'/p'}$$

$$\leq \frac{Cr^{nq'}r^{nq'/p'}}{r^{n(1+(1+\lambda)q')}} \leq C.$$

The proof is completed by applying similar arguments to J_2. $\qquad\square$

4.1.3 *The characterization for $\lambda < 0$*

For the case $-1/p \leq \lambda < 0$, as a concept of highly independent interest, it has received less attention for the characterizations of $\dot{\mathcal{E}}^{p,\lambda}(\mathbb{R}^n)$ by the boundedness of the commutators of Hardy type operators. We settle this problem under the assumption that b satisfies the following mean value inequality. A function f is said to satisfy the well-known mean value inequality if there exists a constant $C > 0$ such that for any ball $B \subset \mathbb{R}^n$,

$$\sup_{B \ni x} |f(x) - f_B| \leq \frac{C}{|B|} \int_B |f(x) - f_B| dx. \tag{4.9}$$

The first main result of this section can be formulated as follows.

Theorem 4.5. *Let $1 < p < \infty$, $-1/p \leq \lambda < 0$, $-1/p_i \leq \lambda_i < 0$, $i = 1,2$, $1/p = \sum_{i=1}^2 1/p_i$, $\lambda = \sum_{i=1}^2 \lambda_i$ and let b satisfy (4.9). Then the following* *two statements are equivalent:*

(a) $b \in \dot{\mathcal{E}}^{p_1,\lambda_1}(\mathbb{R}^n)$;

(b) *Both* $[b, \mathcal{H}]$ *and* $[b, \mathcal{H}^*]$ *are bounded operators from* $\dot{B}^{p_2,\lambda_2}(\mathbb{R}^n)$ *to* $\dot{B}^{p,\lambda}(\mathbb{R}^n)$.

Proof. $(a) \Rightarrow (b)$. For a fixed ball $B = B(\mathbf{0}, r) \subset \mathbb{R}^n$. There is no loss of generality in assuming $B(\mathbf{0}, r) = B(\mathbf{0}, 2^{k_0}) = B_{k_0}$ with $k_0 \in \mathbb{Z}$. The task is now to show that there exists a constant $C > 0$ such that

$$\frac{1}{|B_{k_0}|^{\lambda}} \left(\frac{1}{|B_{k_0}|} \int_{B_{k_0}} |[b, \mathcal{H}]f(x)|^p dx \right)^{1/p} \leq C \|f\|_{\dot{B}^{p_2,\lambda_2}(\mathbb{R}^n)} \tag{4.10}$$

and

$$\frac{1}{|B_{k_0}|^{\lambda}} \left(\frac{1}{|B_{k_0}|} \int_{B_{k_0}} |[b, \mathcal{H}^*]f(x)|^p dx \right)^{1/p} \leq C \|f\|_{\dot{B}^{p_2,\lambda_2}(\mathbb{R}^n)}. \tag{4.11}$$

The definition of $[b, \mathcal{H}]$ gives

$$\int_{B_{k_0}} |[b, \mathcal{H}]f(x)|^p dx = \int_{B_{k_0}} \left| \frac{1}{|x|^n} \int_{|y|<|x|} (b(x) - b(y)) f(y) dy \right|^p dx$$

$$\leq \sum_{k=-\infty}^{k_0} \int_{C_k} \left| \frac{1}{|x|^n} \int_{B_k} (b(x) - b(y)) f(y) dy \right|^p dx$$

$$\leq C \sum_{k=-\infty}^{k_0} \int_{C_k} \left(\frac{1}{|x|^n} \sum_{i=-\infty}^{k} \int_{C_i} |b(x) - b_{B_k}| |f(y)| dy \right)^p dx$$

$$+ C \sum_{k=-\infty}^{k_0} \int_{C_k} \left(\frac{1}{|x|^n} \sum_{i=-\infty}^{k} \int_{C_i} |b(y) - b_{B_k}| |f(y)| dy \right)^p dx$$

$$=: I + II.$$

Applying Hölder's inequality to p_1/p and $(p_1/p)'$, the term I can be estimated as

$$I \leq C \sum_{k=-\infty}^{k_0} 2^{-knp} \int_{C_k} |b(x) - b_{B_k}|^p dx \left(\sum_{i=-\infty}^{k} \int_{C_i} |f(y)| dy \right)^p$$

$$\leq C \sum_{k=-\infty}^{k_0} 2^{-knp} \left(\int_{B_k} |b(x) - b_{B_k}|^{p_1} dx \right)^{p/p_1} |B_k|^{1/(p_1/p)'}$$

$$\times \left| \sum_{i=-\infty}^{k} \left(\int_{B_i} |f(y)|^{p_2} dy \right)^{1/p_2} |B_i|^{1/p_2'} \right|^p$$

$$\leq C\|b\|_{\dot{\mathcal{E}}^{p_1,\lambda_1}(\mathbb{R}^n)}^p \|f\|_{\dot{B}^{p_2,\lambda_2}(\mathbb{R}^n)}^p \sum_{k=-\infty}^{k_0} 2^{-knp}|B_k|^{1+p\lambda_1} \left(\sum_{i=-\infty}^{k} |B_i|^{1+\lambda_2}\right)^p$$

$$\leq C\|b\|_{\dot{\mathcal{E}}^{p_1,\lambda_1}(\mathbb{R}^n)}^p \|f\|_{\dot{B}^{p_2,\lambda_2}(\mathbb{R}^n)}^p \sum_{k=-\infty}^{k_0} 2^{kn(1+p\lambda)}$$

$$\leq C\|b\|_{\dot{\mathcal{E}}^{p_1,\lambda_1}(\mathbb{R}^n)}^p \|f\|_{\dot{B}^{p_2,\lambda_2}(\mathbb{R}^n)}^p |B_{k_0}|^{1+p\lambda}.$$

The fact $1/p = 1/p_1 + 1/p_2$ allows us to estimate the term II as

$$II \leq C \sum_{k=-\infty}^{k_0} 2^{-knp} \int_{C_k} \left(\sum_{i=-\infty}^{k} \int_{C_i} |b(y) - b_{B_k}||f(y)|dy\right)^p dx$$

$$\leq C \sum_{k=-\infty}^{k_0} 2^{-knp} \int_{C_k} \left(\sum_{i=-\infty}^{k} \left(\int_{B_i} (|b(y) - b_{B_k}||f(y)|)^p dy\right)^{1/p} |B_i|^{1/p'}\right)^p dx$$

$$\leq C\|b\|_{\dot{\mathcal{E}}^{p_1,\lambda_1}(\mathbb{R}^n)}^p \|f\|_{\dot{B}^{p_2,\lambda_2}(\mathbb{R}^n)}^p \sum_{k=-\infty}^{k_0} 2^{-knp}|B_k|^{\frac{p\lambda_1 p_1 + p}{p_1}}$$

$$\times \int_{C_k} \left(\sum_{i=-\infty}^{k} |B_i|^{\frac{p_2 p' \lambda_2 + p_2 + p'}{p_2 p'}}\right)^p dx$$

$$\leq C\|b\|_{\dot{\mathcal{E}}^{p_1,\lambda_1}(\mathbb{R}^n)} \|f\|_{\dot{B}^{p_2,\lambda_2}(\mathbb{R}^n)} \sum_{k=-\infty}^{k_0} 2^{kn(1+p\lambda)}$$

$$\leq C\|b\|_{\dot{\mathcal{E}}^{p_1,\lambda_1}(\mathbb{R}^n)} \|f\|_{\dot{B}^{p_2,\lambda_2}(\mathbb{R}^n)} |B_{k_0}|^{1+p\lambda}.$$

On account of the above estimates for I and II, (4.10) is obtained.

We are now in a position to show (4.11). We note that

$$\int_{B_{k_0}} |[b, \mathcal{H}^*]f(x)|^p dx = \int_{B_{k_0}} \left|\int_{|y| \geq |x|} \frac{(b(x) - b(y))}{|y|^n} f(y)dy\right|^p dx$$

$$\leq \int_{B_{k_0}} \left(\int_{2^{k_0 n} \geq |y| \geq |x|} \frac{|b(x) - b(y)|}{|y|^n} |f(y)|dy\right)^p dx$$

$$+ \int_{B_{k_0}} \left(\int_{|y| > 2^{k_0 n}} \frac{|b(x) - b(y)|}{|y|^n} |f(y)|dy\right)^p dx$$

$$=: I' + II'.$$

The term I' can be handled in much the same way as that of (4.10), the only difference being in the analysis of the term II'. Analysis similar to that of $[b, \mathcal{H}]$ shows

$$I' \leq \int_{B_{k_0}} \left(\frac{1}{|x|^n} \int_{|y| \leq 2^{k_0} n} |b(x) - b(y)||f(y)|dy \right)^p dx$$

$$\leq C \sum_{k=-\infty}^{k_0} 2^{-knp} \int_{B_k} \left(\sum_{i=-\infty}^{k} \int_{B_i} |(b(x) - b(y))f(y)|dy \right)^p dx$$

$$\leq C \|b\|_{\dot{\mathcal{E}}^{p_1, \lambda_1}(\mathbb{R}^n)}^p \|f\|_{\dot{B}^{p_2, \lambda_2}(\mathbb{R}^n)}^p |B_{k_0}|^{1+p\lambda}.$$

For the term II', we proceed to show that

$$II' \leq \int_{B_{k_0}} \left(\sum_{k=k_0}^{\infty} \int_{C_k} \frac{|b(x) - b_{B_{k_0}}|}{|y|^n} |f(y)|dy \right)^p dx$$

$$+ \int_{B_{k_0}} \left(\sum_{k=k_0}^{\infty} \int_{C_k} \frac{|b(y) - b_{B_{k_0}}|}{|y|^n} |f(y)|dy \right)^p dx$$

$$=: II'_1 + II'_2.$$

By the Hölder inequality, we have

$$II'_1 \leq \int_{B_{k_0}} |b(x) - b_{B_{k_0}}|^p dx \left(\sum_{k=k_0}^{\infty} \int_{C_k} \frac{|f(y)|}{|y|^n} dy \right)^p$$

$$\leq \left(\int_{B_{k_0}} |b(x) - b_{B_{k_0}}|^{p_1} dx \right)^{p/p_1} |B_{k_0}|^{1/(p_1/p)'}$$

$$\times \left(\sum_{k=k_0}^{\infty} \left(\int_{C_k} \left(\frac{|f(y)|}{|y|^n} \right)^{p_2} dy \right)^{1/p_2} |B_k|^{1/p_2'} \right)^p$$

$$\leq C \|b\|_{\dot{\mathcal{E}}^{p_1, \lambda_1}(\mathbb{R}^n)}^p \|f\|_{\dot{B}^{p_2, \lambda_2}(\mathbb{R}^n)}^p |B_{k_0}|^{1+p\lambda_1} \left(\sum_{k=k_0}^{\infty} |B_k|^{\lambda_2} \right)^p$$

$$\leq C \|b\|_{\dot{\mathcal{E}}^{p_1, \lambda_1}(\mathbb{R}^n)}^p \|f\|_{\dot{B}^{p_2, \lambda_2}(\mathbb{R}^n)}^p |B_{k_0}|^{1+p\lambda}.$$

To get the boundedness for the term II'_2, we need the following decomposition

$$II'_2 \leq \int_{B_{k_0}} \left(\sum_{k=k_0}^{\infty} \int_{C_k} \frac{|b(y) - b_{B_k}||f(y)|}{|y|^n} dy \right)^p dx$$

$$+ \int_{B_{k_0}} \left(\sum_{k=k_0}^{\infty} \int_{C_k} \frac{|b_{B_{k_0}} - b_{B_k}||f(y)|}{|y|^n} dy \right)^p dx$$

$$=: II'_{21} + II'_{22},$$

We first calculate II'_{21}. To do this, the Hölder inequality and $1/p = 1/p_1 + 1/p_2$ show that

$$II'_{21} \leq \int_{B_{k_0}} \left(\sum_{k=k_0}^{\infty} \left(\int_{C_k} \left| \frac{|b(y) - b_{B_k}|}{|y|^n} |f(y)| \right|^p dy \right)^{1/p} |B_k|^{1/p'} \right)^p dx$$

$$\leq \int_{B_{k_0}} \left(\sum_{k=k_0}^{\infty} |B_k|^{1/p'} \left(\int_{B_k} |b(y) - b_{B_k}|^{p_1} dy \right)^{1/p_1} \right.$$

$$\times \left. \left(\int_{B_k} \left| \frac{f(y)}{|y|^n} \right|^{p_2} dy \right)^{1/p_2} \right)^p dx$$

$$\leq C\|b\|^p_{\dot{\mathcal{E}}^{p_1,\lambda_1}(\mathbb{R}^n)} \|f\|^p_{\dot{B}^{p_2,\lambda_2}(\mathbb{R}^n)} \int_{B_{k_0}} \left(\sum_{k=k_0}^{\infty} |B_k|^{\lambda} \right)^p dx$$

$$\leq C\|b\|^p_{\dot{\mathcal{E}}^{p_1,\lambda_1}(\mathbb{R}^n)} \|f\|^p_{\dot{B}^{p_2,\lambda_2}(\mathbb{R}^n)} |B_{k_0}|^{1+p\lambda}.$$

For the term II'_{22}, we claim first that for $k > k_0$

$$|b_{B_{k_0}} - b_{B_k}| \leq C 2^{(k_0+1)n\lambda_1} \|b\|_{\dot{\mathcal{E}}^{p_1,\lambda_1}(\mathbb{R}^n)}.$$

In fact

$$|b_{B_{k_0}} - b_{B_k}| \leq \sum_{j=k_0}^{k-1} |b_{B_j} - b_{B_{j+1}}| \leq \sum_{j=k_0}^{k-1} \frac{1}{|B_j|} \int_{B_{j+1}} |b(y) - b_{B_{j+1}}| dy$$

$$\leq \sum_{j=k_0}^{k-1} \frac{1}{|B_j|} \left(\int_{B_{j+1}} |b(y) - b_{B_{j+1}}|^{p_1} dy \right)^{1/p_1} |B_{j+1}|^{1/p'_1}$$

$$\leq C\|b\|_{\dot{\mathcal{E}}^{p_1,\lambda_1}(\mathbb{R}^n)} \sum_{j=k_0}^{k-1} \frac{|B_{j+1}|^{1+\lambda_1}}{|B_j|}$$

$$\leq C\|b\|_{\dot{\mathcal{E}}^{p_1,\lambda_1}(\mathbb{R}^n)} \sum_{j=k_0}^{k-1} 2^{(j+1)n\lambda_1}$$

$$\leq C 2^{(k_0+1)n\lambda_1} \|b\|_{\dot{\mathcal{E}}^{p_1,\lambda_1}(\mathbb{R}^n)}.$$

Therefore

$$II'_{22} \leq C\|b\|^p_{\dot{\mathcal{E}}^{p_1,\lambda_1}(\mathbb{R}^n)} \int_{B_{k_0}} \left(\sum_{k=k_0}^{\infty} \int_{C_k} \frac{2^{k_0 n\lambda_1}}{|y|^n} |f(y)| dy \right)^p dx$$

$$\leq C\|b\|^p_{\dot{\mathcal{E}}^{p_1,\lambda_1}(\mathbb{R}^n)}$$

$$\times \int_{B_{k_0}} 2^{k_0 n \lambda_1 p} \left(\sum_{k=k_0}^{\infty} \frac{1}{2^{kn}} \left(\int_{C_k} |f(y)|^{p_2} dy \right)^{1/p_2} |B_k|^{1/p_2'} \right)^p dx$$

$$\leq C \|b\|_{\dot{\mathcal{E}}^{p_1,\lambda_1}(\mathbb{R}^n)}^p \|f\|_{\dot{B}^{p_2,\lambda_2}(\mathbb{R}^n)}^p \int_{B_{k_0}} 2^{k_0 n \lambda_1 p} \left(\sum_{k=k_0}^{\infty} |B_k|^{\lambda_2} \right)^p dx$$

$$\leq C \|b\|_{\dot{\mathcal{E}}^{p_1,\lambda_1}(\mathbb{R}^n)}^p \|f\|_{\dot{B}^{p_2,\lambda_2}(\mathbb{R}^n)}^p |B_{k_0}|^{1+p\lambda}.$$

Summarizing, we have

$$II' \leq C \|b\|_{\dot{\mathcal{E}}^{p_1,\lambda_1}(\mathbb{R}^n)}^p \|f\|_{\dot{B}^{p_2,\lambda_2}(\mathbb{R}^n)}^p |B_{k_0}|^{1+p\lambda},$$

which implies (4.11). This is the desired result.

$(b) \Rightarrow (a)$ In this case, the proof consists of the construction of a proper commutator. We are reduced to prove that for a fixed ball $B = B(0, r)$,

$$\frac{1}{|B|^{1+p_1\lambda_1}} \int_B |b(y) - b_B|^{p_1} dy \leq C.$$

We conclude from (4.9) and Hölder's inequality that

$$\frac{1}{|B|^{1+p_1\lambda_1}} \int_B |b(y) - b_B|^{p_1} dy \leq \frac{1}{|B|^{p_1\lambda_1}} \sup_{y \in B} |b(y) - b_B|^{p_1}$$

$$\leq \frac{C}{|B|^{p_1\lambda_1}} \left(\frac{1}{|B|} \int_B |b(y) - b_B| dy \right)^{p_1}$$

$$\leq \frac{C}{|B|^{p_1\lambda_1}} \left(\frac{1}{|B|} \int_B |b(y) - b_B|^p dy \right)^{p_1/p}.$$

To deal with the above term, we note that

$$\int_B |b(y) - b_B|^p dy \leq \frac{1}{|B|^p} \int_B \left(\int_B |b(y) - b(z)| dz \right)^p dy$$

$$\leq \frac{1}{|B|^p} \int_B \left(\int_{\{z \in B, |z| < |y|\}} |b(y) - b(z)| \chi_B(z) dz \right)^p dy$$

$$+ \frac{1}{|B|^p} \int_B \left(\int_{\{z \in B, |z| \geq |y|\}} |b(y) - b(z)| \chi_B(z) dz \right)^p dy$$

$$=: J + JJ.$$

The $(\dot{B}^{p_2,\lambda_2}(\mathbb{R}^n), \dot{B}^{p,\lambda}(\mathbb{R}^n))$ boundedness of $[b, \mathcal{H}]$ allows us to estimate J as

$$J \leq \frac{1}{|B|^p} \int_B |y|^{np} |([b, \mathcal{H}] \chi_B)(y)|^p dy$$

$$\leq |B|^{1+p\lambda}\|[b,\mathcal{H}]\chi_B\|^p_{\dot{B}^{p,\lambda}(\mathbb{R}^n)}$$

$$\leq C|B|^{1+p\lambda}\|\chi_B\|^p_{\dot{B}^{p_2,\lambda_2}(\mathbb{R}^n)}$$

$$\leq C|B|^{1+p\lambda_1}.$$

By the $(\dot{B}^{p_2,\lambda_2}(\mathbb{R}^n), \dot{B}^{p,\lambda}(\mathbb{R}^n))$ boundedness of $[b,\mathcal{H}^*]$, it is easy to check that

$$JJ \leq \frac{1}{|B|^p}\int_B\left(\int_{|z|\geq|y|}|z|^n\frac{|b(y)-b(z)|}{|z|^n}\chi_B(z)dz\right)^p dy$$

$$\leq C\int_B|([b,\mathcal{H}^*]\chi_B)(y)|^p\,dy \leq |B|^{1+p\lambda}\|[b,\mathcal{H}^*]\chi_B\|^p_{\dot{B}^{p,\lambda}(\mathbb{R}^n)}$$

$$\leq C|B|^{1+p\lambda}\|\chi_B\|^p_{\dot{B}^{p_2,\lambda_2}(\mathbb{R}^n)}$$

$$\leq C|B|^{1+p\lambda_1}.$$

We thus have established the following inequality if we combine the above estimates for J and JJ,

$$\frac{1}{|B|^{1+p_1\lambda_1}}\int_B|b(y)-b_B|^{p_1} \leq \frac{C}{|B|^{p_1\lambda_1}}\left(\frac{|B|^{1+p\lambda_1}}{|B|}\right)^{p_1/p} \leq C.$$

The proof is completed. □

Unlike the case $\lambda \geq 0$, more difficulties are caused for the case $\lambda < 0$. The methods currently available for $\lambda < 0$ depend heavily on the structures of the Hardy type operators and the space $\dot{\mathcal{E}}^{p,\lambda}(\mathbb{R}^n)$. Therefore, we need the condition (4.9) to set up Theorem 4.5. This condition is essential to the characterization of $\dot{\mathcal{E}}^{p_1,\lambda_1}(\mathbb{R}^n)$ and can not be weakened in our arguments.

Lemma 4.8. *Let* $1 < p < \infty$, $-1/p \leq \lambda < 0$, $i,k \in \mathbb{Z}$ *and* $b \in \dot{\mathcal{E}}^{p,\lambda}(\mathbb{R}^n)$. *Then*

$$|b(y)-b_{B_k}| \leq |b(y)-b_{B_i}| + C\max\{|B_k|^\lambda,|B_i|^\lambda\}\|b\|_{\dot{\mathcal{E}}^{p,\lambda}(\mathbb{R}^n)}.$$

Proof. Using Hölder's inequality to p and p', one has

$$|b_{B_j}-b_{B_{j+1}}| \leq \frac{1}{|B_j|}\int_{B_{j+1}}|b(y)-b_{B_{j+1}}|dy$$

$$\leq \frac{1}{|B_j|^{1/p}}\left(\int_{B_{j+1}}|b(y)-b_{B_{j+1}}|^p dy\right)^{1/p}$$

$$\leq C|B_{j+1}|^\lambda\|b\|_{\dot{\mathcal{E}}^{p,\lambda}(\mathbb{R}^n)}.$$

The proof falls naturally into two cases. In the case $k < i$, we note that

$$|b(y) - b_{B_k}| \leq |b(y) - b_{B_i}| + \sum_{j=k}^{i-1} |b_{B_j} - b_{B_{j+1}}|$$

$$\leq |b(y) - b_{B_i}| + C\sum_{j=k}^{i-1} |B_{j+1}|^\lambda \|b\|_{\dot{\mathcal{E}}^{p,\lambda}(\mathbb{R}^n)}$$

$$\leq |b(y) - b_{B_i}| + C|B_k|^\lambda \|b\|_{\dot{\mathcal{E}}^{p,\lambda}(\mathbb{R}^n)}.$$

For the case $k > i$, it is easy to obtain

$$|b(y) - b_{B_k}| \leq |b(y) - b_{B_i}| + \sum_{j=i}^{k-1} |b_{B_j} - b_{B_{j+1}}|$$

$$\leq |b(y) - b_{B_i}| + C\sum_{j=i}^{k-1} |B_{j+1}|^\lambda \|b\|_{\dot{\mathcal{E}}^{p,\lambda}(\mathbb{R}^n)}$$

$$\leq |b(y) - b_{B_i}| + C|B_i|^\lambda \|b\|_{\dot{\mathcal{E}}^{p,\lambda}(\mathbb{R}^n)}.$$

$\square$

Under some stronger conditions on λ and p, the following result can be deduced if we drop the assumption that b satisfies (4.9).

Theorem 4.6. *Let $2 < p < \infty$ and $-1/(2p) < \lambda < 0$. Then the following statements are equivalent:*
(a) $b \in \dot{\mathcal{E}}^{p,\lambda}(\mathbb{R}^n)$;
(b) Both $[b, \mathcal{H}]$ and $[b, \mathcal{H}^*]$ are bounded from $\dot{B}^{p,\lambda}(\mathbb{R}^n)$ to $\dot{B}^{p,2\lambda}(\mathbb{R}^n)$.

Proof. $(a) \Rightarrow (b)$ The task is now to find a constant $C > 0$ such that for a fixed ball B_{k_0} with $k_0 \in \mathbb{Z}$, the following inequalities are true

$$\frac{1}{|B_{k_0}|^{2\lambda}} \left(\frac{1}{|B_{k_0}|} \int_{B_{k_0}} |[b, \mathcal{H}]f(x)|^p dx \right)^{1/p} \leq C\|f\|_{\dot{B}^{p,\lambda}(\mathbb{R}^n)} \qquad (4.12)$$

and

$$\frac{1}{|B_{k_0}|^{2\lambda}} \left(\frac{1}{|B_{k_0}|} \int_{B_{k_0}} |[b, \mathcal{H}^*]f(x)|^p dx \right)^{1/p} \leq C\|f\|_{\dot{B}^{p,\lambda}(\mathbb{R}^n)}. \qquad (4.13)$$

To deal with (4.12), we note that

$$\int_{B_{k_0}} |[b, \mathcal{H}]f(x)|^p dx \leq \int_{B_{k_0}} \left| \frac{1}{|x|^n} \int_{|y|<|x|} |b(x) - b(y)| f(y) dy \right|^p dx$$

$$\leq C \sum_{k=-\infty}^{k_0} \frac{1}{|B_k|^p} \int_{B_k} \left(\sum_{i=-\infty}^{k} \int_{B_i} |b(x) - b_{B_k}| |f(y)| dy \right)^p dx$$

$$+ C \sum_{k=-\infty}^{k_0} \frac{1}{|B_k|^p} \int_{B_k} \left(\sum_{i=-\infty}^{k} \int_{B_i} |b(y) - b_{B_k}| |f(y)| dy \right)^p dx$$

$$=: K + KK.$$

Using Hölder's inequality to p_1/p and $(p_1/p)'$, K can be controlled by

$$K \leq C \sum_{k=-\infty}^{k_0} \frac{1}{|B_k|^p} \int_{B_k} |b(x) - b_{B_k}|^p dx \left(\sum_{i=-\infty}^{k} \int_{B_i} |f(y)| dy \right)^p$$

$$\leq C \|b\|_{\dot{\mathcal{E}}^{p,\lambda}(\mathbb{R}^n)}^p \sum_{k=-\infty}^{k_0} |B_k|^{1+p\lambda-p} \left(\sum_{i=-\infty}^{k} \left(\int_{B_i} |f(y)|^p dy \right)^{1/p} |B_i|^{1/p'} \right)^p$$

$$\leq C \|b\|_{\dot{\mathcal{E}}^{p,\lambda}(\mathbb{R}^n)}^p \|f\|_{\dot{B}^{p,\lambda}(\mathbb{R}^n)}^p \sum_{k=-\infty}^{k_0} |B_k|^{1+p\lambda-p} \left(\sum_{i=-\infty}^{k} |B_i|^{1+\lambda} \right)^p$$

$$\leq C \|b\|_{\dot{\mathcal{E}}^{p,\lambda}(\mathbb{R}^n)}^p \|f\|_{\dot{B}^{p,\lambda}(\mathbb{R}^n)}^p \sum_{k=-\infty}^{k_0} 2^{kn(1+2p\lambda)}$$

$$\leq C \|b\|_{\dot{\mathcal{E}}^{p,\lambda}(\mathbb{R}^n)}^p \|f\|_{\dot{B}^{p,\lambda}(\mathbb{R}^n)}^p |B_{k_0}|^{1+2p\lambda},$$

where the fact $\lambda > -1/(2p)$ has been used in the last inequality. Applying Lemma 4.8 to the term KK can produce that

$$KK \leq C \sum_{k=-\infty}^{k_0} \frac{1}{|B_k|^p} \int_{B_k} \left(\sum_{i=-\infty}^{k} \int_{B_i} |b(y) - b_{B_i}| |f(y)| dy \right)^p dx$$

$$+ C \|b\|_{\dot{\mathcal{E}}^{p,\lambda}(\mathbb{R}^n)}^p \sum_{k=-\infty}^{k_0} \frac{1}{|B_k|^p} \int_{B_k} \left(\sum_{i=-\infty}^{k} \int_{B_i} |B_i|^\lambda |f(y)| dy \right)^p dx$$

$$=: KK_1 + KK_2.$$

Repeated application of Hölder's inequality shows that KK_1 is not bigger than

$$C \sum_{k=-\infty}^{k_0} \frac{1}{|B_k|^p} \int_{B_k} \left(\sum_{i=-\infty}^{k} \left(\int_{B_i} |b(y) - b_{B_i}|^{p'} dy \right)^{\frac{1}{p'}} \left(\int_{B_i} |f(y)|^p dy \right)^{\frac{1}{p}} \right)^p dx$$

$$\leq C \|f\|_{\dot{B}^{p,\lambda}(\mathbb{R}^n)}^p$$

$$\times \sum_{k=-\infty}^{k_0} \frac{1}{|B_k|^p} \int_{B_k} \left(\sum_{i=-\infty}^{k} \left(\int_{B_i} |b(y) - b_{B_i}|^p dy \right)^{1/p} |B_i|^{\lambda + \frac{1}{p'}} \right)^p dx$$

$$\leq C\|b\|_{\dot{\mathcal{E}}^{p,\lambda}(\mathbb{R}^n)}^p \|f\|_{\dot{B}^{p,\lambda}(\mathbb{R}^n)}^p \sum_{k=-\infty}^{k_0} \frac{1}{|B_k|^p} \int_{B_k} \left(\sum_{i=-\infty}^{k} |B_i|^{1+2\lambda} \right)^p dx$$

$$\leq C\|b\|_{\dot{\mathcal{E}}^{p,\lambda}(\mathbb{R}^n)}^p \|f\|_{\dot{B}^{p,\lambda}(\mathbb{R}^n)}^p \sum_{k=-\infty}^{k_0} \frac{1}{|B_k|^p} \int_{B_k} |B_k|^{(1+2\lambda)p} dx$$

$$\leq C\|b\|_{\dot{\mathcal{E}}^{p,\lambda}(\mathbb{R}^n)}^p \|f\|_{\dot{B}^{p,\lambda}(\mathbb{R}^n)}^p |B_{k_0}|^{1+2p\lambda}.$$

The term KK_2 can be bounded by

$$C\|b\|_{\dot{\mathcal{E}}^{p,\lambda}(\mathbb{R}^n)}^p \sum_{k=-\infty}^{k_0} \frac{1}{|B_k|^p} \int_{B_k} \left(\sum_{i=-\infty}^{k} \int_{B_i} |B_i|^{\lambda} |f(y)| dy \right)^p dx$$

$$\leq C\|b\|_{\dot{\mathcal{E}}^{p,\lambda}(\mathbb{R}^n)}^p \sum_{k=-\infty}^{k_0} \frac{1}{|B_k|^p} \int_{B_k} \left(\sum_{i=-\infty}^{k} |B_i|^{\frac{1}{p'}+\lambda} \left(\int_{B_i} |f(y)|^p dy \right)^{\frac{1}{p}} \right)^p dx$$

$$\leq C\|b\|_{\dot{\mathcal{E}}^{p,\lambda}(\mathbb{R}^n)}^p \|f\|_{\dot{B}^{p,\lambda}(\mathbb{R}^n)}^p \sum_{k=-\infty}^{k_0} \frac{1}{|B_k|^p} \int_{B_k} \left(\sum_{i=-\infty}^{k} |B_i|^{1+2\lambda} \right)^p dx$$

$$\leq C\|b\|_{\dot{\mathcal{E}}^{p,\lambda}(\mathbb{R}^n)}^p \|f\|_{\dot{B}^{p,\lambda}(\mathbb{R}^n)}^p \sum_{k=-\infty}^{k_0} |B_k|^{1+2\lambda p}$$

$$\leq C\|b\|_{\dot{\mathcal{E}}^{p,\lambda}(\mathbb{R}^n)}^p \|f\|_{\dot{B}^{p,\lambda}(\mathbb{R}^n)}^p |B_{k_0}|^{1+2p\lambda}.$$

We have thus proved the estimate for KK. (4.12) is based on the above estimates for K and KK.

With a slight modification of the proofs for (4.11) and (4.12), (4.13) can be obtained easily, we omit its proof here for the similarity.

$(b) \Rightarrow (a)$ We only need to show that for a fixed ball $B = B(\mathbf{0}, r)$,

$$\frac{1}{|B|^{1+p\lambda}} \int_B |b(y) - b_B|^p \leq C. \tag{4.14}$$

To do this, it will be necessary to note that

$$\frac{1}{|B|^{1+p\lambda}} \int_B |b(y) - b_B|^p dy$$

$$\leq \frac{1}{|B|^{1+p+p\lambda}} \int_B \left| \int_{|z|<|y|} (b(y) - b(z))\chi_B(z) dz \right|^p dy$$

$$+ \frac{1}{|B|^{1+p+p\lambda}} \int_B \left| \int_{|z|\geq|y|} (b(y) - b(z))\chi_B(z) dz \right|^p dy$$

$$=: L + LL.$$

The rest of the proof runs as that of Theorem 4.5, we now apply that argument again. The $(\dot{B}^{p,\lambda}(\mathbb{R}^n), \dot{B}^{p,2\lambda}(\mathbb{R}^n))$ boundedness of $[b, \mathcal{H}]$ produces the following estimate for the term L,

$$
\begin{aligned}
L &\leq \frac{C}{|B|^{1+p+p\lambda}} \int_B |y|^{np} \, |([b, \mathcal{H}]\chi_B)(y)|^p \, dy \\
&\leq C|B|^{p\lambda} \|[b, \mathcal{H}]\chi_B\|^p_{\dot{B}^{p,2\lambda}(\mathbb{R}^n)} \\
&\leq C|B|^{p\lambda} \|\chi_B\|^p_{\dot{B}^{p,\lambda}(\mathbb{R}^n)} \\
&\leq C.
\end{aligned}
$$

Similar arguments apply to the term LL, the following can be confirmed easily

$$
\begin{aligned}
LL &\leq \frac{C}{|B|^{1+p+p\lambda}} \int_B \left(\int_{|z| \geq |y|} |z|^n \frac{|b(y) - b(z)|}{|z|^n} \chi_B(z) dz \right)^p dy \\
&\leq \frac{C}{|B|^{1+p\lambda}} \int_B |([b, \mathcal{H}^*]\chi_B)(y)|^p \, dy \\
&\leq C|B|^{p\lambda} \|[b, \mathcal{H}^*]\chi_B\|^p_{\dot{B}^{p,2\lambda}(\mathbb{R}^n)} \\
&\leq C|B|^{p\lambda} \|\chi_B\|^p_{\dot{B}^{p,\lambda}(\mathbb{R}^n)} \\
&\leq C.
\end{aligned}
$$

After noticing the above estimates for L and LL, (4.14) can then be proved. $\qquad\square$

Next, we give some characterizations of $\dot{\mathcal{E}}^{p,\lambda}(\mathbb{R}^n)$ with $\lambda < 0$ via the boundedness of $[b, \mathcal{H}_\beta]$ and $[b, \mathcal{H}_\beta^*]$ on $\dot{B}^{p,\lambda}(\mathbb{R}^n)$.

Theorem 4.7. *Let p, λ, $p_i, \lambda_i, i = 1, 2$, b as in Theorem 4.5, $0 < \beta < \min\{n(1 - 1/p), n(\lambda_2 + 1/p_2)\}$ and let $\alpha = \lambda_2 - \beta/n$. Then the following statements are equivalent:*
(a) $b \in \dot{\mathcal{E}}^{p_1, \lambda_1}(\mathbb{R}^n)$;
(b) Both $[b, \mathcal{H}_\beta]$ and $[b, \mathcal{H}_\beta^*]$ are bounded operators from $\dot{B}^{p_2, \alpha}(\mathbb{R}^n)$ to $\dot{B}^{p,\lambda}(\mathbb{R}^n)$.

Theorem 4.8. *Let $2 < p < \infty$, $-1/(2p) < \lambda < 0$, $0 < \beta < \min\{n(1 - 1/p), n(\lambda + 1/p)\}$ and let $\alpha = \lambda - \beta/n$. Then the following statements are equivalent:*
(a) $b \in \dot{\mathcal{E}}^{p,\lambda}(\mathbb{R}^n)$;
(b) Both $[b, \mathcal{H}_\beta]$ and $[b, \mathcal{H}_\beta^*]$ are bounded operators from $\dot{B}^{p,\alpha}(\mathbb{R}^n)$ to $\dot{B}^{p,2\lambda}(\mathbb{R}^n)$.

The methods used in the proofs of Theorem 4.5 and Theorem 4.6 remain valid for that of Theorem 4.7 and Theorem 4.8 with only a slight modification. We omit their proofs here for their similarity.

It is also interesting to know that the corresponding results on bounded domains gain independent interest since many problems for PDEs are about the local behavior of solutions and discussed on bounded domains. For instance, the original form of Morrey space was first introduced by Morrey [Morrey (1938)] to investigate the local behavior of solutions to the second order elliptic PDEs. Under the condition (4.1), we are thus led to the following version of Theorems 4.5–4.8 on any bounded domain Ω as a result of (4.2).

Corollary 4.1. *Let p, λ, $p_i, \lambda_i, i = 1, 2$, and let b be as in Theorem 4.5. Then the following statements are equivalent:*
(a) $b \in \dot{B}^{p_1, \lambda_1}(\Omega)$;
(b) Both $[b, \mathcal{H}]$ and $[b, \mathcal{H}^*]$ are bounded operators from $\dot{B}^{p_2, \lambda_2}(\Omega)$ to $\dot{B}^{p, \lambda}(\Omega)$.

Corollary 4.2. *Let p, λ be as in Theorem 4.5. Then the following statements are equivalent:*
(a) $b \in \dot{B}^{p, \lambda}(\Omega)$;
(b) Both $[b, \mathcal{H}]$ and $[b, \mathcal{H}^*]$ are bounded operators from $\dot{B}^{p, \lambda}(\Omega)$ to $\dot{B}^{p, 2\lambda}(\Omega)$.

Corollary 4.3. *Let p, λ, b, β, $p_i, \lambda_i, i = 1, 2$, and let α be as in Theorem 4.7. Then the following statements are equivalent:*
(a) $b \in \dot{B}^{p_1, \lambda_1}(\Omega)$;
(b) Both $[b, \mathcal{H}_\beta]$ and $[b, \mathcal{H}_\beta^*]$ are bounded operators from $\dot{B}^{p_2, \alpha}(\Omega)$ to $\dot{B}^{p, \lambda}(\Omega)$.

Corollary 4.4. *Let p, λ, β and let α be as in Theorem 4.8. Then the following statements are equivalent:*
(a) $b \in \dot{B}^{p, \lambda}(\Omega)$;
(b) Both $[b, \mathcal{H}_\beta]$ and $[b, \mathcal{H}_\beta^*]$ are bounded operators from $\dot{B}^{p, \alpha}(\Omega)$ to $\dot{B}^{p, 2\lambda}(\Omega)$.

The methods used in this subsection are quite different from that of [Shi and Lu (2013, 2014)], which depend heavily on the smoothness of the kernel functions of the corresponding integral operators. It is also worth pointing out that the ideas used in [Fu et al. (2007)] and [Zhao and Lu (2013)], which deal with the case $\lambda \geq 0$, can not be adopted directly to the case $\lambda < 0$, especially in the estimates of $[b, \mathcal{H}^*]$ and $[b, \mathcal{H}_\beta^*]$. Our theorems provide a natural and intrinsic characterization of central Campanato space via the boundedness of commutator operators on central Morrey space. This work is intended as an attempt to some further characterizations of central function spaces.

4.2 The compactness characterizations

The existing results for the compactness of operators are all related to singular integral operators. For the average operators, one of the three most important operators in harmonic analysis due to Stein [Stein (1993)], as a concept of highly independent interest, has received little attention to the best of our knowledge. The second aim of this chapter is to explore the compactness of $[b, \mathcal{H}]$ and $[b, \mathcal{H}^*]$ on the $L^p(\mathbb{R}^n)$ space. It is easy to see that the structure of a Hardy operator is centrally symmetric, being quite different from that of other operators (including the singular integral operator, Riesz potential and bilinear operator). This might be the reason why the existing methods (for example, the John-Nirenberg inequality) used to address the singular integral operator can not be applied when considering the Hardy operator. In this section, we prove the main result by some new ideas via exploiting the center symmetry of the Hardy operator and function space deeply. To do so, some lemmas are needed. The first pertains to the properties of the $C\dot{M}O(\mathbb{R}^n)$ space.

Lemma 4.9. *The following properties of the space $C\dot{M}O(\mathbb{R}^n)$ are valid:*

(a) $L^\infty(\mathbb{R}^n) \subset C\dot{M}O(\mathbb{R}^n)$ and $\|b\|_{C\dot{M}O(\mathbb{R}^n)} \leq 2\|b\|_{L^\infty(\mathbb{R}^n)}$.

(b) Assume that there exists $C > 0$ such that for all balls $B(\mathbf{0}, r) \subset \mathbb{R}^n$, there exists a constant c satisfying

$$\sup_r \frac{1}{|B(\mathbf{0}, r)|} \int_{B(\mathbf{0}, r)} |b(x) - c| dx \leq C.$$

Then, $b \in C\dot{M}O(\mathbb{R}^n)$ and $\|b\|_{C\dot{M}O(\mathbb{R}^n)} \leq 2C$.

(c) $\|b\|_{C\dot{M}O(\mathbb{R}^n)} \sim \sup_r \inf_{c \in \mathbb{R}} \frac{1}{|B(\mathbf{0}, r)|} \int_{B(\mathbf{0}, r)} |b(x) - c| dx.$

(d) If $b \in C\dot{M}O(\mathbb{R}^n)$, then $b \in CVMO(\mathbb{R}^n)$ (central $VMO(\mathbb{R}^n)$ space, i.e. the $C\dot{M}O(\mathbb{R}^n)$ closure of $C_c^\infty(\mathbb{R}^n)$) if and only if b satisfies the following two conditions

$$\lim_{r \to 0} \sup_r N(b, B(\mathbf{0}, r)) = 0; \tag{4.15}$$

$$\lim_{r \to \infty} \sup_r N(b, B(\mathbf{0}, r)) = 0, \tag{4.16}$$

where

$$N(b, B(\boldsymbol{0}, r)) := \inf_{c \in \mathbb{R}} \frac{1}{|B(\boldsymbol{0}, r)|} \int_{B(\boldsymbol{0}, r)} |b(x) - c| dx.$$

Proof. We prove (a)–(c) by a slight modification of [Grafakos (2009), Proposition 7.1.2]. It is easy to check (a) by the following observation,

$$\frac{1}{|B(\boldsymbol{0}, r)|} \int_{B(\boldsymbol{0}, r)} |b(x) - b_{B(\boldsymbol{0}, r)}| dx \le 2\|b\|_{L^\infty(\mathbb{R}^n)}.$$

For (b), we first note that

$$|b - b_{B(\boldsymbol{0}, r)}| \le |b - c| + |b_{B(\boldsymbol{0}, r)} - c| \le |b - c| + \frac{1}{|B(\boldsymbol{0}, r)|} \int_{B(\boldsymbol{0}, r)} |b(x) - c| dx,$$

whence finding

$$\frac{1}{|B(\boldsymbol{0}, r)|} \int_{B(\boldsymbol{0}, r)} |b(x) - b_{B(\boldsymbol{0}, r)}| dx \le \frac{1}{|B(\boldsymbol{0}, r)|} \int_{B(\boldsymbol{0}, r)} |b(x) - c| dx$$

$$+ \frac{1}{|B(\boldsymbol{0}, r)|} \int_{B(\boldsymbol{0}, r)} \frac{1}{|B(\boldsymbol{0}, r)|} \int_{B(\boldsymbol{0}, r)} |b(x) - c| dx dy.$$

Hence $\|b\|_{C\dot{M}O(\mathbb{R}^n)} \le 2C$ is desired. The proof of (c) is equivalent to the following inequality

$$\frac{1}{2} \|b\|_{C\dot{M}O(\mathbb{R}^n)} \le \sup_r \inf_{c \in \mathbb{R}} \frac{1}{|B(\boldsymbol{0}, r)|} \int_{B(\boldsymbol{0}, r)} |b(x) - c| dx \le \|b\|_{C\dot{M}O(\mathbb{R}^n)},$$

where the lower inequality follows from (b), while the upper one is trivial.

(d) is a characterization of the $CVMO$ space and can be seen as a central version of [Uchiyama (1978), Lemma], which was presented in [Neri (1975)] without proof. Next, we show (d), which is partly inspired by the proof of [Uchiyama (1978), Lemma]. As we shall see, it requires a large modification since the $C\dot{M}O(\mathbb{R}^n)$ space is central space.

The sufficiency of (d) is trivial by the definition of the $CVMO(\mathbb{R}^n)$ space. Next, we prove the necessity. To do this, it is sufficient to show that if b satisfies (4.15) and (4.16), then for any $\varepsilon > 0$, there exists $b_\varepsilon \in C\dot{M}O(\mathbb{R}^n)$ such that

$$\inf_{h \in C_c^\infty(\mathbb{R}^n)} \|b_\varepsilon - h\|_{C\dot{M}O(\mathbb{R}^n)} \le C\varepsilon \tag{4.17}$$

and

$$\|b_\varepsilon - b\|_{C\dot{M}O(\mathbb{R}^n)} \le C\varepsilon. \tag{4.18}$$

By (4.15) and (4.16), there exist $K_1, K_2 \in \mathbb{Z}$ such that

$$\sup_k N(b, B_k) < \varepsilon \quad \text{for} \quad k \le K_1 \tag{4.19}$$

and

$$\sup_{k} N(b, B_k) < \varepsilon \quad \text{for} \quad k \geq K_2. \tag{4.20}$$

Define $b'_\varepsilon(x) = b_{B^x}$, where

$$B^x = \begin{cases} B_{K_1}, & x \in B_{K_1}; \\ B_{K_2}, & x \in B_{K_2} \backslash B_{K_1}; \\ B_k, & x \in B_k \backslash B_{k-1}, \ k \geq K_2 + 1. \end{cases}$$

By (4.20), there exists $K_3 > K_2$ such that

$$\sup\{|b'_\varepsilon(x) - b'_\varepsilon(y)| : x, y \in B_{K_3} \backslash B_{K_3-1}\} < \varepsilon. \tag{4.21}$$

Without loss of generality, we may assume that $K_3 = K_2 + 1$ in the following analysis.

Set

$$b_\varepsilon(x) = \begin{cases} b'_\varepsilon(x), & x \in B_{K_3}; \\ b_{B_{K_3}}, & x \in B^c_{K_3}. \end{cases}$$

Then, b_ε is our desired function. Namely, b_ε satisfies (4.17) and (4.18). In doing so, we first claim that for any x and y,

$$|b_\varepsilon(x) - b_\varepsilon(y)| < C\varepsilon \tag{4.22}$$

and

$$\frac{1}{|B(\mathbf{0}, r)|} \int_{B(\mathbf{0}, r)} |b(x) - b_\varepsilon(x)| dx < C\varepsilon \quad \text{for} \quad r > 0. \tag{4.23}$$

Indeed, we can show (4.22) by four cases.

CASE 1. $x, y \in B_{K_3}$. This case can be divided into four subcases.

Subcase 1. $x, y \in B_{K_3} \backslash B_{K_2}$. (4.22) is an immediate consequence of (4.21).

Subcase 2. $x, y \in B_{K_2}$. In this subcase, (4.19) and (4.20) can be used to obtain that $|b_\varepsilon(x) - b_\varepsilon(y)|$

$$= \begin{cases} \frac{1}{|B_{K_1}|} \int_{B_{K_1}} |b(z) - b_{B_{K_1}}| dz < C\varepsilon, \ x, y \in B_{K_1}; \\ \frac{1}{|B_{K_2}|} \int_{B_{K_2}} |b(z) - b_{B_{K_2}}| dz < C\varepsilon, \ x, y \in B_{K_2} \backslash B_{K_1}; \\ \frac{1}{|B_{K_1}|} \int_{B_{K_1}} |b(z) - b_{B_{K_2}}| dz < C\varepsilon, \ x \in B_{K_1}, y \in B_{K_2} \backslash B_{K_1}; \\ \frac{1}{|B_{K_2}|} \int_{B_{K_2}} |b(z) - b_{B_{K_1}}| dz < C\varepsilon, \ y \in B_{K_1}, x \in B_{K_2} \backslash B_{K_1}. \end{cases}$$

Subcase 3. $x \in B_{K_2}$, $y \in B_{K_3} \backslash B_{K_2}$. We conclude from (4.20) that

$$|b_\varepsilon(x) - b_\varepsilon(y)| = \frac{1}{|B_{K_2}|} \int_{B_{K_2}} |b(z) - b_{B_{K_3}}| dz < C\varepsilon.$$

Subcase 4. $y \in B_{K_2}$, $x \in B_{K_3} \backslash B_{K_2}$. This subcase can can be handled in much the same way as that of Subcase 3.

CASE 2. $x, y \in B_{K_3}^c$. (4.22) can be deduced from the definition of b_ε.

CASE 3. $x \in B_{K_3}, y \in B_{K_3}^c$. A further use of (4.19) and (4.20), we can get (4.22) since

$$
|b_\varepsilon(x) - b_\varepsilon(y)| =
\begin{cases}
\frac{1}{|B_{K_1}|} \int_{B_{K_1}} |b(z) - b_{B_{K_3}}| dz < C\varepsilon, & x \in B_{K_1}; \\
\frac{1}{|B_{K_2}|} \int_{B_{K_2}} |b(z) - b_{B_{K_3}}| dz < C\varepsilon, & x \in B_{K_2} \backslash B_{K_1}; \\
\frac{1}{|B_{K_3}|} \int_{B_{K_3}} |b(z) - b_{B_{K_3}}| dz < C\varepsilon, & x \in B_{K_3} \backslash B_{K_2}.
\end{cases}
$$

CASE 4. $x \in B_{K_3}^c, y \in B_{K_3}$. CASE (iii) still works for this case.

It is concluded from (4.22) that $b_\varepsilon \in C\dot{M}O(\mathbb{R}^n)$ and (4.17) is obvious.

Now, we are in a position to show (4.23) which can be divided into three cases.

CASE 1. $B(\mathbf{0}, r) \subseteq B_{K_1}$. Using (4.19), we have

$$
\frac{1}{|B(\mathbf{0}, r)|} \int_{B(\mathbf{0}, r)} |b(y) - b_\varepsilon(y)| dy = \frac{1}{|B(\mathbf{0}, r)|} \int_{B(\mathbf{0}, r)} |b(y) - b_{B_{K_1}}| dy < C\varepsilon.
$$

CASE 2. $B(\mathbf{0}, r) \subseteq B_{K_2}$ with $r > 2^{K_1}$. In this case, Minkowski's inequality and (4.20) are used to obtain

$$
\frac{1}{|B(\mathbf{0}, r)|} \int_{B(\mathbf{0}, r)} |b(y) - b_\varepsilon(y)| dy \leq \frac{1}{|B_{K_2}|} \int_{B_{K_2}} |b(y) - b_{B(\mathbf{0}, r)}| dy < C\varepsilon.
$$

CASE 3. $B(\mathbf{0}, r) \subseteq B_{K_2}^c$. By applying (4.20) to $B(\mathbf{0}, r)$, we obtain

$$
\frac{1}{|B(\mathbf{0}, r)|} \int_{B(\mathbf{0}, r)} |b(y) - b_\varepsilon(y)| dy \leq \frac{1}{|B(\mathbf{0}, r)|} \int_{B(\mathbf{0}, r)} |b(y) - b_{B_{K_3}}| dy < C\varepsilon,
$$

thus finding (4.23). This shows that for any $B(\mathbf{0}, r) \subset \mathbb{R}^n$ with $r > 0$, $N(b - b_\varepsilon, B(\mathbf{0}, r)) < C\varepsilon$. This result together with (c) gives (4.18). $\quad\square$

The second lemma is the Frechet-Kolmogorov theorem [Yosida (1995)], which gives a characterization of compact sets.

Lemma 4.10. *Let $G \subset L^p(\mathbb{R}^n)$ and $E_\alpha = \{x \in \mathbb{R}^n : |x| > \alpha\}$. Then G is strongly precompact, if and only if,*

$$
\sup_{f \in G} \|f\|_{L^p(\mathbb{R}^n)} < \infty; \tag{4.24}
$$

$$
\lim_{|y| \to 0} \|f(\cdot + y) - f(\cdot)\|_{L^p(\mathbb{R}^n)} = 0 \quad \text{uniformly in} \quad f \in G; \tag{4.25}
$$

$$
\lim_{\alpha \to \infty} \|f\chi_{E_\alpha}\|_{L^p(\mathbb{R}^n)} = 0 \quad \text{uniformly in} \quad f \in G. \tag{4.26}
$$

The third lemma is the L^p-boundedness of $\mathcal{H}$, which was obtained by Christ and Grafakos [Christ and Grafakos (1995)].

Lemma 4.11. *Let $1 < p < \infty$ and f be a locally integral function on $\mathbb{R}^n$. Then there exists a constant $C > 0$ such that*

$$\|\mathcal{H}f\|_{L^p(\mathbb{R}^n)} \le C\|f\|_{L^p(\mathbb{R}^n)}.$$

Let us now formulate the main results of this section.

Theorem 4.9. *Let $1 < p < \infty$. Then, $b \in CVMO(\mathbb{R}^n)$ if and only if both $[b, \mathcal{H}]$ and $[b, \mathcal{H}^*]$ are compact on $L^p(\mathbb{R}^n)$.*

Theorem 4.9 is contained in the following two theorems.

Theorem 4.10. *Let $[b, \mathcal{H}]$ and $[b, \mathcal{H}^*]$ be compact operators on $L^p(\mathbb{R}^n)$ with $1 < p < \infty$. Then, $b \in CVMO(\mathbb{R}^n)$.*

Theorem 4.11. *If $b \in CVMO(\mathbb{R}^n)$, then both $[b, \mathcal{H}]$ and $[b, \mathcal{H}^*]$ are compact operators on $L^p(\mathbb{R}^n)$ with $1 < p < \infty$.*

Proof of Theorem 4.10. Since $[b, \mathcal{H}]$ and $[b, \mathcal{H}^*]$ are compact operators on $L^p(\mathbb{R}^n)$, $b \in C\dot{M}O^{\max\{p,p'\}}(\mathbb{R}^n)$ [Fu et al. (2007), Corollary 2.1], and hence $b \in C\dot{M}O(\mathbb{R}^n)$.

Without loss of generality, we may assume that $\|b\|_{C\dot{M}O^{\max\{p,p'\}}(\mathbb{R}^n)} = 1$. Our task is to show (4.15)–(4.16) of Lemma 4.9. This can be done by contradiction.

We begin with the assumption that b does not satisfy (4.15). Then, it is immediately that there is a $\delta > 0$ and a sequence of balls $\{B(\mathbf{0}, r_i)\}_{i=1}^{\infty}$ with $\lim_{i \to \infty} r_i = 0$ satisfying

$$\frac{1}{|B(\mathbf{0}, r_i)|} \int_{B(\mathbf{0}, r_i)} |b(y) - b_{B(\mathbf{0}, r_i)}| dy > \delta. \tag{4.27}$$

Letting

$$\varepsilon_0 = \frac{1}{|B(\mathbf{0}, r_i)|} \int_{B(\mathbf{0}, r_i)} sgn(b(y) - b_{B(\mathbf{0}, r_i)}) dy,$$

we choose the function

$$g_i(y) = \frac{\chi_{B(\mathbf{0}, r_i)}(y)}{|B(\mathbf{0}, r_i)|^{\frac{1}{p}}} \left(sgn(b(y) - b_{B(\mathbf{0}, r_i)}) - \varepsilon_0 \right), \quad i = 1, 2, \ldots \tag{4.28}$$

to achieve

$$\begin{cases} g_i \in L^p(\mathbb{R}^n); \\ \text{supp } g_i \subset B(\mathbf{0}, r_i); \\ g_i(y)(b(y) - b_{B(\mathbf{0}, r_i)}) > 0; \\ |g_i(y)| \leq 2|B(\mathbf{0}, r_i)|^{-\frac{1}{p}}, \text{ with } y \in B(\mathbf{0}, r_i); \\ \int_{\mathbb{R}^n} g_i(y) dy = 0. \end{cases} \tag{4.29}$$

Hence, $\{[b, \mathcal{H}]g_i\}_{i=1}^{\infty}$ is a bounded set in $L^p(\mathbb{R}^n)$. Next, we pick a subsequence $\{[b, \mathcal{H}]g_{i_m}\}_{m=1}^{\infty}$ from $\{[b, \mathcal{H}]g_i\}_{i=1}^{\infty}$ such that $\{[b, \mathcal{H}]g_{i_m}\}_{m=1}^{\infty}$ has no convergence subsequence in $L^p(\mathbb{R}^n)$ to show that $[b, \mathcal{H}]$ is not a compact operator on $L^p(\mathbb{R}^n)$. This contradiction will show that b must satisfy (4.15). To do so, we first need the following estimates for $\{[b, \mathcal{H}]g_i\}_{i=1}^{\infty}$.

Lemma 4.12. *Let* $b \in C\dot{M}O(\mathbb{R}^n)$, $1 < p < \infty$ *and* δ, g_i *be defined as in* (4.27) *and* (4.28), *respectively. Then there exist constants* $C_2 > C_1 > 2$, $C_3 > 0$ *and* $0 < \varepsilon \ll C_2$, *which are dependent only on* n, p, δ *and* b *such that*

$$\left(\int_{S_1} |[b, \mathcal{H}]g_i(x)|^p dx \right)^{1/p} \geq C_3 \tag{4.30}$$

$$\left(\int_{S_2} |[b, \mathcal{H}]g_i(x)|^p dx \right)^{1/p} \leq C_3/4 \tag{4.31}$$

and

$$\left(\int_{S_3} |[b, \mathcal{H}]g_i(x)|^p dx \right)^{1/p} \leq C_3/4, \tag{4.32}$$

where

$$\begin{cases} S_1 = \{x : C_1 r_i < |x| < C_2 r_i\}; \\ S_2 = \{x : |x| > C_2 r_i\}; \\ S_3 \subset S_1 \text{ with } |S_3|/|B_{r_i}| < \varepsilon^n. \end{cases}$$

Proof. For fixed r_i and $x \in (\alpha B(\mathbf{0}, r_i))^c$ with $\alpha > 2$, one gets from (4.29) and the Hölder inequality that

$$|\mathcal{H}((b - b_{B(\mathbf{0}, r_i)})g_i)(x)| \leq \frac{1}{|x|^n} \int_{B_{r_i}} |b(y) - b_{B(\mathbf{0}, r_i)}||g_i(y)| dy$$

$$\leq \frac{1}{|x|^n} \left(\int_{B(\mathbf{0}, r_i)} |b(y) - b_{B(\mathbf{0}, r_i)}|^{p'} dy \right)^{1/p'} \left(\int_{B(\mathbf{0}, r_i)} |g_i(y)|^p dy \right)^{1/p}$$

$$\leq C\|b\|_{C\dot{B}MO^{p'}(\mathbb{R}^n)} \frac{|B(\mathbf{0}, r_i)|^{1/p'}}{|x|^n} \left(\int_{B(\mathbf{0}, r_i)} |B(\mathbf{0}, r_i)|^{-1} dy \right)^{1/p}.$$

Namely

$$\left|\mathcal{H}((b-b_{B(\mathbf{0},r_i)})g_i)(x)\right| \leq \frac{C|B(\mathbf{0},r_i)|^{1/p'}}{|x|^n}. \tag{4.33}$$

On the other hand, (4.27) and (4.29) are used to obtain

$$\left|\mathcal{H}((b-b_{B(\mathbf{0},r_i)})g_i)(x)\right| \geq \frac{C\delta|B(\mathbf{0},r_i)|^{1/p'}}{|x|^n}. \tag{4.34}$$

Furthermore, noting that $\int_{\mathbb{R}^n} g_i(y)dy = 0$, one has

$$\left|(b(x)-b_{B(\mathbf{0},r_i)})\mathcal{H}g_i(x)\right| = (b(x)-b_{B(\mathbf{0},r_i)})\frac{1}{|x|^n}\int_{B(\mathbf{0},r_i)}g_i(y)dy$$

$$= 0. \tag{4.35}$$

Hence, we have for $a > \max\{\alpha, 8\}$

$$\left(\int_{\{|x|>ar_i\}}\left|(b(x)-b_{B(\mathbf{0},r_i)})\mathcal{H}(g_i)(x)\right|^p dx\right)^{1/p} = 0. \tag{4.36}$$

Upon setting $S_4 = \{x : ar_i < |x| < dr_i\}$ for $d > a$, (4.34) and (4.36) show that

$$\left(\int_{S_4}|[b,\mathcal{H}]g_i(x)|^p dx\right)^{1/p} \geq \left(\int_{S_4}|\mathcal{H}(b-b_{B(\mathbf{0},r_i)})g_i(x)|^p dx\right)^{1/p}$$

$$-\left(\int_{\{|x|>ar_i\}}|(b(x)-b_{B(\mathbf{0},r_i)})\mathcal{H}g_i(x)|^p dx\right)^{1/p}$$

$$\geq C\delta\left(a^{n(1-p)} - d^{n(1-p)}\right)^{1/p}. \tag{4.37}$$

On the other hand, combining (4.33) with (4.36), we can assert that

$$\left(\int_{\{|x|>dr_i\}}|[b,\mathcal{H}]g_i(x)|^p dx\right)^{1/p}$$

$$\leq \left(\int_{\{|x|>dr_i\}}|\mathcal{H}(b-b_{B(\mathbf{0},r_i)})g_i(x)|^p dx\right)^{1/p}$$

$$+\left(\int_{\{|x|>dr_i\}}|(b(x)-b_{B(\mathbf{0},r_i)})\mathcal{H}g_i(x)|^p dx\right)^{1/p}$$

$$\leq C|B(\mathbf{0},r_i)|^{1/p'}\left(\int_{\{|x|>dr_i\}}|x|^{-np}dx\right)^{1/p}.$$

Namely,

$$\left(\int_{\{|x|>dr_i\}} |[b,\mathcal{H}]g_i(x)|^p dx\right)^{1/p} \leq Cd^{-n/p'}. \tag{4.38}$$

It is easy to check that there exist constants $C_2 > C_1 > 2$ and $C_3 > 0$, which are dependent only on n, p, δ and b, such that the required inequalities (4.30) and (4.31) are true.

Now it remains to prove (4.32). Let $S_3 \subset S_1$ be an arbitrary measurable set. Combining (4.33), (4.35) with the Minkowski inequality, one has

$$\left(\int_{S_3} |[b,\mathcal{H}]g_i(x)|^p dx\right)^{1/p} \leq \left(\int_{S_3} |\mathcal{H}(b - b_{B(\mathbf{0},r_i)})g_i(x)|^p dx\right)^{1/p}$$

$$+ \left(\int_{S_3} |(b(x) - b_{B(\mathbf{0},r_i)})\mathcal{H}g_i(x)|^p dx\right)^{1/p}$$

$$\leq C|B(\mathbf{0},r_i)|^{1/p'} \left(\int_{S_3} |x|^{-np} dx\right)^{1/p}$$

$$\leq C \left(\frac{|S_3|}{|B(\mathbf{0},r_i)|}\right)^{1/p}$$

$$\leq C\varepsilon^{n/p}, \tag{4.39}$$

where in the last inequality we have used $|S_3| < |B(\mathbf{0},r_i)|\varepsilon^n$, and thus (4.32) holds by taking $\varepsilon \ll C_2$. And hence, the proof of Lemma 4.12 is completed. $\qquad\square$

Now, we are in a position to use Lemma 4.12 to choose a subsequence $\{[b,\mathcal{H}]g_{i_m}\}_{m=1}^\infty$ such that $\{[b,\mathcal{H}]g_{i_m}\}_{m=1}^\infty$ has no convergence subsequence in $L^p(\mathbb{R}^n)$, thus reaching (4.15). In doing so, it is sufficient to verify that there exists a subsequence $\{[b,\mathcal{H}]g_{i_m}\}_{m=1}^\infty$ and $\tau > 0$, independent of g_{i_m}, such that

$$\|[b,\mathcal{H}]g_{i_m} - [b,\mathcal{H}]g_{i_{m+k}}\|_{L^p(\mathbb{R}^n)} \geq \tau. \tag{4.40}$$

Since $\lim_{i\to\infty} r_i = 0$, we may choose a subsequence $B(\mathbf{0},r_{i_m})$ from $B(\mathbf{0},r_i)$ such that

$$r_{i_{m+1}}/r_{i_m} < \varepsilon/C_2.$$

Here ε, C_2 are defined as that of in Lemma 4.12. Next, we claim that if the function g_{i_m} is defined relative to $B(\mathbf{0},r_{i_m})$ as in (4.28), then $\{[b,\mathcal{H}]g_{i_m}\}_{m=1}^\infty$ is our desired subsequence. Indeed, for $k, m \in \mathbb{N}$,

$$\|[b,\mathcal{H}]g_{i_m} - [b,\mathcal{H}]g_{i_{m+k}}\|_{L^p(\mathbb{R}^n)}$$

$$\geq \left(\int_{S_5} |[b,\mathcal{H}]g_{i_m}(x)|^p dx\right)^{1/p} - \left(\int_{S_6} |[b,\mathcal{H}]g_{i_{m+k}}(x)|^p dx\right)^{1/p},$$

where $S_5 = \{x : C_1 r_{i_m} < |x| < C_2 r_{i_m}\}$, $S_6 = \{x : |x| > C_2 r_{i_{m+k}}\}$, and C_1, C_2 are defined in Lemma 4.12. Then a further use of (4.30) and (4.31) derives

$$\|[b, \mathcal{H}]g_{i_m} - [b, \mathcal{H}]g_{i_{m+k}}\|_{L^p(\mathbb{R}^n)}$$
$$\geq \left(\int_{S_5} |[b, \mathcal{H}]g_{i_m}(x)|^p dx\right)^{1/p} - \left(\int_{S_6} |[b, \mathcal{H}]g_{i_{m+k}}(x)|^p dx\right)^{1/p}$$
$$\geq C_3 - \frac{C_3}{4} = \frac{3C_3}{4},$$

whence finding (4.40).

The proof of Theorem 4.10 is completed since the proof of (4.15) still works for (4.16). □

To prove Theorem 4.11, we first recall the following lemma which can help us to simplify the proof by considering only $b \in C_c^\infty(\mathbb{R}^n)$.

Lemma 4.13. *Let $[b, T]$ be the commutator of operator T and $1 < p < \infty$. If $[b, T]$ is a compact operator on $L^p(\mathbb{R}^n)$ with $b \in C_c^\infty(\mathbb{R}^n)$, then $[b, T]$ is also a compact operator on $L^p(\mathbb{R}^n)$ with $b \in CVMO(\mathbb{R}^n)$.*

Proof. The proof of Lemma 4.13 follows by a modification of [Chen and Ding (2010), p. 2645]. Note first that for $b \in C\dot{M}O(\mathbb{R}^n)$, there exists $\{b_\varepsilon\} \subset C_c^\infty(\mathbb{R}^n)$ with $\varepsilon > 0$ satisfying $\|b - b_\varepsilon\|_{C\dot{M}O(\mathbb{R}^n)} < \varepsilon$. Then by the L^p boundedness of $[b, T]$, one has

$$\|[b, T] - [b_\varepsilon, T]\|_{L^p(\mathbb{R}^n)} = \|[b - b_\varepsilon, T]\|_{L^p(\mathbb{R}^n)}$$
$$\leq C\|b - b_\varepsilon\|_{C\dot{M}O(\mathbb{R}^n)} \leq C\varepsilon. \tag{4.41}$$

Let

$$Q = \{f : f \in L^p(\mathbb{R}^n) \ \& \ \|f\|_{L^p(\mathbb{R}^n)} \leq C\}.$$

Then the L^p compactness of $[b_\varepsilon, T]$ implies that the set $S = \{[b_\varepsilon, T]f : f \in Q\}$ is strongly precompact. Now, it is sufficient to show that $S = \{[b, T]f : f \in Q\}$ satisfies (4.24)–(4.26). (4.24) and (4.41) are used to produce that

$$\sup_{f \in Q} \|[b, T]f\|_{L^p(\mathbb{R}^n)} \leq \sup_{f \in Q} \|[b_\varepsilon, T]f\|_{L^p(\mathbb{R}^n)} + C\varepsilon < \infty.$$

Moreover, we have

$$\lim_{|y| \to 0} \|[b, T]f(\cdot + y) - [b, T]f(\cdot)\|_{L^p(\mathbb{R}^n)}$$
$$= \lim_{|y| \to 0} \|[b, T]f(\cdot + y) - [b_\varepsilon, T]f(\cdot + y)$$

$$+ [b_\varepsilon, T]f(\cdot + y) - [b_\varepsilon, T]f(\cdot) + [b_\varepsilon, T]f(\cdot) - [b, T]f(\cdot)\|_{L^p(\mathbb{R}^n)}$$

$$= \lim_{|y| \to 0} \|[b_\varepsilon, T]f(\cdot + y) - [b_\varepsilon, T]f(\cdot)$$

$$+ [b - b_\varepsilon, T]f(\cdot + y) + [b - b_\varepsilon, T]f(\cdot)\|_{L^p(\mathbb{R}^n)}$$

$$\leq \lim_{|y| \to 0} \|[b_\varepsilon, T]f(\cdot + y) - [b_\varepsilon, T]f(\cdot)\|_{L^p(\mathbb{R}^n)} + 2\|[b - b_\varepsilon, T]f\|_{L^p(\mathbb{R}^n)}$$

$$\leq 2C\varepsilon \mapsto 0 \quad \text{uniformly for } f \in Q \text{ with } \varepsilon \to 0.$$

A similar argument gives

$$\lim_{\alpha \to \infty} \|[b, T]\chi_{E_\alpha}\|_{L^p(\mathbb{R}^n)} \leq \lim_{\alpha \to \infty} \|[b_\varepsilon, T]f\chi_{E_\alpha}\|_{L^p(\mathbb{R}^n)} + \|[b - b_\varepsilon, T]f\|_{L^p(\mathbb{R}^n)}$$

$$\leq C\varepsilon \mapsto 0 \quad \text{uniformly for } f \in Q \text{ with } \varepsilon \to 0.$$

According to Lemma 4.10, S is a strongly precompact set in $L^p(\mathbb{R}^n)$, and hence, $[b, T]$ is a compact operator on $L^p(\mathbb{R}^n)$. $\qquad \square$

By Lemma 4.13, we only need to consider $b \in C_c^\infty(\mathbb{R}^n)$. Namely, we shall prove that the set

$$G = \{[b, \mathcal{H}]f : f \in \mathcal{F}\} \text{ with } \mathcal{F} = \{f : f \in L^p(\mathbb{R}^n) \ \& \ \|f\|_{L^p(\mathbb{R}^n)} \leq C\}$$

and

$$G^* = \{[b, \mathcal{H}^*]f : f \in \mathcal{F}\} \text{ with } \mathcal{F} = \{f : f \in L^p(\mathbb{R}^n) \ \& \ \|f\|_{L^p(\mathbb{R}^n)} \leq C\}$$

are strongly precompact on $L^p(\mathbb{R}^n)$. By Lemma 4.10, it is sufficient to show (4.24)–(4.26) for G and G^*. We begin with the proof for G.

First, for $f \in \mathcal{F}$ and $b \in C_c^\infty(\mathbb{R}^n)$, the boundedness of $[b, \mathcal{H}]$ on $L^p(\mathbb{R}^n)$ shows

$$\sup_{f \in \mathcal{F}} \|[b, \mathcal{H}]f\|_{L^p(\mathbb{R}^n)} \leq C\|b\|_{C\dot{M}O(\mathbb{R}^n)} \sup_{f \in \mathcal{F}} \|f\|_{L^p(\mathbb{R}^n)} < \infty,$$

whence finding (4.24).

Second, suppose that supp $b \subset \{x : |x| \leq \xi\}$. Then, for $0 < \varepsilon < 1$, we choose $\alpha > \xi + 1$ such that $\left(\frac{\xi}{\alpha^n}\right)^{n/p'} < \varepsilon$ to verify

$$\|[b, \mathcal{H}]f\chi_{E_\alpha}\|_{L^p(\mathbb{R}^n)} < C\varepsilon \quad \text{for} \quad E_\alpha = \{x : |x| > \alpha\}.$$

Indeed

$$\left(\int_{|x|>\alpha} |[b, \mathcal{H}]f(x)|^p dx\right)^{1/p} \leq C \left(\int_{|x|>\alpha} \left(|x|^{-n} \int_{|y|\leq\xi} |f(y)|dy\right)^p dx\right)^{1/p}$$

$$\leq \left(\int_\alpha^\infty t^{n-1-np}dt\right)^{1/p} \left(\int_{|y|\leq\xi} dy\right)^{1/p'} \left(\int_{|y|\leq\xi} |f(y)|^p dy\right)^{1/p}$$

$$\leq C \frac{\xi^{n/p'}}{\alpha^{n/p'}} \left(\int_{|y| \leq \xi} |f(y)|^p dy \right)^{1/p}$$

$$\leq C\varepsilon.$$

This means that (4.26) holds for $[b, \mathcal{H}]$ in G uniformly.

Finally, we proceed the proof by showing (4.25). To do this, it is sufficient to prove that for any $\varepsilon > 0$ and $|z|$ sufficiently small dependent only on ε, one has

$$\|[b, \mathcal{H}]f(\cdot + z) - [b, \mathcal{H}]f(\cdot)\|_{L^p(\mathbb{R}^n)} \leq C\varepsilon, \quad \forall f \in \mathcal{F}. \tag{4.42}$$

Let $0 < \varepsilon < 1/2$ and $z \in \mathbb{R}^n$. We first rewrite $[b, \mathcal{H}]f(x + z) - [b, \mathcal{H}]f(x)$ as

$$
\begin{aligned}
|[b, \mathcal{H}]f(x + z) - [b, \mathcal{H}]f(x)| &= \frac{1}{|x + z|^n} \int_{U_1} [(b(x + z) - b(y))]f(y)dy \\
&+ \frac{1}{|x + z|^n} \int_{U_2} [(b(x + z) - b(y))]f(y)dy \\
&- \frac{1}{|x|^n} \int_{V_1} [(b(x) - b(y))]f(y)dy \\
&- \frac{1}{|x|^n} \int_{V_2} [(b(x) - b(y))]f(y)dy \\
&:= I_1 + I_2 + I_3 - I_4.
\end{aligned}
$$

Here

$$
\begin{cases}
U_1 = \{y : |y| < |x + z| \ \& \ |x| > \varepsilon^{-1}|z|\}; \\
U_2 = \{y : |y| < |x + z| \ \& \ |x| \leq \varepsilon^{-1}|z|\}; \\
V_1 = \{y : |y| < |x| \ \& \ |x| > \varepsilon^{-1}|z|\}; \\
V_2 = \{y : |y| < |x| \leq \varepsilon^{-1}|z|\}
\end{cases}
$$

and

$$
\begin{cases}
I_1 = \frac{1}{|x|^n} \int_{V_1} [(b(x + z) - b(x))]f(y)dy; \\
I_2 = \frac{1}{|x|^n} \int_{V_1} [(b(y) - b(x + z))]f(y)dy \\
\qquad - \frac{1}{|x+z|^n} \int_{U_1} [(b(y) - b(x + z))]f(y)dy; \\
I_3 = \frac{1}{|x|^n} \int_{V_2} [(b(y) - b(x))]f(y)dy; \\
I_4 = \frac{1}{|x+z|^n} \int_{U_2} [(b(y) - b(x + z))]f(y)dy.
\end{cases}
$$

Therefore, (4.42) follows from the following L^p-estimates for I_i, $i = 1, 2, 3, 4$. Note first that $b \in C_c^\infty(\mathbb{R}^n)$, $|b(x + z) - b(x)| \leq C|z|$. So we have

$$|I_1| \leq C|z|\mathcal{H}(|f|)(x)\chi_{\{|x| > \varepsilon^{-1}|z|\}}(x).$$

This together with Lemma 4.11 implies that for $f \in \mathcal{F}$,

$$\|I_1\|_{L^p(\mathbb{R}^n)} \leq C|z| \, \|\mathcal{H}(f)\|_{L^p(\mathbb{R}^n)} \leq C|z| \|f\|_{L^p(\mathbb{R}^n)} \leq C|z|. \qquad (4.43)$$

The fact $|b(x+z) - b(y)| \leq 2\|b\|_{L^\infty(\mathbb{R}^n)} \leq C$ and $|x+z| \sim |z|$ for $|z|$ small enough allow us to obtain the following estimate for I_2,

$$|I_2| \leq \frac{C|z|}{|x|^{n+1}} \int_{V_1} |f(y)| dy \leq \frac{C\varepsilon}{|x|^n} \int_{|y|<|x|} |f(y)| dy \chi_{\{|x|>\varepsilon^{-1}|z|\}}(x)$$

$$\leq C\varepsilon \mathcal{H}(|f|)(x)\chi_{\{|x|>\varepsilon^{-1}|z|\}}(x).$$

A further use of Lemma 4.11, we obtain

$$\|I_2\|_{L^p(\mathbb{R}^n)} \leq C\varepsilon\|\mathcal{H}(|f|)\|_{L^p(\mathbb{R}^n)} \leq C\varepsilon\|f\|_{L^p(\mathbb{R}^n)} \leq C\varepsilon. \qquad (4.44)$$

For I_3, we note the fact that $|b(x) - b(y)| \leq C|x - y|$ since $b \in C_c^\infty(\mathbb{R}^n)$ and that $|x - y| < 2|x|$ since $y \in V_2$; therefore,

$$|I_3| \leq \frac{C}{|x|^{n-1}} \int_{V_2} |f(y)| dy \leq \frac{C\varepsilon^{-1}|z|}{|x|^n} \int_{V_2} |f(y)| dy$$

$$\leq C\varepsilon^{-1}|z|\mathcal{H}(|f|)(x)\chi_{\{|x|<\varepsilon^{-1}|z|\}}(x).$$

Hence

$$\|I_3\|_{L^p(\mathbb{R}^n)} \leq C\varepsilon^{-1}|z|\|f\|_{L^p(\mathbb{R}^n)} \leq C\varepsilon^{-1}|z|. \qquad (4.45)$$

To address the term I_4, we use the fact that $|b(x+z) - b(y)| \leq C|x + z - y| < C|x+z|$ since $y \in U_2$ to obtain

$$|I_4| \leq C\frac{\varepsilon^{-1}|z| + |z|}{|x+z|^n} \int_{\{|y|<|x+z|\}} |f(y)| dy \chi_{\{|x|<\varepsilon^{-1}|z|\}}(x)$$

$$\leq C(\varepsilon^{-1}|z| + |z|)\mathcal{H}(|f|)(x+z)\chi_{\{|x|<\varepsilon^{-1}|z|\}}(x).$$

Therefore

$$\|I_4\|_{L^p(\mathbb{R}^n)} \leq C(\varepsilon^{-1}|z| + |z|)\|f\|_{L^p(\mathbb{R}^n)} \leq C(\varepsilon^{-1}|z| + |z|). \qquad (4.46)$$

The desired estimate (4.42) can be obtained by (4.43)–(4.46) and by taking $|z|$ to be sufficiently small. We proceed the proof of Theorem 4.11 to show that (4.24)–(4.26) hold for G^*. In fact, similar arguments for G can be used to deal with G^* by recalling the L^p-boundedness of $[b, \mathcal{H}^*]$. We omit its proof here due to the similarity.

Theorem 4.9 can be seen as a first work on the problem of central function space characterization via the compactness of the commutator of the classical Hardy operator. This theorem makes up for the compactness results of average integral operators and enriches the characterization theory of central function spaces via the compactness of operators.

For the fractional Hardy-Littlewood maximal operator M_β, we have the following result (see [Lu et al. (2007), Theorem 3.2]).

Lemma 4.14. *Let* $0 < \beta < n$, $1 < p \leq n/\beta$ *and* $1/q = 1/p - \beta/n$. *Then there exists a constant* $C > 0$ *such that*

$$\|M_\beta f\|_{L^q(\mathbb{R}^n)} \leq C\|f\|_{L^p(\mathbb{R}^n)}.$$

For the boundedness of $[b, \mathcal{H}_\beta]$ on the Lebesgue space, the known result is from [Fu et al. (2007), Theorem 2.1].

Lemma 4.15. *Let* $0 < \beta < n$, $1 < p \leq n/\beta$ *and* $1/q = 1/p - \beta/n$. *Then* $[b, \mathcal{H}_\beta] : L^p(\mathbb{R}^n) \to L^q(\mathbb{R}^n)$ *if and only if* $b \in C\dot{M}O^{p'}(\mathbb{R}^n)$.

We end this chapter with the compactness characterization of $[b, \mathcal{H}_\beta]$ on the Lebesgue space.

Theorem 4.12. *Suppose that* $0 < \beta < n$, $1 < p \leq n/\beta$ *and* $1/q = 1/p - \beta/n$. *Then* $[b, \mathcal{H}_\beta]$ *is compact from* $L^p(\mathbb{R}^n)$ *to* $L^q(\mathbb{R}^n)$ *if and only if* $b \in CVMO(\mathbb{R}^n)$.

Theorem 4.12 will be proved by the following two results.

Theorem 4.13. *Assume that* $0 < \beta < n$, $1 < p \leq n/\beta$, $1/q = 1/p - \beta/n$ *and* $[b, \mathcal{H}_\beta]$ *is compact from* $L^p(\mathbb{R}^n)$ *to* $L^q(\mathbb{R}^n)$. *Then* $b \in CVMO(\mathbb{R}^n)$.

Theorem 4.14. *Let* β, p, q *be defined as that of Theorem 4.13 and* $b \in CVMO(\mathbb{R}^n)$. *Then* $[b, \mathcal{H}_\beta]$ *is compact from* $L^p(\mathbb{R}^n)$ *to* $L^q(\mathbb{R}^n)$.

The proofs of Theorems 4.13 and 4.14 can be modified directly by that of Theorems 4.10 and 4.11 thanks to Lemmas 4.14 and 4.15. We omit their proofs here for their similarity.

4.3 Notes

The pioneer work on $[b, T]$ when T belongs to a class of nonconvolution operators and $b \in BMO(\mathbb{R}^n)$ can be traced to Coifman, Rochberg and Weiss [Coifman et al. (1976)], the well known result of which is a new characterization of $BMO(\mathbb{R}^n)$ via the boundedness of $[b, T]$. Since 1920, a considerable amount of research has been done to estimate $\mathcal{H}$ and $[b, \mathcal{H}]$; see, for example, [Andersen and Muckenhoupt (1982); Christ and Grafakos (1995); Golubov (1997); Hardy et al. (1952); Komori (2003); Sawyer (1984); Stein and Weiss (1971)] and the references therein. The function class that satisfies (4.9) is also called the reverse Hölder class which contains many kinds of functions, such as polynomial functions [Fefferman (1983)] and harmonic functions [Gilbarg and Trudinger (1983)]. It is interesting to note that solutions to a large class of elliptic PDEs of the second order satisfy (4.9) (cf. [Fazio and Ragusa (1993); Gilbarg and Trudinger (1983); Lemarié-Rieusset (2007)]). More information about the reverse Hölder classes, see also [Cruz-Uribe and Neugebauer (1995); Harboure et al. (1998)] for example. Theorem 4.1 was given by Fu, Liu, Lu and Wang in [Fu et al. (2007), Corollary 2.1] as a special case of [Fu et al. (2007), Theorem 2.1] by choosing $\alpha = \beta = 0$, $p_1 = p_2 = q_1 = q_2$. Theorem 4.4 was proved by Zhao and Lu in [Zhao and Lu (2013), Theorem 1] for given λ. Theorem 4.5 and Theorem 4.6 were just [Shi and Lu (2015), Theorem 1.1] and [Shi and Lu (2015), Theorem 1.2] due to Shi and Lu. The fractional case, Theorem 4.7 and Theorem 4.8, were [Shi and Lu (2015), Theorem 1.3] and [Shi and Lu (2015), Theorem 1.4]. Theorem 4.3 was proved by Shi and Lu in [Shi and Lu (2015)].

The earliest study on the compactness of operators can be traced to Uchiyama [Uchiyama (1978)], where the author obtained the characterization of the L^p-compactness of $[b, T]$ when T is the classical Calderón-Zygmund singular integral operator and $b \in VMO(\mathbb{R}^n)$, the $BMO(\mathbb{R}^n)$ closure of $C_c^\infty(\mathbb{R}^n)$ (the space of all functions being infinite-times continuously differential in $\mathbb{R}^n$ with compact support). Since then, the study of compactness for commutators on different function spaces and their applications (for example, the application in PDEs [Iwaniec and Sbordone (1998); Palagachev and Softova (2004)]) has been a basic component of harmonic analysis. If T is the multiplication operator, then Beatrous and Li [Beatrous and Li (1993)] obtained the compactness of $[b, T]$ on $L^p(\mathbb{R}^n)(1 < p < \infty)$ when b is in an appropriately BMO space, and applications to Hankel-type operators on Begman spaces have been given. In [Chen and Ding (2010)], Chen and Ding proved that $[b, T]$ is a compact operator on $L^p(\mathbb{R}^n)(1 < p < \infty)$ if and only if $b \in VMO(\mathbb{R}^n)$ with T being the parabolic singular integral. For generalized Toeplitz operators (including the singular integral operator and multiplication operator), Krantz and Li developed the compactness theory on $L^p(\mathbb{R}^n)$ [Krantz and Li (2001)]. Furthermore, as applications, they formulated some characterization theorems for the compactness of $[b, T]$ on holomorphic Hardy spaces. For Morrey spaces, Chen and Ding considered the compactness of $[b, T]$ when T is the Riesz potential [Chen et al. (2009)] and T is the singular integral operator [Chen et al. (2012)], respectively, and gave some characterizations of $VMO(\mathbb{R}^n)$ via the compactness of $[b, T]$. Ding and Mei [Ding and Mei (2015)] showed the compactness of

$[b, T]$ for bilinear Calderón-Zygmund operators, bilinear fractional integrals and bilinear pesudodifferential operators on Morrey spaces. Theorems 4.9 and 4.12 were obtained by Shi, Fu and Lu in [Shi et al. (2020)], which was first published in Pacific Journal of Mathematics in [Vol. 307, No. 1, 2020], published by Mathematical Sciences Publishers. Compared with the boundedness characterization of commutators, the compactness characterization only deals with the BMO type space, i.e. the case of $\lambda = 0$. For $\lambda \neq 0$, there is no corresponding characterization results, to the best of our knowledge and will be the direction of our future research.

Chapter 5

Hardy operators on Heisenberg groups and p-adic fields

In this chapter, we first investigate sharp estimates for the Hardy operator in the setting of the Heisenberg group $\mathbb{H}^n$, which plays an important role in several branches of mathematics, such as representation theory, harmonic analysis, several complex variables, partial differential equations and quantum mechanics; see [Stein (1993)] for more details. It will be shown that the norm of the Hardy operator on $L^p(\mathbb{H}^n)$ is still $p/(p-1)$. This goes some way to imply that L^p norms of the Hardy operator are the same despite the domains are intervals on $\mathbb{R}$, balls in $\mathbb{R}^n$ or "ellipsoids" on the Heisenberg group. In other words, the shape of domain does not influence the L^p norm of the Hardy operator. As we know, the real dimension of the Heisenberg group is at least 3. Therefore, if we want to study the Hardy operator on the Heisenberg group, we should consider the multi-dimensional case. For the other type of Hardy inequalities on the Heisenberg group one can refer to [Niu (2001)].

The Heisenberg group $\mathbb{H}^n$ is a non-commutative nilpotent Lie group, with the underlying manifold $\mathbb{R}^{2n} \times \mathbb{R}$ and the group law

$$(x_1, \ldots, x_{2n}, x_{2n+1})(x_1', \ldots, x_{2n}', x_{2n+1}')$$
$$= \left(x_1 + x_1', \ldots, x_{2n} + x_{2n}', x_{2n+1} + x_{2n+1}' + 2 \sum_{j=1}^{n} (x_j' x_{n+j} - x_j x_{n+j}') \right).$$

By the definition, we can see that the identity element on $\mathbb{H}^n$ is $\mathbf{0} \in \mathbb{R}^{2n+1}$, while the element x^{-1} inverse to x is $-x$. The corresponding Lie algebra is generated by the left-invariant vector fields:

$$X_j = \frac{\partial}{\partial x_j} + 2x_{n+j} \frac{\partial}{\partial x_{2n+1}}, \quad j = 1, \ldots, n,$$

$$X_{n+j} = \frac{\partial}{\partial x_{n+j}} - 2x_j \frac{\partial}{\partial x_{2n+1}}, \quad j = 1, \ldots, n,$$

$$X_{2n+1} = \frac{\partial}{\partial x_{2n+1}}.$$

The only non-trivial commutator relations are

$$[X_j, X_{n+j}] = -4X_{2n+1}, \quad j = 1, \dots, n.$$

$\mathbb{H}^n$ is a homogeneous group with dilations

$$\delta_r(x_1, x_2, \dots, x_{2n}, x_{2n+1}) = (rx_1, rx_2, \dots, rx_{2n}, r^2 x_{2n+1}), \quad r > 0.$$

The Haar measure on $\mathbb{H}^n$ coincides with the usual Lebesgue measure on $\mathbb{R}^{2n} \times \mathbb{R}$. We denote by $|E|$ the measure of any measurable set $E \subset \mathbb{H}^n$. Then

$$|\delta_r(E)| = r^Q |E|, \quad d(\delta_r x) = r^Q dx,$$

where $Q = 2n + 2$ is called the homogeneous dimension of $\mathbb{H}^n$.

The Heisenberg distance derived from the norm

$$|x|_h = \left(\left(\sum_{i=1}^{2n} x_i^2 \right)^2 + x_{2n+1}^2 \right)^{1/4}$$

is given by

$$d(p, q) = d(q^{-1}p, \mathbf{0}) = |q^{-1}p|_h.$$

The distance d is left-invariant in the sense that $d(p, q)$ remains unchanged when p and q are both left-translated by some fixed vector on $\mathbb{H}^n$. Furthermore, d satisfies the triangular inequality ([Korányi and Reimann (1985), p. 320])

$$d(p, q) \le d(p, x) + d(x, q), \quad p, x, q \in \mathbb{H}^n.$$

For $r > 0$ and $x \in \mathbb{H}^n$, the ball and sphere with center x and radius r on $\mathbb{H}^n$ are given by

$$B(x, r) = \{y \in \mathbb{H}^n : d(x, y) < r\},$$

and

$$S(x, r) = \{y \in \mathbb{H}^n : d(x, y) = r\},$$

respectively. And we have

$$|B(x, r)| = |B(\mathbf{0}, r)| = \Omega_Q r^Q,$$

where

$$\Omega_Q = \frac{2\pi^{n+\frac{1}{2}} \Gamma(\frac{n}{2})}{(n+1)\Gamma(n)\Gamma(\frac{n+1}{2})}.$$

is the volume of the unit ball $B(\mathbf{0}, 1)$ on $\mathbb{H}^n$. The area of $S(\mathbf{0}, 1)$ on $\mathbb{H}^n$ is $\omega_Q = Q\Omega_Q$. For more details about Heisenberg group, one can refer to [Coulhon et al. (1996); Folland and Stein (1982)].

Next, we shall provide the definition of Hardy operator on the Heisenberg group.

Definition 5.1. Let f be a locally integrable function on $\mathbb{H}^n$. The Hardy operator on $\mathbb{H}^n$ is defined by

$$\mathsf{H}f(x) := \frac{1}{|B(\mathbf{0}, |x|_h)|} \int_{B(\mathbf{0}, |x|_h)} f(y)dy, \quad x \in \mathbb{H}^n \backslash \{\mathbf{0}\}.$$

It is clear that $|\mathsf{H}f(x)| \le C\mathsf{M}f(x)$, where M is the centered Hardy-Littlewood maximal operator on $\mathbb{H}^n$ which is defined by

$$\mathsf{M}f(x) = \sup_{r>0} \frac{1}{|B(x, r)|} \int_{B(x, r)} |f(y)|dy.$$

Li in [Li (2009)] has shown that M is of weak type $(1, 1)$, i.e.

$$\|\mathsf{M}f\|_{L^{1,\infty}(\mathbb{H}^n)} \le A_n \|f\|_{L^1(\mathbb{H}^n)},$$

where the weak $L^1(\mathbb{H}^n)$ space $L^{1,\infty}(\mathbb{H}^n)$ is defined as the set of all measurable functions f on $\mathbb{H}^n$ satisfying

$$\|f\|_{L^{1,\infty}(\mathbb{H}^n)} := \sup_{\lambda>0} \lambda|\{x \in \mathbb{H}^n : |f(x)| > \lambda\}| < \infty.$$

Zienkiewicz in [Zienkiewicz (2005)] found that the constant of (p, p) boundedness of M is dimension free, where $p > 1$. For the weak $(1, 1)$ boundedness of M on Heisenberg type group, we refer to [Li and Qian (2014)]. However, it seems difficult to obtain the sharp bounds of these inequalities. Since $|\mathsf{H}f| \le C\mathsf{M}f$, it is easy to see that H is of weak type $(1, 1)$ and strong type (p, p) (we say (p, p) for short). This naturally raises the question as to whether the best constants in these inequalities for the Hardy operator on the Heisenberg group can be obtained. It turns out that the answer is yes. In Section 5.2 we shall establish the sharp bounds for the (p, p) and weak type $(1, 1)$ inequalities. In dealing with the (p, p) boundedness we shall use the "rotation method" constructed in [Fu et al. (2012)]. As we know, the Hardy-Littlewood maximal operator is not bounded from Hardy space $H^1(\mathbb{R}^n)$ to $L^1(\mathbb{R}^n)$. However, by translation, we obtain that H is bounded from $H^1(\mathbb{H}^n)$ to $L^1(\mathbb{H}^n)$. In Section 5.3, we shall introduce the M_p weights on the Heisenberg group, which is different from the A_p weights. Also we shall give the characterization of such weights by the weighted Hardy inequality on the Heisenberg group.

On the other hand, we shall establish sharp estimates of p-adic Hardy operators. For a prime number p, let $\mathbb{Q}_p$ be the field of p-adic numbers. It is defined as the completion of the field of rational numbers $\mathbb{Q}$ with respect to the non-Archimedean p-adic norm $|\cdot|_p$. This norm is defined as follows: $|0|_p = 0$; if any non-zero rational number x is represented as $x = p^\gamma \frac{m}{n}$, where m and n are integers which are not divisible by p, and γ is an integer, then $|x|_p = p^{-\gamma}$. It is not difficult to show that the norm satisfies the following properties:

$$|xy|_p = |x|_p |y|_p, \qquad |x + y|_p \leq \max\{|x|_p, |y|_p\}.$$

It follows from the second property that when $|x|_p \neq |y|_p$, then $|x + y|_p = \max\{|x|_p, |y|_p\}$. From the standard p-adic analysis [Vladimirov et al. (1994)], we see that any non-zero p-adic number $x \in \mathbb{Q}_p$ can be uniquely represented in the canonical series

$$x = p^\gamma \sum_{j=0}^{\infty} a_j p^j, \qquad \gamma = \gamma(x) \in \mathbb{Z}, \tag{5.1}$$

where a_j are integers, $0 \leq a_j \leq p - 1$, $a_0 \neq 0$. The series (5.1) converges in the p-adic norm because $|a_j p^j|_p = p^{-j}$. We set $\mathbb{Q}_p^* = \mathbb{Q}_p \backslash \{0\}$.

The space $\mathbb{Q}_p^n$ consists of points $x = (x_1, x_2, \ldots, x_n)$, where $x_j \in \mathbb{Q}_p$, $j = 1, 2, \ldots, n$. The p-adic norm on $\mathbb{Q}_p^n$ is

$$|x|_p := \max_{1 \leq j \leq n} |x_j|_p, \qquad x \in \mathbb{Q}_p^n.$$

Denote by

$$B_\gamma(a) = \{x \in \mathbb{Q}_p^n : |x - a|_p \leq p^\gamma\}$$

the ball with center at $a \in \mathbb{Q}_p^n$ and radius p^γ, and its boundary

$$S_\gamma(a) = \{x \in \mathbb{Q}_p^n : |x - a|_p = p^\gamma\} = B_\gamma(a) \backslash B_{\gamma-1}(a).$$

Since $\mathbb{Q}_p^n$ is a locally compact commutative group under addition, it follows from the standard analysis that there exists a Haar measure dx on $\mathbb{Q}_p^n$, which is unique up to positive constant multiple and is translation invariant. We normalize the measure dx by the equality

$$\int_{B_0(0)} dx = |B_0(0)|_H = 1,$$

where $|E|_H$ denotes the Haar measure of a measurable subset E of $\mathbb{Q}_p^n$. By a simple calculation, we can obtain that

$$|B_\gamma(a)|_H = p^{\gamma n} \quad \text{and} \quad |S_\gamma(a)|_H = p^{\gamma n}(1 - p^{-n})$$

for any $a \in \mathbb{Q}_p^n$. For a more complete introduction to the p-adic field, see [Vladimirov et al. (1994)] or [Talenti (1969)].

Next, we introduce the definition of the p-adic Hardy operator.

Definition 5.2. For a function f on $\mathbb{Q}_p^n$, we define the p-adic Hardy operator as follows

$$\mathcal{H}^p f(x) = \frac{1}{|x|_p^n} \int_{B(0,|x|_p)} f(t)dt, \quad x \in \mathbb{Q}_p^n \backslash \{0\}, \quad x \in \mathbb{Q}_p^n \backslash \{0\},$$

where $B(0, |x|_p)$ is a ball in $\mathbb{Q}_p^n$ with center at $0 \in \mathbb{Q}_p^n$ and radius $|x|_p$.

Let $f \in L^q(\mathbb{Q}_p^n)$ and $g \in L^{q'}(\mathbb{Q}_p^n)$, $1/q + 1/q' = 1$ for $1 < q < \infty$. It follows from

$$\int_{\mathbb{Q}_p^n} g(x)\mathcal{H}^p f(x)dx = \int_{\mathbb{Q}_p^n} f(x)\mathcal{H}^{p,*} g(x)dx,$$

that the adjoint operator of the p-adic Hardy operator $\mathcal{H}^{p,*}$ is

$$\mathcal{H}^{p,*} f(x) = \int_{\mathbb{Q}_p^n \backslash B(0,|x|_p)} \frac{f(t)}{|t|_p^n} dt, \quad x \in \mathbb{Q}_p^n \backslash \{0\}.$$

It is obvious that $|\mathcal{H}^p f(x)| \leq C\mathcal{M}^p f(x)$, where $\mathcal{M}^p$ is the Hardy-Littlewood maximal operator [Kim (2009)] defined by

$$\mathcal{M}^p f(x) = \sup_{\gamma \in \mathbb{Z}} \frac{1}{|B_\gamma(x)|_H} \int_{B_\gamma(x)} |f(y)|dy, \quad f \in L^1_{\mathrm{loc}}(\mathbb{Q}_p^n).$$

The boundedness of $\mathcal{M}^p$ on $L^q(\mathbb{Q}_p^n)$ has been solved (see for instance [Talenti (1969)]). But the best estimate of $\mathcal{M}^p$ on $L^q(\mathbb{Q}_p^n)$, $q > 1$, even that of the Hardy-Littlewood maximal operator on Euclidean spaces $\mathbb{R}^n$ is very difficult to obtain. Instead, in Section 5.3.1, we obtain the sharp estimate of $\mathcal{H}^p$ (and the p-adic Hardy-Littlewood-Pólya operator), and the norm of $\mathcal{M}^p$ should be no less than that of $\mathcal{H}^p$. In Section 5.3.2, we shall study the boundedness of the commutators of the p-adic Hardy and Hardy-Littlewood-Pólya operators. One of the main innovative points is that we estimate the commutator of Hardy-Littlewood-Pólya operator by those of p-adic Hardy operators.

5.1   Bounds for Hardy operators on $\mathbb{H}^n$

It is known that $H^1(\mathbb{H}^n) \subset L^1(\mathbb{H}^n) \subset L^{1,\infty}(\mathbb{H}^n)$. In this section, we shall demonstrate the following results:

$$\text{H:} \begin{cases} L^p(\mathbb{H}^n) \to L^p(\mathbb{H}^n), & 1 < p < \infty; & \text{(true)} \\ L^1(\mathbb{H}^n) \to L^1(\mathbb{H}^n); & & \text{(false)} \\ L^1(\mathbb{H}^n) \to L^{1,\infty}(\mathbb{H}^n); & & \text{(true)} \\ H^1(\mathbb{H}^n) \to H^1(\mathbb{H}^n); & & \text{(false)} \\ H^1(\mathbb{H}^n) \to L^1(\mathbb{H}^n). & & \text{(true)} \end{cases}$$

For convenience, let us recall the Hardy space $H^1(\mathbb{H}^n)$ on the Heisenberg group.

Definition 5.3. [Folland and Stein (1982)] A $(1, \infty, 0)$-atom is a compactly supported $L^\infty(\mathbb{H}^n)$ function a such that
(i) there is a ball B whose closure contains supp a such that $\|a\|_{L^\infty} \leq |B|^{-1}$;
(ii) $\displaystyle\int_{\mathbb{H}^n} a(x)dx = 0$.

From the results of [Folland and Stein (1982), Proposition 2.15 and Theorem 3.28], we can see that the Hardy space $H^1(\mathbb{H}^n)$ can be defined by

$$H^1(\mathbb{H}^n) := \left\{ f \in L^1(\mathbb{H}^n) : f(x) = \sum_{k=1}^\infty \lambda_k a_k(x), \ \sum_{k=1}^\infty |\lambda_k| < \infty \right\},$$

where each a_k is a $(1, \infty, 0)$-atom, and the H^1 norm of f can be defined by

$$\|f\|_{H^1(\mathbb{H}^n)} := \inf \sum_{k=1}^\infty |\lambda_k|,$$

where the infimum is taken over all the decompositions of $f = \sum_k \lambda_k a_k$ as above.

Theorem 5.1. (1) H *is bounded from* $L^p(\mathbb{H}^n)$ *to* $L^p(\mathbb{H}^n)$, $1 < p \leq \infty$. *Moreover,*

$$\|H\|_{L^p(\mathbb{H}^n) \to L^p(\mathbb{H}^n)} = \frac{p}{p-1}, \quad 1 < p < \infty,$$

$$\|H\|_{L^\infty(\mathbb{H}^n) \to L^\infty(\mathbb{H}^n)} = 1.$$

(2) H *is bounded from* $L^1(\mathbb{H}^n)$ *to* $L^{1,\infty}(\mathbb{H}^n)$. *Moreover,*

$$\|H\|_{L^1(\mathbb{H}^n) \to L^{1,\infty}(\mathbb{H}^n)} = 1.$$

Proof. (1) It is trivial to get the norm by taking $f_0(x) \equiv 1$ for $p = \infty$. Therefore, we shall only consider the case $1 < p < \infty$. In this proof we shall use the "rotation method" constructed in [Fu et al. (2012)] and [Zhao et al. (2012)].

Firstly, we claim that the operator H and its restriction to the radial functions have the same operator norm on $L^p(\mathbb{H}^n)$. In fact, for any $f \in L^p(\mathbb{H}^n)$, we set

$$g_f(x) = \frac{1}{\omega_Q} \int_{|\xi|_h=1} f(\delta_{|x|_h}\xi)d\xi, \qquad x \in \mathbb{H}^n.$$

Then g_f is a radial function. By making a change of variables, we have

$$\mathsf{H}(g_f)(x) = \frac{1}{|B(0,|x|_h)|} \int_{B(0,|x|_h)} \left(\frac{1}{\omega_Q} \int_{|\xi|_h=1} f(\delta_{|y|_h}\xi)d\xi \right) dy$$

$$= \frac{1}{|B(0,|x|_h)|} \int_0^{|x|_h} \int_{|y'|_h=1} \left(\frac{1}{\omega_Q} \int_{|\xi|_h=1} f(\delta_r\xi)d\xi \right) r^{Q-1}dy'dr$$

$$= \frac{1}{|B(0,|x|_h)|} \int_0^{|x|_h} \int_{|\xi|_h=1} f(\delta_r\xi)r^{Q-1}d\xi dr$$

$$= \mathsf{H}f(x).$$

Using Hölder's inequality, we have

$$\|g_f\|_{L^p(\mathbb{H}^n)} = \frac{1}{\omega_Q} \left(\int_{\mathbb{H}^n} \left| \int_{|\xi|_h=1} f(\delta_{|x|_h}\xi)d\xi \right|^p dx \right)^{1/p}$$

$$\leq \frac{1}{\omega_Q} \left(\int_{\mathbb{H}^n} \left(\int_{|\xi|_h=1} |f(\delta_{|x|_h}\xi)|^p d\xi \right) \left(\int_{|\xi|_h=1} d\xi \right)^{p/p'} dx \right)^{1/p}$$

$$= \omega_Q^{-1/p} \left(\int_0^\infty \int_{|x'|_h=1} \left(\int_{|\xi|_h=1} |f(\delta_r\xi)|^p d\xi \right) r^{Q-1}dx'dr \right)^{1/p}$$

$$= \|f\|_{L^p(\mathbb{H}^n)}.$$

Thus we have

$$\frac{\|\mathsf{H}f\|_{L^p(\mathbb{H}^n)}}{\|f\|_{L^p(\mathbb{H}^n)}} \leq \frac{\|\mathsf{H}(g_f)\|_{L^p(\mathbb{H}^n)}}{\|g_f\|_{L^p(\mathbb{H}^n)}},$$

which implies the claim. In the following, without loss of generality, we may assume that $f \in L^p(\mathbb{H}^n)$ is a radial function. The Minkowski inequality yields that

$$\|Hf\|_{L^p(\mathbb{H}^n)} = \left(\int_{\mathbb{H}^n} \left| \frac{1}{|B(\mathbf{0}, |x|_h)|} \int_{B(\mathbf{0}, |x|_h)} f(y)dy \right|^p dx \right)^{1/p}$$

$$= \frac{1}{\Omega_Q} \left(\int_{\mathbb{H}^n} \left| \int_{B(\mathbf{0},1)} f(\delta_{|x|_h} z)dz \right|^p dx \right)^{1/p}$$

$$\leq \frac{1}{\Omega_Q} \int_{B(\mathbf{0},1)} \left(\int_{\mathbb{H}^n} |f(\delta_{|x|_h} z)|^p dx \right)^{1/p} dz$$

$$= \frac{1}{\Omega_Q} \int_{B(\mathbf{0},1)} \left(\int_{\mathbb{H}^n} |f(\delta_{|z|_h} x)|^p dx \right)^{1/p} dz$$

$$= \frac{1}{\Omega_Q} \int_{B(\mathbf{0},1)} \left(\int_{\mathbb{H}^n} |f(t)|^p |z|_h^{-Q} dt \right)^{1/p} dz$$

$$= \frac{1}{\Omega_Q} \|f\|_{L^p(\mathbb{H}^n)} \int_{B(\mathbf{0},1)} |z|_h^{-\frac{Q}{p}} dz$$

$$= \frac{p}{p-1} \|f\|_{L^p(\mathbb{H}^n)}.$$

Therefore, we have

$$\|Hf\|_{L^p(\mathbb{H}^n) \to L^p(\mathbb{H}^n)} \leq \frac{p}{p-1}, \quad 1 < p < \infty.$$

On the other hand, for $0 < \varepsilon < 1$, take

$$f_\varepsilon = \begin{cases} 0 & |x|_h \leq 1, \\ |x|_h^{-\frac{Q}{p}-\varepsilon} & |x|_h > 1. \end{cases}$$

Then $\|f_\varepsilon\|_{L^p(\mathbb{H}^n)}^p = \frac{\omega_Q}{\varepsilon p}$, and

$$Hf_\varepsilon(x) = \begin{cases} 0 & |x|_h \leq 1, \\ \Omega_Q^{-1} |x|_h^{-\frac{Q}{p}-\varepsilon} \int_{|x|_h^{-1} < |y|_h < 1} |y|_h^{-\frac{Q}{p}-\varepsilon} dy & |x|_h > 1. \end{cases}$$

Consequently, we obtain that $\|Hf_\varepsilon\|_{L^p(\mathbb{H}^n)}$ is equal to

$$\left(\int_{|x|_h > 1} \left(\frac{1}{\Omega_Q} |x|_h^{-\frac{Q}{p}-\varepsilon} \int_{|x|_h^{-1} < |y|_h < 1} |y|_h^{-\frac{Q}{p}-\varepsilon} dy \right)^p dx \right)^{1/p}$$

$$\geq \frac{1}{\Omega_Q} \left(\int_{|x|_h > \varepsilon^{-1}} \left(|x|_h^{-\frac{Q}{p}-\varepsilon} \int_{\varepsilon < |y|_h < 1} |y|_h^{-\frac{Q}{p}-\varepsilon} dy \right)^p dx \right)^{1/p}$$

$$= \varepsilon^{\varepsilon p} \frac{1 - \varepsilon^{Q - \frac{Q}{p} - \varepsilon}}{1 - \frac{1}{p} - \frac{\varepsilon}{Q}} \|f_\varepsilon\|_{L^p(\mathbb{H}^n)}.$$

Thus

$$\|\mathsf{H}\|_{L^p(\mathbb{H}^n) \to L^p(\mathbb{H}^n)} \geq \varepsilon^{\varepsilon p} \frac{1 - \varepsilon^{Q - \frac{Q}{p} - \varepsilon}}{1 - \frac{1}{p} - \frac{\varepsilon}{Q}}.$$

Since $\varepsilon^\varepsilon \to 1$ as $\varepsilon \to 0^+$, by letting $\varepsilon \to 0^+$, we obtain that

$$\|\mathsf{H}f\|_{L^p(\mathbb{H}^n) \to L^p(\mathbb{H}^n)} \geq \frac{p}{p - 1}.$$

This completes the proof of (1) of Theorem 5.1.

(2) For $0 < \lambda < \infty$, we have

$$|\{x \in \mathbb{H}^n : |\mathsf{H}f(x)| > \lambda\}|$$

$$= \left|\left\{ x \in \mathbb{H}^n : \frac{1}{|B(\mathbf{0}, |x|_h)|} \left| \int_{B(\mathbf{0}, |x|_h)} f(y) dy \right| > \lambda \right\}\right|$$

$$\leq \left|\left\{ x \in \mathbb{H}^n : \frac{\|f\|_{L^1(\mathbb{H}^n)}}{\Omega_Q |x|_h^Q} > \lambda \right\}\right|$$

$$= \left|\left\{ x \in \mathbb{H}^n : |x|_h^Q < \frac{\|f\|_{L^1(\mathbb{H}^n)}}{\lambda \Omega_Q} \right\}\right|$$

$$= \left| B\left(\mathbf{0}, \left(\frac{\|f\|_{L^1(\mathbb{H}^n)}}{\lambda \Omega_Q} \right)^{\frac{1}{Q}} \right) \right| = \frac{\|f\|_{L^1(\mathbb{H}^n)}}{\lambda}.$$

Therefore

$$\|\mathsf{H}f\|_{L^{1,\infty}(\mathbb{H}^n)} \leq \|f\|_{L^1(\mathbb{H}^n)}. \tag{5.2}$$

On the other hand, take $f_0(x) = \chi_{B(\mathbf{0},r)}(x)$ for $r > 0$. Then we have

$$\|f_0\|_{L^1(\mathbb{H}^n)} = \Omega_Q r^Q.$$

And

$$\mathsf{H}f_0(x) = \frac{1}{|B(\mathbf{0}, |x|_h)|} \int_{B(\mathbf{0}, |x|_h)} \chi_{B(\mathbf{0},r)}(y) dy$$

$$\leq \frac{1}{|B(\mathbf{0}, |x|_h)|} \int_{B(\mathbf{0}, |x|_h)} dy = 1.$$

Thus, we can only consider the case $0 < \lambda < 1$.

(I) If $x \in B(\mathbf{0}, r)$, then $|x|_h < r$ and

$$\mathsf{H}f_0(x) = \frac{1}{|B(\mathbf{0}, |x|_h)|} \int_{B(\mathbf{0}, |x|_h)} dy = 1.$$

In this case

$$|\{x \in B(\mathbf{0}, r) : |\mathsf{H}f_0(x)| > \lambda\}| = |B(\mathbf{0}, r)| = \Omega_Q r^Q.$$

(II) If $x \in \mathbb{H}^n \backslash B(\mathbf{0}, r)$, then $|x|_h \geq r$ and

$$\mathsf{H}f_0(x) = \frac{1}{|B(\mathbf{0}, |x|_h)|} \int_{B(\mathbf{0}, r)} dy = \frac{r^Q}{|x|_h^Q}.$$

In this case

$$|\{x \in \mathbb{H}^n \backslash B(\mathbf{0}, r) : |\mathsf{H}f_0(x)| > \lambda\}|$$

$$= \left|\left\{x \in \mathbb{H}^n \backslash B(\mathbf{0}, r) : \frac{r^Q}{|x|_h^Q} > \lambda\right\}\right|$$

$$= \left|\left\{x \in \mathbb{H}^n : r < |x|_h < \frac{r}{\lambda^{\frac{1}{Q}}}\right\}\right|$$

$$= \Omega_Q r^Q \left(\frac{1}{\lambda} - 1\right).$$

From the above discussion we can see that

$$|\{x \in \mathbb{H}^n : |\mathsf{H}f_0(x)| > \lambda\}|$$
$$= |\{x \in B(\mathbf{0}, r) : |\mathsf{H}f_0(x)| > \lambda\}|$$
$$+ |\{x \in \mathbb{H}^n \backslash B(\mathbf{0}, r) : |\mathsf{H}f_0(x)| > \lambda\}|$$
$$= \frac{1}{\lambda} \Omega_Q r^Q = \frac{1}{\lambda} \|f_0\|_{L^1(\mathbb{H}^n)}.$$

We therefore conclude that

$$\lambda |\{x \in \mathbb{H}^n : |\mathsf{H}f_0(x)| > \lambda\}| = \|f_0\|_{L^1(\mathbb{H}^n)},$$

which implies that

$$\|\mathsf{H}\|_{L^1(\mathbb{H}^n) \to L^{1,\infty}(\mathbb{H}^n)} \geq 1. \tag{5.3}$$

At last, (2) of Theorem 5.1 follows from (5.2) and (5.3). $\qquad\qquad \square$

Remark 5.1. H is not bounded from $L^1(\mathbb{H}^n)$ to $L^1(\mathbb{H}^n)$. For example, we can take

$$f_0(x) = |x|_h^\alpha \chi_{B(\mathbf{0}, R)}(x), \quad \alpha > -Q.$$

It is easy to see that $f_0 \in L^1(\mathbb{H}^n)$, and $\|f_0\|_{L^1(\mathbb{H}^n)} = \omega_Q R^{\alpha+Q}/(\alpha + Q)$. But

$$\mathsf{H}f_0(x) = \begin{cases} Q|x|_h^\alpha/(Q+\alpha) & |x|_h \leq R, \\ QR^{Q+\alpha}|x|_h^{-Q}/(Q+\alpha) & |x|_h > R, \end{cases}$$

does not belong to $L^1(\mathbb{H}^n)$.

Remark 5.2. H is not bounded from $H^1(\mathbb{H}^n)$ to $H^1(\mathbb{H}^n)$. To illustrate this, we take

$$f_0(x) = \frac{1 - 2^Q}{\Omega_Q 2^{Q+1}} \chi_{\{|x|_h \leq 1\}}(x) + \frac{1}{\Omega_Q 2^{Q+1}} \chi_{\{1 < |x|_h \leq 2\}}(x).$$

Then f_0 is a $(1, \infty, 0)$-atom of $H^1(\mathbb{H}^n)$, and

$$\mathsf{H}f_0(x) = \frac{1 - 2^Q}{\Omega_Q 2^{Q+1}} \chi_{\{|x|_h \leq 1\}}(x) + \frac{1 - 2^Q |x|_h^{-Q}}{\Omega_Q 2^{Q+1}} \chi_{\{1 < |x|_h \leq 2\}}(x).$$

It is clear that

$$\int_{\mathbb{H}^n} \mathsf{H}f_0(x) dx = -\frac{Q}{2} \ln 2 \neq 0.$$

Thus $\mathsf{H}f_0 \notin H^1(\mathbb{H}^n)$.

Theorem 5.2. H *is bounded from* $H^1(\mathbb{H}^n)$ *to* $L^1(\mathbb{H}^n)$.

Proof. Assume that f is an atom of $H^1(\mathbb{H}^n)$ and satisfies the following conditions: (i) supp $f \subset \overline{B(x_0, r)}$, (ii) $\|f\|_{L^\infty} \leq |B(x_0, r)|^{-1}$, (iii) $\int_{\mathbb{H}^n} f(x) dx = 0$. Now take $\tilde{f} = f(x_0^{-1}x)$. Then $\tilde{f}$ satisfies (i) supp $\tilde{f} \subset \overline{B(0, r)}$, (ii) $\|\tilde{f}\|_{L^\infty} \leq |B(0, r)|^{-1}$, (iii) $\int_{\mathbb{H}^n} \tilde{f}(x) dx = 0$.

It is sufficient to show that $\int_{\mathbb{H}^n} |\mathsf{H}\tilde{f}(x)| dx < C$, where C is independent of $\tilde{f}$. By the definition, we have

$$\int_{\mathbb{H}^n} |\mathsf{H}\tilde{f}(x)| dx = \int_{B(0,2r)} |\mathsf{H}\tilde{f}(x)| dx + \int_{\mathbb{H}^n \backslash B(0,2r)} |\mathsf{H}\tilde{f}(x)| dx$$

$$= \int_{B(0,2r)} \left| \frac{1}{|B(0, |x|_h)|} \int_{B(0,|x|_h)} \tilde{f}(y) dy \right| dx$$

$$+ \int_{\mathbb{H}^n \backslash B(0,2r)} \left| \frac{1}{|B(0, |x|_h)|} \int_{B(0,|x|_h)} \tilde{f}(y) dy \right| dx$$

$$= I_1 + I_2.$$

Since supp $\tilde{f} \subset \overline{B(0, r)}$ for $r > 0$, we have

$$I_1 \leq \int_{B(0,2r)} \frac{1}{|B(0, |x|_h)|} \int_{B(0,|x|_h) \cap \overline{B(0,r)}} \left| \tilde{f}(y) \right| dy dx$$

$$\leq \int_{B(0,2r)} \frac{1}{|B(0, |x|_h)|} |B(0, r)|^{-1} \int_{B(0,|x|_h)} dy dx$$

$$= \frac{1}{|B(0, r)|} \int_{B(0,2r)} dx = 2^Q.$$

For I_2, since $x \in \mathbb{H}^n \backslash B(\mathbf{0}, 2r)$, we have $B(\mathbf{0}, |x|_h) \cap \overline{B(\mathbf{0}, r)} = \overline{B(\mathbf{0}, r)}$. Thus

$$I_2 = \int_{\mathbb{H}^n \backslash B(\mathbf{0}, 2r)} \frac{1}{|B(\mathbf{0}, |x|_h)|} \left| \int_{B(\mathbf{0}, |x|_h) \cap \overline{B(\mathbf{0}, r)}} \tilde{f}(y) dy \right| dx$$

$$\leq \int_{\mathbb{H}^n \backslash B(\mathbf{0}, 2r)} \frac{1}{|B(\mathbf{0}, |x|_h)|} \left| \int_{\overline{B(\mathbf{0}, r)}} \tilde{f}(y) dy \right| dx$$

$$= 0.$$

This completes the proof. $\qquad\qquad\qquad\qquad\qquad\qquad\qquad\qquad\qquad\square$

5.2 A weight characterization of Hardy operators on $\mathbb{H}^n$

In this section, we shall study the properties of M_p (resp. M^p) weights on the Heisenberg group.

Definition 5.4. Assume that (w, v) is a pair of nonnegative functions.

(1) (w, v) is called an $M_1(\mathbb{H}^n)$ weight if for almost all $x \in \mathbb{H}^n$,

$$\int_{|x|_h > r} |x|_h^{-Q} w(x) dx \leq C \operatorname{essinf}_{|x|_h < r} v(x), \quad r > 0,$$

for some constant C.

(2) (w, v) is called an $M_p(\mathbb{H}^n)$ $(1 < p < \infty)$ weight if

$$\sup_{0 < r < \infty} \left(\int_{|x|_h > r} |x|_h^{-Qp} w(x) dx \right)^{1/p} \left(\int_{|x|_h < r} v(x)^{1-p'} dx \right)^{1/p'} < \infty,$$

where $1/p + 1/p' = 1$.

To maintain the integrity of the whole paper, we shall list some properties of M_p weights. With slight modifications, the proofs of these properties are analogous to those in Propositions 2.3 and 2.5 in [Zhao et al. (2014)] and so they have been omitted.

Proposition 5.1. *Let $w \in M_p(\mathbb{H}^n)$ for some $1 \leq p < \infty$. Then*
(i) $M_p(\mathbb{H}^n) \subsetneq M_q(\mathbb{H}^n)$, *for $1 \leq p < q$.*
(ii) *For $1 \leq p < \infty$, $0 < \alpha, \varepsilon < 1$. If $w \in M_p$, then $w^\alpha \in M_{\alpha p + 1 - \varepsilon p}(\mathbb{H}^n)$.*
(iii) *For $1 \leq p < \infty$, $0 < \alpha < 1$. If $w_1, w_2 \in M_p$, then $w_1^\alpha w_2^{1-\alpha} \in M_p(\mathbb{H}^n)$.*
(iv) *If $w_1, w_2 \in M_p$, then $w_1 + w_2 \in M_p(\mathbb{H}^n)$.*
(v) *For $\lambda > 0$, then $\lambda w \in M_p(\mathbb{H}^n)$.*
(vi) *For $\lambda > 0$, then $w \circ \delta_\lambda \in M_p(\mathbb{H}^n)$, where*

$$w \circ \delta_\lambda(x) = w(\lambda x_1, \ldots, \lambda x_{2n}, \lambda^2 x_{2n+1}).$$

Proposition 5.2. *Let $x \in \mathbb{H}^n$. Then*
(i) $|x|_h^\alpha \in M_1$ *if and only if $\alpha < 0$.*
(ii) $|x|_h^\alpha \in M_p$, $1 < p < \infty$, *if and only if $\alpha \le Q(p-1)$.*

The following result, together with Proposition 5.2, implies the difference between the classes of M_p weights and A_p weights. The A_p weights in the setting of Heisenberg group are similar to that in the setting of Euclidean spaces. For more details we refer the reader to [Guliev (1994)] and [Hytönen et al. (2012)].

Proposition 5.3. *Let $x \in \mathbb{H}^n$. Then*
(i) $|x|_h^\alpha \in A_1$ *if and only if $-Q < \alpha \le 0$.*
(ii) $|x|_h^\alpha \in A_p$, $1 < p < \infty$, *if and only if $-Q < \alpha \le Q(p-1)$.*

The proof of this proposition can follow the standard proofs of Propositions 1.4.3 and 1.4.4 in [Lu et al. (2007), p. 35]. It is easy to see from these two propositions that a constant is an $A_1(\mathbb{H}^n)$ weight but not an $M_1(\mathbb{H}^n)$ weight.

The weighted Hardy inequalities on the Heisenberg group can also characterize M_p weights.

Theorem 5.3. *Let w and v be nonnegative weight functions on $\mathbb{H}^n$. For $1 < p \le q < \infty$, the inequality*

$$\left(\int_{\mathbb{H}^n} w(x) \, |\mathsf{H}f(x)|^q \, dx \right)^{1/q} \le C \left(\int_{\mathbb{H}^n} v(x) |f(x)|^p dx \right)^{1/p} \tag{5.4}$$

holds for $f \ge 0$ if and only if

$$A := \sup_{0 < r < \infty} \left(\int_{|x|_h > r} |x|_h^{-Qq} w(x) dx \right)^{1/q} \left(\int_{|x|_h < r} v(x)^{1-p'} dx \right)^{1/p'} < \infty. \tag{5.5}$$

Moreover, if C is the smallest constant for which (5.4) holds, then

$$A \le \Omega_Q C \le A p'^{1/p'} p^{1/q}. \tag{5.6}$$

Corollary 5.1. *Let w and v be nonnegative weight functions on $\mathbb{H}^n$. For $1 < p < \infty$, the inequality*

$$\left(\int_{\mathbb{H}^n} w(x) \, |\mathsf{H}f(x)|^p \, dx \right)^{1/p} \le C \left(\int_{\mathbb{H}^n} v(x) |f(x)|^p dx \right)^{1/p} \tag{5.7}$$

holds for $f \ge 0$ if and only if $(w, v) \in M_p$. Moreover, if C is the smallest constant for which (5.7) holds, then

$$A \le \Omega_Q C \le A p'^{1/p'} p^{1/p}.$$

To prove Theorem 5.3, we need the following version of Minkowski's inequality in [Drábek et al. (1995)].

Lemma 5.1. *Suppose that $f(x) \geq 0$, $g(x) \geq 0$ on $(0, \infty)$ and $p \geq 1$, then*

$$\int_0^\infty f(x) \left(\int_0^x g(y) dy \right)^p dx \leq \left(\int_0^\infty g(y) \left(\int_y^\infty f(x) dx \right)^{1/p} dy \right)^p.$$

Proof of Theorem 5.3. We proceed as in [Drábek et al. (1995)]. For any $r > 0$, (5.5) can be written in the polar coordinate form:

$$\left(\int_r^\infty \rho^{-Qq+Q-1} d\rho \int_{S(0,1)} w(\rho x') dx' \right)^{1/q}$$

$$\times \left(\int_0^r \rho^{Q-1} d\rho \int_{S(0,1)} v(\rho x')^{1-p'} dx' \right)^{1/p'} \leq A. \tag{5.8}$$

The left-hand side of (5.4) has the form

$$\left(\int_{\mathbb{H}^n} w(x) \left(Hf(x) \right)^q dx \right)^{1/q}$$

$$= \left(\int_{\mathbb{H}^n} w(x) \left(\frac{1}{|B(0, |x|_h)|} \int_{B(0, |x|_h)} f(y) dy \right)^q dx \right)^{1/q}$$

$$= \frac{1}{\Omega_Q} \left(\int_0^\infty \int_{S(0,1)} \rho^{Q-1-Qq} w(\rho x') \right.$$

$$\times \left. \left(\int_0^\rho \int_{S(0,1)} f(ry') r^{Q-1} dy' dr \right)^q dx' d\rho \right)^{1/q}. \tag{5.9}$$

Let

$$h(\rho) = \left(\int_0^\rho \int_{S(0,1)} r^{Q-1} v(ry')^{1-p'} dy' dr \right)^{\frac{1}{pp'}}.$$

Then by Hölder's inequality, we have

$$\int_0^\rho \int_{S(0,1)} f(ry') r^{Q-1} dy' dr$$

$$= \int_0^\rho \int_{S(0,1)} r^{\frac{Q-1}{p}} f(ry') v(ry')^{\frac{1}{p}} h(r) r^{\frac{Q-1}{p'}} \left(v(ry')^{\frac{1}{p}} h(r) \right)^{-1} dy' dr$$

$$\leq \left(\int_0^\rho \int_{S(0,1)} r^{Q-1} \left(f(ry') v(ry')^{\frac{1}{p}} h(r) \right)^p dy' dr \right)^{1/p}$$

$$\times \left(\int_0^\rho \int_{S(0,1)} r^{Q-1} \left(v(ry')^{\frac{1}{p}} h(r) \right)^{-p'} dy' dr \right)^{1/p'}.$$

If we define

$$W(\rho) = \int_{S(0,1)} \rho^{Q-1-Qq} w(\rho x') dx',$$

$$F(r) = \int_{S(0,1)} r^{Q-1} \left(f(ry') v(ry')^{\frac{1}{p}} h(r) \right)^p dy',$$

$$G(\rho) = \int_0^\rho \int_{S(0,1)} r^{Q-1} \left(v(ry')^{\frac{1}{p}} h(r) \right)^{-p'} dy' dr,$$

then by (5.9) and Minkowski's inequality, we obtain that

$$\left(\int_{\mathbb{H}^n} w(x) \left(\mathsf{H}f(x) \right)^q dx \right)^{1/q}$$

$$\leq \frac{1}{\Omega_Q} \left(\int_0^\infty W(\rho) \left(\int_0^\rho F(r) dr \right)^{q/p} G(\rho)^{q/p'} d\rho \right)^{1/q}$$

$$\leq \frac{1}{\Omega_Q} \left(\int_0^\infty F(r) \left(\int_r^\infty W(\rho) G(\rho)^{q/p'} d\rho \right)^{p/q} dr \right)^{1/p}. \tag{5.10}$$

By the definition, we have

$$G(\rho) = \int_0^\rho \int_{S(0,1)} r^{Q-1} \left(v(ry')^{\frac{1}{p}} h(r) \right)^{-p'} dy' dr$$

$$= \int_0^\rho \int_{S(0,1)} r^{Q-1} v(ry')^{1-p'} \left(\int_0^r \int_{S(0,1)} t^{Q-1} v(tz)^{1-p} dz dt \right)^{-1/p} dy' dr.$$

If we set

$$U(r) = \int_{S(0,1)} r^{Q-1} v(ry')^{1-p'} dy',$$

then by (5.8), we obtain

$$G(\rho) = \int_0^\rho U(r) \left(\int_0^r U(t) dt \right)^{-1/p} dr$$

$$= p' \int_0^\rho \frac{d}{dr} \left(\int_0^r U(t) dt \right)^{1/p'} dr = p' \left(\int_0^\rho U(t) dt \right)^{1/p'}$$

$$= p' \left(\int_0^\rho \int_{S(0,1)} t^{Q-1} v(tz)^{1-p'} dz dt \right)^{1/p'}$$

$$\leq p'A \left(\int_\rho^\infty \int_{S(0,1)} t^{-Qq+Q-1} w(tz) \, dz \, dt \right)^{-1/q}$$

$$= p'A \left(\int_\rho^\infty W(t) \, dt \right)^{-1/q}.$$

Similarly, using (5.8), we show that

$$\int_r^\infty W(\rho) \left(\int_\rho^\infty W(t) \, dt \right)^{-1/p'} d\rho$$

$$= p \left(\int_r^\infty W(\rho) \, d\rho \right)^{\frac{1}{p}}$$

$$= p \left(\int_r^\infty \int_{S(0,1)} \rho^{Q-1-Qq} w(\rho x') \, dx' \, d\rho \right)^{1/p}$$

$$\leq p A^{q/p} \left(\int_0^r \int_{S(0,1)} \rho^{Q-1} v(\rho x')^{1-p'} \, dx' \, d\rho \right)^{-\frac{q}{p'p}}$$

$$= p A^{q/p} h(r)^{-q}.$$

Hence (5.10) can be estimated from above by

$$\left(\int_{\mathbb{H}^n} w(x) \left(\mathsf{H} f(x) \right)^q dx \right)^{1/q}$$

$$\leq \frac{1}{\Omega_Q} \left(\int_0^\infty F(r) \left(\int_r^\infty W(\rho) G(\rho)^{q/p'} d\rho \right)^{p/q} dr \right)^{1/q}$$

$$\leq \frac{(p'A)^{1/p'}}{\Omega_Q} \left(\int_0^\infty F(r) \left(\int_r^\infty W(\rho) \left(\int_\rho^\infty W(t) \, dt \right)^{-1/p'} d\rho \right)^{p/q} dr \right)^{1/p}$$

$$\leq \frac{1}{\Omega_Q} p'^{1/p'} p^{1/q} A \left(\int_0^\infty F(r) h(r)^{-p} dr \right)^{1/p}$$

$$= \frac{1}{\Omega_Q} p'^{1/p'} p^{1/q} A \left(\int_0^\infty \int_{S(0,1)} r^{Q-1} f(ry')^p v(ry') \, dy' \, dr \right)^{1/p}$$

$$= \frac{1}{\Omega_Q} p'^{1/p'} p^{1/q} A \left(\int_{\mathbb{H}^n} f(x)^p v(x) \, dx \right)^{1/p}. \tag{5.11}$$

On the other hand, if (5.4) holds, by taking $f_0(x) = v(x)^{1-p'} \chi_{B(0,r)}(x)$

with $r > 0$, we have

$$C \left(\int_{\mathbb{H}^n} v(x) f(x)^p dx \right)^{1/p}$$

$$\geq \left(\int_{|x|_h \geq r} w(x) \left(\frac{1}{|B(0, |x|_h)|} \int_{B(0,r)} v(y)^{1-p'} dy \right)^q dx \right)^{1/q}$$

$$= \frac{1}{\Omega_Q} \left(\int_{|x|_h \geq r} w(x) |x|_h^{-Qq} dx \right)^{1/q} \left(\int_{|y|_h \leq r} v(y)^{1-p'} dy \right).$$

Consequently,

$$\left(\int_{|x|_h \geq r} w(x) |x|_h^{-Qq} dx \right)^{1/q} \left(\int_{|y|_h \leq r} v^{1-p'}(y) dy \right)^{1/p'} \leq C\Omega_Q.$$

Thus by taking the supremum over $r > 0$, we have

$$A \leq C\Omega_Q. \tag{5.12}$$

Moreover, (5.11) and (5.12) imply (5.6). □

The adjoint operator of Hardy operator is defined by

$$H^* f(x) := \frac{1}{\Omega_Q} \int_{|y|_h > |x|_h} \frac{f(y)}{|y|_h^Q} dy, \quad x \in \mathbb{H}^n \backslash \{0\}.$$

Obviously, the operators H and H^* are adjoint mutually:

$$\int_{\mathbb{H}^n} g(x) Hf(x) dx = \int_{\mathbb{H}^n} f(x) H^* g(x) dx,$$

where $f \in L^p(\mathbb{H}^n)$ and $g \in L^{p'}(\mathbb{H}^n)$, $1/p + 1/p' = 1$ for $1 < p < \infty$. By duality,

$$\|H^*\|_{L^{p'}(\mathbb{H}^n) \to L^{p'}(\mathbb{H}^n)} = \|H\|_{L^p(\mathbb{H}^n) \to L^p(\mathbb{H}^n)} = \frac{p}{p-1}$$

and H^* is bounded from $L^\infty(\mathbb{H}^n)$ to $BMO(\mathbb{H}^n)$. However, the dual space of $L^{1,\infty}(\mathbb{H}^n)$, which is the special case of Lorentz space $L^{p,q}(\mathbb{H}^n)$, $0 < p < \infty$, $0 < q \leq \infty$, is unknown. Therefore, the corresponding result for H^* can not be obtained by duality. But it can be got by the similar method as above. The proof is left to the reader. For H^*, there also exist M^p weights. A pair of nonnegative functions (w, v) is said to be of class $M^p(\mathbb{H}^n)$ if for $1 < p < \infty$,

$$\sup_{0 < r < \infty} \left(\int_{|x|_h < r} w(x) dx \right)^{1/p} \left(\int_{|x|_h > r} v(x)^{1-p'} |x|_h^{-Qp'} dx \right)^{1/p'} < \infty$$

and for $p = 1$,

$$\int_{|x|_h < r} w(x) dx \leq C \operatorname{essinf}_{|x|_h > r} \{ |x|_h^{-Q} v(x) \}, \quad r > 0.$$

Analogue to Corollary 5.1, one can also show that $H^* : L^p(v) \to L^p(w)$ if and only if $(w, v) \in M^p(\mathbb{H}^n)$, $1 < p < \infty$.

5.3 p-adic Hardy operators and their commutators

5.3.1   *Sharp estimates of p-adic Hardy and Hardy-Littlewood-Pólya operators*

We get the following sharp estimate of $\mathcal{H}^p$ on $L^q(|x|_p^\alpha dx)$.

Theorem 5.4. *Let $1 < q < \infty$ and $\alpha < n(q-1)$. Then*

$$\|\mathcal{H}^p\|_{L^q(|x|_p^\alpha dx)\to L^q(|x|_p^\alpha dx)} = \frac{1 - p^{-n}}{1 - p^{\frac{\alpha}{q} - \frac{n}{q'}}},$$

where $1/q + 1/q' = 1$.

When $\alpha = 0$, we get the following corollary.

Corollary 5.2. *Let $1 < q < \infty$. Then*

$$\|\mathcal{H}^p\|_{L^q(\mathbb{Q}_p^n)\to L^q(\mathbb{Q}_p^n)} = \frac{1 - p^{-n}}{1 - p^{-\frac{n}{q'}}},$$

where $1/q + 1/q' = 1$.

Remark 5.3. Obviously, the L^q norm of $\mathcal{H}^p$ on $\mathbb{Q}_p^n$ depends on n, however, the L^q norms of $\mathcal{H}$ on $\mathbb{R}^n$ and $\mathbb{H}^n$ are independent of the dimensions of underlying spaces n and $2n+1$, respectively.

The Hardy-Littlewood-Pólya operator [Bényi and Oh (2006)] is defined by

$$\mathbf{P}f(x) = \int_0^\infty \frac{f(y)}{\max\{x, y\}} dy.$$

In [Bényi and Oh (2006)], the authors obtained that the norm of Hardy-Littlewood-Pólya's operator on $L^q(\mathbb{R}_+)$ (see also in [Hardy et al. (1952), p. 254]), $1 < q < \infty$, is

$$\|\mathbf{P}\|_{L^q(\mathbb{R}_+)\to L^q(\mathbb{R}_+)} = \frac{q^2}{q-1}.$$

Next, we consider the p-adic version of Hardy-Littlewood-Pólya operator. We define the p-adic Hardy-Littlewood-Pólya operator as

$$T^p f(x) = \int_{\mathbb{Q}_p^*} \frac{f(y)}{\max\{|x|_p, |y|_p\}} dy, \quad x \in \mathbb{Q}_p.$$

By the similar method to the proof of Theorem 5.4, we obtain the norm of p-adic Hardy-Littlewood-Pólya operator from $L^q(|x|_p^\alpha dx)$ to $L^q(|x|_p^\alpha dx)$.

Theorem 5.5. *Let* $1 < q < \infty$ *and* $-1 < \alpha < q - 1$. *Then for any* $f \in L^q(|x|_p^\alpha dx)$

$$\|T^p f\|_{L^q(|x|_p^\alpha dx)} \leq C \|f\|_{L^q(|x|_p^\alpha dx)}. \tag{5.13}$$

Moreover,

$$\|T^p\|_{L^q(|x|_p^\alpha dx) \to L^q(|x|_p^\alpha dx)} = C, \tag{5.14}$$

where

$$C = \left(1 - \frac{1}{p}\right) \left(\frac{1}{1 - p^{\frac{\alpha+1}{q} - 1}} + \frac{p^{-\frac{\alpha+1}{q}}}{1 - p^{-\frac{\alpha+1}{q}}}\right).$$

When $\alpha = 0$, we get the norm of T^p on $L^q(\mathbb{Q}_p)$ as follows.

Corollary 5.3. *Let* $1 < q < \infty$. *Then*

$$\|T^p\|_{L^q(\mathbb{Q}_p) \to L^q(\mathbb{Q}_p)} = \left(1 - \frac{1}{p}\right) \left(\frac{1}{1 - p^{1/q-1}} + \frac{p^{-1/q}}{1 - p^{-1/q}}\right).$$

Proof of Theorem 5.4. Firstly, we claim that the operator $\mathcal{H}^p$ and its restriction to the functions g satisfying $g(x) = g(|x|_p^{-1})$ have the same operator norm on $L^q(|x|_p^\alpha dx)$. In fact, set

$$g_f(x) = \frac{1}{1 - p^{-n}} \int_{|\xi|_p=1} f(|x|_p^{-1}\xi) d\xi, \qquad x \in \mathbb{Q}_p^n.$$

It is easy to see that g_f satisfies that $g_f(x) = g_f(|x|_p^{-1})$ and $\mathcal{H}^p(g_f) = \mathcal{H}^p f$. By Hölder's inequality, we get

$$\|g_f\|_{L^q(|x|_p^\alpha dx)} = \left(\int_{\mathbb{Q}_p^n} \left|\frac{1}{1-p^{-n}} \int_{|\xi|_p=1} f(|x|_p^{-1}\xi) d\xi\right|^q |x|_p^\alpha dx\right)^{1/q}$$

$$\leq \left(\int_{\mathbb{Q}_p^n} \frac{1}{(1-p^{-n})^q} \left(\int_{|\xi|_p=1} |f(|x|_p^{-1}\xi)|^q d\xi\right) \left(\int_{|\xi|_p=1} d\xi\right)^{q/q'} |x|_p^\alpha dx\right)^{1/q}$$

$$= \left(\int_{\mathbb{Q}_p^n} \frac{1}{1-p^{-n}} \int_{|\xi|_p=1} |f(|x|_p^{-1}\xi)|^q d\xi |x|_p^\alpha dx\right)^{1/q}$$

$$= \frac{1}{(1-p^{-n})^{1/q}} \left(\int_{\mathbb{Q}_p^n} \int_{|y|_p=|x|_p} |f(y)|^q dy |x|_p^{\alpha-n} dx\right)^{1/q}$$

$$= \frac{1}{(1-p^{-n})^{1/q}} \left(\int_{\mathbb{Q}_p^n} \int_{|x|_p=|y|_p} |x|_p^{\alpha-n} dx |f(y)|^q dy\right)^{1/q}$$

$$= \|f\|_{L^q(|x|_p^\alpha dx)}.$$

Therefore, we have

$$\frac{\|\mathcal{H}^p f\|_{L^q(|x|_p^\alpha dx)}}{\|f\|_{L^q(|x|_p^\alpha dx)}} \le \frac{\|\mathcal{H}^p(g_f)\|_{L^q(|x|_p^\alpha dx)}}{\|g_f\|_{L^q(|x|_p^\alpha dx)}},$$

which implies the claim. In the following, without loss of generality, we may assume that $f \in L^q(|x|_p^\alpha dx)$ satisfies $f(x) = f(|x|_p^{-1})$.

Substituting $t = |x|_p^{-1} y$, we have

$$\|\mathcal{H}^p f\|_{L^q(|x|_p^\alpha dx)} = \left(\int_{\mathbb{Q}_p^n} \left| \frac{1}{|x|_p^n} \int_{B(0,|x|_p)} f(t) dt \right|^q |x|_p^\alpha dx \right)^{1/q}$$

$$= \left(\int_{\mathbb{Q}_p^n} \left| \int_{B(0,1)} f(|x|_p^{-1} y) dy \right|^q |x|_p^\alpha dx \right)^{1/q}.$$

Then using Minkowski's integral inequality, we get

$$\|\mathcal{H}^p f\|_{L^q(|x|_p^\alpha dx)} \le \int_{B(0,1)} \left(\int_{\mathbb{Q}_p^n} |f(|y|_p^{-1} x)|^q |x|_p^\alpha dx \right)^{1/q} dy$$

$$\le \left(\int_{B(0,1)} |y|_p^{-\frac{n}{q} - \frac{\alpha}{q}} dy \right) \|f\|_{L^q(|x|_p^\alpha dx)}$$

$$= \left(\sum_{k=0}^{\infty} \int_{|y|_p = p^{-k}} p^{\frac{k(n+\alpha)}{q}} dy \right) \|f\|_{L^q(|x|_p^\alpha dx)}$$

$$= \frac{1 - p^{-n}}{1 - p^{\frac{\alpha}{q} - \frac{n}{q'}}} \|f\|_{L^q(|x|_p^\alpha dx)},$$

where $1/q + 1/q' = 1$. Therefore, we get

$$\|\mathcal{H}^p\|_{L^q(|x|_p^\alpha dx) \to L^q(|x|_p^\alpha dx)} \le \frac{1 - p^{-n}}{1 - p^{\frac{\alpha}{q} - \frac{n}{q'}}}. \tag{5.15}$$

On the other hand, for $0 < \varepsilon < 1$, we take

$$f_\varepsilon = \begin{cases} 0 & |x|_p < 1, \\ |x|_p^{-\frac{n}{q} - \frac{\alpha}{q} - \varepsilon} & |x|_p \ge 1. \end{cases}$$

Then $\|f_\varepsilon\|_{L^q(|x|_p^\alpha dx)}^q = \frac{1 - p^{-n}}{1 - p^{-\varepsilon q}}$, and

$$\mathcal{H}^p f_\varepsilon(x) = \begin{cases} 0 & |x|_p < 1, \\ |x|_p^{-\frac{n}{q} - \frac{\alpha}{q} - \varepsilon} \int_{\frac{1}{|x|_p} \le |t|_p \le 1} |t|_p^{-\frac{n}{q} - \frac{\alpha}{q} - \varepsilon} dt & |x|_p \ge 1. \end{cases}$$

Let $|\varepsilon|_p > 1$. Then we have that $\|\mathcal{H}^p f_\varepsilon\|_{L^q(|x|_p^\alpha dx)}$ is equal to

$$\left(\int_{|x|_p \geq 1} \left(|x|_p^{-\frac{n}{q}-\frac{\alpha}{q}-\varepsilon} \int_{\frac{1}{|x|_p} \leq |t|_p \leq 1} |t|_p^{-\frac{n}{q}-\frac{\alpha}{q}-\varepsilon} dt\right)^q |x|_p^\alpha dx\right)^{1/q}$$

$$\geq \left(\int_{|x|_p \geq |\varepsilon|_p} \left(|x|_p^{-\frac{n}{q}-\frac{\alpha}{q}-\varepsilon} \int_{\frac{1}{|\varepsilon|_p} \leq |t|_p \leq 1} |t|_p^{-\frac{n}{q}-\frac{\alpha}{q}-\varepsilon} dt\right)^q |x|_p^\alpha dx\right)^{1/q}$$

$$= \left(\int_{|x|_p \geq |\varepsilon|_p} |x|_p^{-n-\varepsilon q} dx\right)^{1/q} \int_{\frac{1}{|\varepsilon|_p} \leq |t|_p \leq 1} |t|_p^{-\frac{n}{q}-\frac{\alpha}{q}-\varepsilon} dt$$

$$= \|f_\varepsilon\|_{L^q(|x|_p^\alpha dx)} |\varepsilon|_p^{-\varepsilon} \int_{\frac{1}{|\varepsilon|_p} \leq |t|_p \leq 1} |t|_p^{-\frac{n}{q}-\frac{\alpha}{q}-\varepsilon} dt.$$

Therefore, we have

$$\int_{\frac{1}{|\varepsilon|_p} \leq |t|_p \leq 1} |t|_p^{-\frac{n}{q}-\frac{\alpha}{q}-\varepsilon} dt \leq \|\mathcal{H}^p\|_{L^q(|x|_p^\alpha dx) \to L^q(|x|_p^\alpha dx)} |\varepsilon|_p^\varepsilon.$$

Now take $\varepsilon = p^{-k}$, $k = 1, 2, 3, \ldots$. Then $|\varepsilon|_p = p^k > 1$. Letting k approach to ∞, then ε approaches to 0^+ and $|\varepsilon|_p^\varepsilon = p^{\frac{k}{p^k}}$ approaches to 1. By Fatou's lemma, we obtain

$$\frac{1 - p^{-n}}{1 - p^{\frac{\alpha}{q} - \frac{n}{q'}}} = \int_{0 < |t|_p \leq 1} |t|_p^{-\frac{n}{q}-\frac{\alpha}{q}} dt \leq \|\mathcal{H}^p\|_{L^q(|x|_p^\alpha dx) \to L^q(|x|_p^\alpha dx)}. \tag{5.16}$$

Then (5.15) and (5.16) imply that

$$\|\mathcal{H}^p\|_{L^q(|x|_p^\alpha dx) \to L^q(|x|_p^\alpha dx)} = \frac{1 - p^{-n}}{1 - p^{\frac{\alpha}{q} - \frac{n}{q'}}}.$$

Theorem 5.4 is proved. $\qquad \square$

By similar arguments, we may obtain the sharp bound for the operator $\mathcal{H}^{p,*}$ on $L^q(|x|_p^\alpha dx)$ and we leave it to the interested reader.

Proof of Theorem 5.5. By the definition and the change of variables $y = xt$, we have

$$\|T^p f\|_{L^q(|x|_p^\alpha dx)} = \left(\int_{\mathbb{Q}_p} \left|\int_{\mathbb{Q}_p^*} \frac{f(y)}{\max\{|x|_p, |y|_p\}} dy\right|^q |x|_p^\alpha dx\right)^{1/q}$$

$$\leq \left(\int_{\mathbb{Q}_p} \left(\int_{\mathbb{Q}_p^*} \frac{|f(y)|}{\max\{|x|_p, |y|_p\}} dy\right)^q |x|_p^\alpha dx\right)^{1/q}$$

$$= \left(\int_{\mathbb{Q}_p} \left(\int_{\mathbb{Q}_p^*} \frac{|f(xt)|}{\max\{1, |t|_p\}} dt\right)^q |x|_p^\alpha dx\right)^{1/q}.$$

By Minkowski's integral inequality, we get

$$\|T^p f\|_{L^q(|x|_p^\alpha dx)} \leq \int_{\mathbb{Q}_p^*} \left(\int_{\mathbb{Q}_p} |f(xt)|^q |x|_p^\alpha dx \right)^{1/q} \frac{1}{\max\{1, |t|_p\}} dt$$

$$\leq \int_{\mathbb{Q}_p^*} \left(\int_{\mathbb{Q}_p} |f(x)|^q |x|_p^\alpha dx \right)^{1/q} \frac{|t|_p^{-\frac{\alpha+1}{q}}}{\max\{1, |t|_p\}} dt$$

$$= \|f\|_{L^q(|x|_p^\alpha dx)} \int_{\mathbb{Q}_p^*} \frac{|t|_p^{-\frac{\alpha+1}{q}}}{\max\{1, |t|_p\}} dt. \tag{5.17}$$

Note that

$$\int_{\mathbb{Q}_p^*} \frac{|t|_p^{-\frac{\alpha+1}{q}}}{\max\{1, |t|_p\}} dt = \sum_{k=-\infty}^{0} \int_{S_k} \frac{|t|_p^{-\frac{\alpha+1}{q}}}{\max\{1, |t|_p\}} dt + \sum_{k=1}^{\infty} \int_{S_k} \frac{|t|_p^{-\frac{\alpha+1}{q}}}{\max\{1, |t|_p\}} dt$$

$$= \left(1 - \frac{1}{p} \right) \left(\sum_{k=-\infty}^{0} p^{k(1-\frac{\alpha+1}{q})} + \sum_{k=1}^{\infty} p^{-\frac{k(\alpha+1)}{q}} \right)$$

$$= \left(1 - \frac{1}{p} \right) \left(\frac{1}{1 - p^{\frac{\alpha+1}{q}-1}} + \frac{p^{-\frac{\alpha+1}{q}}}{1 - p^{-\frac{\alpha+1}{q}}} \right). \tag{5.18}$$

Substituting (5.18) into (5.17), we can obtain that (5.13) holds and

$$\|T^p\|_{L^q(|x|_p^\alpha dx) \to L^q(|x|_p^\alpha dx)} \leq \left(1 - \frac{1}{p} \right) \left(\frac{1}{1 - p^{\frac{\alpha+1}{q}-1}} + \frac{p^{-\frac{\alpha+1}{q}}}{1 - p^{-\frac{\alpha+1}{q}}} \right). \tag{5.19}$$

On the other hand, for $0 < \varepsilon < 1$, let

$$f_\varepsilon(x) = \begin{cases} 0 & |x|_p < 1, \\ |x|_p^{\frac{-1-\alpha}{q}-\varepsilon} & |x|_p \geq 1. \end{cases}$$

Then we have

$$\|f_\varepsilon\|_{L^q(|x|_p^\alpha dx)}^q = \frac{1 - p^{-1}}{1 - p^{-\varepsilon q}}$$

and

$$T^p f_\varepsilon(x) = \int_{|y|_p \geq 1} \frac{|y|_p^{\frac{-1-\alpha}{q}-\varepsilon}}{\max\{|x|_p, |y|_p\}} dy.$$

Set $|\varepsilon|_p > 1$. We have

$$
\|T^p f_\varepsilon\|_{L^q(|x|_p^\alpha dx)} = \left(\int_{\mathbb{Q}_p} \left(\int_{|y|_p \geq 1} \frac{|y|_p^{\frac{-1-\alpha}{q}-\varepsilon}}{\max\{|x|_p, |y|_p\}} dy \right)^q |x|_p^\alpha dx \right)^{1/q}
$$

$$
\geq \left(\int_{|x|_p \geq |\varepsilon|_p} \left(\int_{|t|_p \geq \frac{1}{|\varepsilon|_p}} \frac{|t|_p^{\frac{-1-\alpha}{q}-\varepsilon}}{\max\{1, |t|_p\}} dt \right)^q |x|_p^{-1-\varepsilon q} dx \right)^{1/q}
$$

$$
= \|f_\varepsilon\|_{L^q(|x|_p^\alpha dx)}^q |\varepsilon|_p^{-\varepsilon} \int_{|t|_p \geq \frac{1}{|\varepsilon|_p}} \frac{|t|_p^{\frac{-1-\alpha}{q}-\varepsilon}}{\max\{1, |t|_p\}} dt.
$$

Therefore,

$$
\int_{|t|_p \geq \frac{1}{|\varepsilon|_p}} \frac{|t|_p^{\frac{-1-\alpha}{q}-\varepsilon}}{\max\{1, |t|_p\}} dt \leq |\varepsilon|_p^\varepsilon \|T^p\|_{L^q(|x|_p^\alpha dx) \to L^q(|x|_p^\alpha dx)}.
$$

Now take $\varepsilon = p^{-k}$, $k = 1, 2, 3, \ldots$. Then $|\varepsilon|_p = p^k > 1$. Letting $k \to \infty$ and using Fatou's lemma, we obtain that

$$
\left(1 - \frac{1}{p}\right) \left(\frac{1}{1 - p^{\frac{\alpha+1}{q}-1}} + \frac{p^{-\frac{\alpha+1}{q}}}{1 - p^{-\frac{\alpha+1}{q}}} \right) = \int_{\mathbb{Q}_p^*} \frac{|t|_p^{-\frac{\alpha+1}{q}}}{\max\{1, |t|_p\}} dt
$$

$$
\leq \|T^p\|_{L^q(|x|_p^\alpha dx) \to L^q(|x|_p^\alpha dx)}. \tag{5.20}
$$

Then (5.14) follows from (5.19) and (5.20). Theorem 5.5 is proved. □

5.3.2 Boundedness of commutators of p-adic Hardy and Hardy-Littlewood-Pólya operators

In this section, we consider the boundedness of commutators of p-adic Hardy and Hardy-Littlewood-Pólya operators.

Definition 5.5. Let $b \in L_{\text{loc}}(\mathbb{Q}_p^n)$. The commutators of p-adic Hardy operators are defined by

$$
[b, \mathcal{H}^p]f = b\mathcal{H}^p f - \mathcal{H}^p(bf) \text{ and } [b, \mathcal{H}^{p,*}]f = b\mathcal{H}^{p,*} f - \mathcal{H}^{p,*}(bf).
$$

Definition 5.6. Let $b \in L_{\text{loc}}(\mathbb{Q}_p^n)$. The commutator of p-adic Hardy-Littlewood-Pólya operator is defined by

$$
[b, T^p]f = bT^p f - T^p(bf).
$$

Similar to the definition of $C\dot{M}O^q$ spaces in the setting of Euclidean spaces, we define the $C\dot{M}O^q$ spaces on $\mathbb{Q}_p^n$.

Definition 5.7. Let $1 \leq q < \infty$. A function $f \in L_{\text{loc}}^q(\mathbb{Q}_p^n)$ is said to be in $C\dot{M}O^q(\mathbb{Q}_p^n)$, if

$$\|f\|_{C\dot{M}O^q(\mathbb{Q}_p^n)} := \sup_{\gamma \in \mathbb{Z}} \left(\frac{1}{|B_\gamma(0)|_H} \int_{B_\gamma(0)} |f(x) - f_{B_\gamma(0)}|^q dx \right)^{1/q} < \infty,$$

where

$$f_{B_\gamma(0)} = \frac{1}{|B_\gamma(0)|_H} \int_{B_\gamma(0)} f(x) dx.$$

Remark 5.4. It is obvious that $L^\infty(\mathbb{Q}_p^n) \subset BMO(\mathbb{Q}_p^n) \subset C\dot{M}O^q(\mathbb{Q}_p^n)$.

Since Herz space is a natural generalization of Lebesgue space with power weight, we further study the boundedness of commutators of p-adic Hardy and Hardy-Littlewood-Pólya operators on Herz space. Let us firstly give the definition of Herz space.

Let $B_k = B_k(0) = \{x \in \mathbb{Q}_p^n : |x|_p \leq p^k\}$, $S_k = B_k \backslash B_{k-1}$ and let χ_E be the characteristic function of the set E.

Definition 5.8. [Zhu and Zheng (1998)] Suppose $\alpha \in \mathbb{R}$, $0 < q < \infty$ and $0 < r < \infty$. The homogeneous p-adic Herz space $\dot{K}_r^{\alpha,q}(\mathbb{Q}_p^n)$ is defined by

$$\dot{K}_r^{\alpha,q}(\mathbb{Q}_p^n) = \left\{ f \in L_{\text{loc}}^r(\mathbb{Q}_p^n) : \|f\|_{\dot{K}_r^{\alpha,q}(\mathbb{Q}_p^n)} < \infty \right\},$$

where

$$\|f\|_{\dot{K}_r^{\alpha,q}(\mathbb{Q}_p^n)} = \left(\sum_{k=-\infty}^{\infty} p^{k\alpha q} \|f\chi_k\|_{L^r(\mathbb{Q}_p^n)}^q \right)^{1/q}$$

with the usual modifications made when $q = \infty$ or $r = \infty$.

Remark 5.5. $\dot{K}_q^{0,q}(\mathbb{Q}_p^n)$ is the generalization of $L^q(|x|_p^\alpha dx)$, and

$$\dot{K}_q^{\frac{\alpha}{q},q}(\mathbb{Q}_p^n) = L^q(|x|_p^\alpha dx), \quad \dot{K}_q^{0,q}(\mathbb{Q}_p^n) = L^q(\mathbb{Q}_p^n)$$

for all $0 < q \leq \infty$ and $\alpha \in \mathbb{R}$.

Motivated by [Fu et al. (2007)], we get the following results for $0 < q_1 \leq q_2 < \infty$.

Theorem 5.6. Let $1 < r < \infty$ and $b \in C\dot{M}O^{\max\{r',r\}}(\mathbb{Q}_p^n)$. Then
(i) if $\alpha < \frac{n}{r'}$, then $[b, \mathcal{H}^p]$ is bounded from $\dot{K}_r^{\alpha,q_1}(\mathbb{Q}_p^n)$ to $\dot{K}_r^{\alpha,q_2}(\mathbb{Q}_p^n)$;
(ii) if $\alpha > -\frac{n}{r}$, then $[b, \mathcal{H}^{p,*}]$ is bounded from $\dot{K}_r^{\alpha,q_1}(\mathbb{Q}_p^n)$ to $\dot{K}_r^{\alpha,q_2}(\mathbb{Q}_p^n)$.

From Remark 5.5, we get the following two corollaries.

Corollary 5.4. *Suppose that $1 < q < \infty$ and $b \in C\dot{M}O^{\max\{q',q\}}(\mathbb{Q}_p^n)$. Then*
(i) if $\alpha < \frac{nq}{q'}$, then $[b, \mathcal{H}^p]$ is bounded from $L^q(|x|_p^\alpha dx)$ to $L^q(|x|_p^\alpha dx)$;
(ii) if $\alpha > -n$, then $[b, \mathcal{H}^{p,*}]$ is bounded from $L^q(|x|_p^\alpha dx)$ to $L^q(|x|_p^\alpha dx)$.

Corollary 5.5. *Suppose that $1 < q < \infty$ and $b \in C\dot{M}O^{\max\{q',q\}}(\mathbb{Q}_p^n)$. Then both $[b, \mathcal{H}^p]$ and $[b, \mathcal{H}^{p,*}]$ are bounded from $L^q(\mathbb{Q}_p^n)$ to $L^q(\mathbb{Q}_p^n)$.*

Due to Theorem 5.6, we can also obtain the boundedness of commutators generated by p-adic Hardy-Littlewood-Pólya operators and $C\dot{M}O$ functions.

Theorem 5.7. *Suppose that $0 < q_1 \leq q_2 < \infty$, $1 < r < \infty$, $-1/r < \alpha < 1/r'$ and $b \in C\dot{M}O^{\max\{r',r\}}(\mathbb{Q}_p^n)$. Then $[b, T^p]$ is bounded from $\dot{K}_r^{\alpha,q_1}(\mathbb{Q}_p^n)$ to $\dot{K}_r^{\alpha,q_2}(\mathbb{Q}_p^n)$.*

To prove Theorem 5.6, we need the following lemmas.

Lemma 5.2. *Suppose that b is a $C\dot{M}O$ function and $1 \leq q < r < \infty$. Then $C\dot{M}O^r(\mathbb{Q}_p^n) \subset C\dot{M}O^q(\mathbb{Q}_p^n)$ and $\|b\|_{C\dot{M}O^q(\mathbb{Q}_p^n)} \leq \|b\|_{C\dot{M}O^r(\mathbb{Q}_p^n)}$.*

Proof. For any $b \in C\dot{M}O^r(\mathbb{Q}_p^n)$, by Hölder's inequality, we have

$$\left(\frac{1}{|B_\gamma(\mathbf{0})|_H} \int_{B_\gamma(\mathbf{0})} |b(x) - b_{B_\gamma(\mathbf{0})}|^q dx\right)^{1/q}$$

$$\leq \left(\frac{1}{|B_\gamma(\mathbf{0})|_H} \left(\int_{B_\gamma(\mathbf{0})} |b(x) - b_{B_\gamma(\mathbf{0})}|^r dx\right)^{q/r} |B_\gamma(\mathbf{0})|_H^{1-q/r}\right)^{1/q}$$

$$= \left(\frac{1}{|B_\gamma(\mathbf{0})|_H} \int_{B_\gamma(\mathbf{0})} |b(x) - b_{B_\gamma(\mathbf{0})}|^r dx\right)^{1/r}$$

$$\leq \|b\|_{C\dot{M}O^r(\mathbb{Q}_p^n)}.$$

Therefore, $b \in C\dot{M}O^q(\mathbb{Q}_p^n)$ and $\|b\|_{C\dot{M}O^q(\mathbb{Q}_p^n)} \leq \|b\|_{C\dot{M}O^r(\mathbb{Q}_p^n)}$. This completes the proof. $\square$

Lemma 5.3. *Suppose that b is a $C\dot{M}O^1(\mathbb{Q}_p^n)$ function, $j, k \in \mathbb{Z}$. Then*

$$|b(t) - b_{B_k}| \leq |b(t) - b_{B_k}| + p^n |j - k| \|b\|_{C\dot{M}O^1(\mathbb{Q}_p^n)}.$$

Proof. For $i \in \mathbb{Z}$, we have

$$|b_{B_i} - b_{B_{i+1}}| \leq \frac{1}{|B_i|_H} \int_{B_i} |b(t) - b_{B_{i+1}}| dt$$

$$\leq \frac{p^n}{|B_{i+1}|_H} \int_{B_{i+1}} |b(t) - b_{B_{i+1}}| dt$$

$$\leq p^n \|b\|_{C\dot{M}O^1(\mathbb{Q}_p^n)}. \tag{5.21}$$

For j, $k \in \mathbb{Z}$, without loss of generality, we can assume that $j \leq k$. By (5.21), we get

$$|b(t) - b_{B_k}| \leq |b(t) - b_{B_j}| + \sum_{i=k}^{j-1} |b_{B_i} - b_{B_{i+1}}|$$

$$\leq |b(t) - b_{B_j}| + p^n |j - k| \|b\|_{C\dot{M}O^1(\mathbb{Q}_p^n)}.$$

The lemma is proved. $\qquad\square$

Proof of Theorem 5.6. We first prove that (i) holds. Denote by $f(x)\chi_i(x) = f_i(x)$. Then we have

$$\|([b, \mathcal{H}^p]f)\chi_k\|^r_{L^r(\mathbb{Q}_p^n)} = \int_{S_k} |x|_p^{-rn} \left| \int_{B(0,|x|_p)} f(t)(b(x) - b(t)) dt \right|^r dx$$

$$\leq \int_{S_k} p^{-krn} \left(\int_{B(0,p^k)} |f(t)(b(x) - b(t))| \, dt \right)^r dx$$

$$= p^{-krn} \int_{S_k} \left(\sum_{j=-\infty}^{k} \int_{S_j} |f(t)(b(x) - b(t))| \, dt \right)^r dx$$

$$\leq Cp^{-krn} \int_{S_k} \left(\sum_{j=-\infty}^{k} \int_{S_j} |f(t)(b(x) - b_{B_k})| \, dt \right)^r dx$$

$$+ Cp^{-krn} \int_{S_k} \left(\sum_{j=-\infty}^{k} \int_{S_j} |f(t)(b(t) - b_{B_k})| \, dt \right)^r dx$$

$$=: I + II.$$

Next, we turn to estimate I and J, respectively. For I, by Hölder's inequality $(1/r + 1/r' = 1)$, we have

$$I = Cp^{-krn} \left(\int_{S_k} |b(x) - b_{B_k}|^r dx \right) \left(\sum_{j=-\infty}^{k} \int_{S_j} |f(t)| \, dt \right)^r$$

$$\le Cp^{\frac{-krn}{r'}} \left(\frac{1}{|B_k|_H} \int_{B_k} |b(x) - b_{B_k}|^r dx \right)$$

$$\times \left(\sum_{j=-\infty}^{k} \left(\int_{S_j} |f(t)|^r dt \right)^{1/r} \left(\int_{S_j} dt \right)^{1/r'} \right)^r$$

$$\le C\|b\|_{C\dot{M}O^r(\mathbb{Q}_p^n)}^r \left(\sum_{j=-\infty}^{k} p^{\frac{(j-k)n}{r'}} \|f_j\|_{L^r(\mathbb{Q}_p^n)} \right)^r. \tag{5.22}$$

For II, by Lemma 5.3, we get

$$II = Cp^{-krn} \int_{S_k} \left(\sum_{j=-\infty}^{k} \int_{S_j} |f(t)(b(t) - b_{B_k})| \, dt \right)^r dx$$

$$= Cp^{-krn} p^{kn} (1 - p^{-n}) \left(\sum_{j=-\infty}^{k} \int_{S_j} |f(t)(b(t) - b_{B_k})| \, dt \right)^r$$

$$\le Cp^{\frac{-krn}{r'}} \left(\sum_{j=-\infty}^{k} \int_{S_j} \left| f(t)(b(t) - b_{B_j}) \right| dt \right)^r$$

$$+ Cp^{\frac{-krn}{r'}} \|b\|_{C\dot{M}O^1(\mathbb{Q}_p^n)}^r \left(\sum_{j=-\infty}^{k} (k-j) \int_{S_j} |f(t)| \, dt \right)^r$$

$$=: II_1 + II_2.$$

For II_1 and II_2, by Hölder's inequality, we obtain that II_1 is not bigger than

$$Cp^{\frac{-krn}{r'}} \left(\left(\sum_{j=-\infty}^{k} \int_{S_j} |f(t)|^r dt \right)^{\frac{1}{r}} \left(\int_{S_j} |b(t) - b_{B_j}|^{r'} dt \right)^{\frac{1}{r'}} \right)^r$$

$$\le Cp^{\frac{-krn}{r'}} \left(\sum_{j=-\infty}^{k} \|f_j\|_{L^r(\mathbb{Q}_p^n)} p^{\frac{jn}{r'}} \left(\frac{1}{|B_j|_H} \int_{B_j} |b(t) - b_{B_j}|^{r'} dt \right)^{\frac{1}{r'}} \right)^r$$

$$\le C\|b\|_{C\dot{M}O^{r'}(\mathbb{Q}_p^n)}^r \left(\sum_{j=-\infty}^{k} p^{\frac{(j-k)n}{r'}} \|f_j\|_{L^r(\mathbb{Q}_p^n)} \right)^r. \tag{5.23}$$

And II_2 is not greater than

$$Cp^{\frac{-krn}{r'}}\|b\|^r_{C\dot{M}O^1(\mathbb{Q}_p^n)}\left(\sum_{j=-\infty}^{k}(k-j)\left(\int_{S_j}|f(t)|^r\,dt\right)^{\frac{1}{r}}\left(\int_{S_j}dt\right)^{\frac{1}{r'}}\right)^r$$

$$\leq C\|b\|^r_{C\dot{M}O^1(\mathbb{Q}_p^n)}\left(\sum_{j=-\infty}^{k}(k-j)p^{\frac{(j-k)n}{r'}}\|f_j\|_{L^r(\mathbb{Q}_p^n)}\right)^r. \qquad (5.24)$$

Then (5.22)–(5.24) together with Lemma 5.2 imply that

$$\|\mathcal{H}_bf\|_{\dot{K}_r^{\alpha,q_2}(\mathbb{Q}_p^n)}=\left(\sum_{k=-\infty}^{\infty}p^{k\alpha q_2}\|(U_{\beta,b}f)\chi_k\|^{q_2}_{L^r(\mathbb{Q}_p^n)}\right)^{1/q_2}$$

$$\leq\left(\sum_{k=-\infty}^{\infty}p^{k\alpha q_1}\|(U_{\beta,b}f)\chi_k\|^{q_1}_{L^r(\mathbb{Q}_p^n)}\right)^{1/q_1}$$

$$\leq C\left(\sum_{k=-\infty}^{\infty}p^{k\alpha q_1}\|b\|^{q_1}_{C\dot{M}O^r(\mathbb{Q}_p^n)}\left(\sum_{j=-\infty}^{k}p^{\frac{(j-k)n}{r'}}\|f_j\|_{L^r(\mathbb{Q}_p^n)}\right)^{q_1}\right)^{1/q_1}$$

$$+C\left(\sum_{k=-\infty}^{\infty}p^{k\alpha q_1}\|b\|^{q_1}_{C\dot{M}O^{r'}(\mathbb{Q}_p^n)}\left(\sum_{j=-\infty}^{k}p^{\frac{(j-k)n}{r'}}\|f_j\|_{L^r(\mathbb{Q}_p^n)}\right)^{q_1}\right)^{1/q_1}$$

$$+C\|b\|_{C\dot{M}O^1(\mathbb{Q}_p^n)}\left(\sum_{k=-\infty}^{\infty}p^{k\alpha q_1}\left(\sum_{j=-\infty}^{k}(k-j)p^{\frac{(j-k)n}{r'}}\|f_j\|_{L^r(\mathbb{Q}_p^n)}\right)^{q_1}\right)^{1/q_1}$$

$$\leq C\|b\|_{C\dot{M}O^{\max\{r',r\}}(\mathbb{Q}_p^n)}$$

$$\times\left(\sum_{k=-\infty}^{\infty}p^{k\alpha q_1}\left(\sum_{j=-\infty}^{k}(k-j)p^{\frac{(j-k)n}{r'}}\|f_j\|_{L^r(\mathbb{Q}_p^n)}\right)^{q_1}\right)^{1/q_1}$$

$$=:J.$$

For the case $0<q_1\leq 1$, since $\alpha<n/r'$, we have

$$J^{q_1}=C\|b\|^{q_1}_{C\dot{M}O^{\max\{r',r\}}(\mathbb{Q}_p^n)}$$

$$\times\sum_{k=-\infty}^{\infty}p^{k\alpha q_1}\left(\sum_{j=-\infty}^{k}(k-j)p^{\frac{(j-k)n}{r'}}\|f_j\|_{L^r(\mathbb{Q}_p^n)}\right)^{q_1}$$

$$=C\|b\|^{q_1}_{C\dot{M}O^{\max\{r',r\}}(\mathbb{Q}_p^n)}$$

$$\times \sum_{k=-\infty}^{\infty} \left(\sum_{j=-\infty}^{k} p^{j\alpha} \|f_j\|_{L^r(\mathbb{Q}_p^n)}(k-j)p^{(j-k)(\frac{n}{r'}-\alpha)} \right)^{q_1}$$

$$\leq C\|b\|_{C\dot{M}O^{\max\{r',r\}}(\mathbb{Q}_p^n)}^{q_1}$$

$$\times \sum_{k=-\infty}^{\infty} \sum_{j=-\infty}^{k} p^{j\alpha q_1} \|f_j\|_{L^r(\mathbb{Q}_p^n)}^{q_1}(k-j)^{q_1} p^{(j-k)(\frac{n}{r'}-\alpha)q_1}$$

$$= C\|b\|_{C\dot{M}O^{\max\{r',r\}}(\mathbb{Q}_p^n)}^{q_1}$$

$$\times \sum_{j=-\infty}^{\infty} p^{j\alpha q_1} \|f_j\|_{L^r(\mathbb{Q}_p^n)}^{q_1} \sum_{k=j}^{\infty} (k-j)^{q_1} p^{(j-k)(\frac{n}{r'}-\alpha)q_1}$$

$$= C\|b\|_{C\dot{M}O^{\max\{r',r\}}(\mathbb{Q}_p^n)}^{q_1} \|f\|_{\dot{K}_r^{\alpha,r}(\mathbb{Q}_p^n)}^{q_1}.$$

For the case $q_1 > 1$, by Hölder's inequality, we have

$$J^{q_1} \leq C\|b\|_{C\dot{M}O^{\max\{r',r\}}(\mathbb{Q}_p^n)}^{q_1}$$

$$\times \sum_{k=-\infty}^{\infty} \left(\sum_{j=-\infty}^{k} p^{j\alpha q_1} \|f_j\|_{L^r(\mathbb{Q}_p^n)}^{q_1} p^{\frac{(j-k)}{2}(\frac{n}{r'}-\alpha)q_1} \right)$$

$$\times \left(\sum_{j=-\infty}^{k} (k-j)^{q_1'} p^{\frac{(j-k)}{2}(\frac{n}{r'}-\alpha)q_1'} \right)^{q_1/q_1'}$$

$$= C\|b\|_{C\dot{M}O^{\max\{r',r\}}(\mathbb{Q}_p^n)}^{q_1}$$

$$\times \sum_{j=-\infty}^{\infty} p^{j\alpha q_1} \|f_j\|_{L^r(\mathbb{Q}_p^n)}^{q_1} \sum_{k=j}^{\infty} p^{\frac{(j-k)}{2}(\frac{n}{r'}-\alpha)q_1}$$

$$= C\|b\|_{C\dot{M}O^{\max\{r',r\}}(\mathbb{Q}_p^n)}^{q_1} \|f\|_{\dot{K}_r^{\alpha,r}(\mathbb{Q}_p^n)}^{q_1}.$$

By the similar method, we can obtain (ii). Theorem 5.6 is proved. $\qquad\square$

Proof of Theorem 5.7. It follows from that

$$|[b, T^p]f(x)| = \left| \int_{\mathbb{Q}_p^*} \frac{f(y)}{\max\{|x|_p, |y|_p\}} (b(x) - b(y)) \, dy \right|$$

$$\leq \left| \int_{B(0,|x|_p)} \frac{f(y)(b(x) - b(y))}{|x|_p} dy \right|$$

$$+ \left| \int_{\mathbb{Q}_p^n \setminus B(0,|x|_p)} \frac{f(y)(b(x) - b(y))}{|y|_p} dy \right|$$

$$= |[b, \mathcal{H}^p]f(x)| + |[b, \mathcal{H}^{p,*}]f(x)| \, .$$

By Minkowski's inequality, we have

$$\|[b, T^p]f\|_{\dot{K}_r^{\alpha, q_2}(\mathbb{Q}_p^n)} \le C\|[b, \mathcal{H}^p]f\|_{K_r^{\alpha, q_2}(\mathbb{Q}_p^n)} + C\|[b, \mathcal{H}^{p,*}]f\|_{K_r^{\alpha, q_2}(\mathbb{Q}_p^n)}.$$

Then Theorem 5.7 holds with the help of Theorem 5.6. $\qquad\qquad\square$

5.4   Notes

Geller [Geller (1980)] studied the characterizations of Hardy operator through Riesz transforms on the Heisenberg group $\mathbb{H}^n$. Wu and Fu in [Wu and Fu (2016)] first proved a sharp estimate for the n-dimensional Hardy operator on the Heisenberg group (Theorems 5.1, 5.2 and 5.3). Carton-Lebrun and Fosset [García-Cuerva and Rubio de Francia (1985)] defined the weighted Hardy operators.

In recent years, p-adic analysis has received a lot of attention due to its application in Mathematical Physics ([Albeverio and Karwowski (1994); Avetisov et al. (2002); Khrennikov (1994, 1997); Varadarajan (1997); Vladimirov et al. (1994)]). There are numerous papers on p-adic analysis, such as [Haran (1990, 1993)] about Riesz potentials, [Albeverio et al. (2006); Chuong and Co (2008); Chuong et al. (2007); Kochubei (2008); Zuniga-Galindo (2004)] about p-adic pseudo differential equations, etc. The Harmonic Analysis on p-adic field has been drawing more and more concern ([Kim (2009, 2008); Rim and Lee (2006); Rogers (2005, 2004)] and references therein). Fu et al. in [Fu et al. (2013)] established the sharp estimates of p-adic Hardy operators on p-adic weighted Lebesgue spaces (Theorems 5.4 and 5.5). Wu et al. [Wu et al. (2013)] obtained the sharp bounds of p-adic Hardy operators on p-adic central Morrey spaces and p-adic λ-central BMO spaces (Theorems 5.4 and 5.5). They also obtained the boundedness for commutators of p-adic Hardy operators on these spaces (Theorems 5.6 and 5.7). The authors in [Liu and Zhou (2017)] computed the sharp estimates for the p-adic Hardy type operators on higher-dimensional product spaces. More information about p-adic Hardy operators, see [He et al. (2019); Chuong et al. (2018); Rim and Lee (2006)] and the references therein.

Chapter 6

Sharp constants for Hausdorff q-inequalities

The classical Hausdorff operator becomes an attractive research topic in recent years (see [Liflyand (2007); Lerner and Liflyand (2007); Liflyand and Miyachi (2009)]). For a locally integrable function ϕ on $(0, \infty)$, the one-dimensional Hausdorff operator is defined in the integral form by

$$h_\phi(f)(x) = \int_0^\infty \frac{\phi(t)}{t} f\left(\frac{x}{t}\right) dt.$$

The operator h_ϕ has an important feature: many classical positive integral operators actually are special cases of the Hausdorff operator if one chooses suitable functions ϕ. Let $x > 0$. Then one obtains:

(i) the one-dimensional Hardy operator

$$Hf(x) = \frac{1}{x} \int_0^x f(t)dt, \text{ if } \phi(t) = \frac{\chi_{(1,\infty)}(t)}{t};$$

(ii) the one-dimensional adjoint Hardy operator

$$H^* f(x) = \int_x^\infty \frac{f(t)}{t} dt, \text{ if } \phi(t) = \chi_{(0,1)}(t);$$

(iii) the Hilbert operator

$$T^1 f(x) = \int_0^\infty \frac{f(t)}{x+t} dt, \text{ if } \phi(t) = \frac{1}{1+t} \chi_{[0,\infty)}(t)$$

and

(iv) the Hardy-Littlewood-Pólya operator

$$\mathbf{P}f(x) = \int_0^\infty \frac{f(t)}{\max\{t, x\}} dt = Hf(x) + H^* f(x),$$

if $\phi(t) = \frac{\chi_{(1,\infty)}(t)}{t} + \chi_{(0,1)}(t)$.

For the Hausdorff operator, the classical Hausdorff inequality (see [Fefferman (1983)] for instance)

$$\int_0^\infty \left(\int_0^\infty \frac{\phi(t)}{t} f(\frac{x}{t}) dt \right)^p dx \le \widetilde{C_\phi} \int_0^\infty f^p(t) dt \tag{6.1}$$

is well known, where $f \ge 0$, $\phi(t)$ is a nonnegative valued function and the constant $\widetilde{C_\phi}$ is

$$\widetilde{C_\phi} = \left(\int_0^\infty \frac{\phi(t)}{t} t^{\frac{1}{p}} dt \right)^p .$$

By taking the special functions ϕ, (6.1) becomes the classical Hardy type inequalities in [Hardy et al. (1952)], for $1 < p < \infty$,

$$\int_0^\infty \left(\frac{1}{x} \int_0^x f(t) dt \right)^p dx \le \left(\frac{p}{p-1} \right)^p \int_0^\infty f^p(t) dt$$

and

$$\int_0^\infty \left(\int_x^\infty \frac{f(t)}{t} dt \right)^{p'} dx \le \left(\frac{p}{p-1} \right)^{p'} \int_0^\infty f^{p'}(t) dt,$$

where $1/p + 1/p' = 1$.

The quantum calculus (q-calculus) initially introduced by Jackson [Jackson (1910b)], but this kind of calculus had already been studied by Euler who started analytic number theory by inventing the Euler-Maclaurin summation formula [Euler (1732, 1741)], and Jacobi in [Jacobi (1829)]. Kac and Cheung in [Kac and Cheung (2002)] had written a book about many of the fundamental aspects of quantum calculus. The q-calculus has plenty of applications in different mathematical areas such as number theory, combinatorics, orthogonal polynomials, basic hyper-geometric functions and other sciences quantum theory, mechanics and the theory of relativity (see [Andersen and Muckenhoupt (1982); Bangerezako (2005); Ernst (1999, 2012); Exton (1983)]). As one of interesting topics, q-analogues of many inequalities from the classical analysis have been established, such as [Baiarystanov et al. (2016); Guo and Zhao (2017); Maligranda et al. (2014); Miao and Qi (2009); Sulaiman (2011)]. These integral inequalities can be used for the study of qualitative and quantitative properties of integrals, see [Andersen (2003); Mitrinović (1993); Wu and Debnath (2007)].

This chapter is aimed to study the Hausdorff operators in the frame of quantum calculus.

6.1   Some q-inequalities for Hausdorff operators

Recently, Maligranda, Oinarov and Persson in [Maligranda et al. (2014)] studied the q-analysis variants of Hardy type inequalities and obtained the sharp bound constant. Thus, it is interesting to extend the result of [Maligranda et al. (2014)] to the more general Hausdorff operator. We shall establish the following q-analogue of the Hausdorff inequality

$$\int_0^\infty \left(\int_0^\infty \frac{\phi(t)}{t} f(\frac{x}{t}) d_q t \right)^p d_q x \le C_\Phi \int_0^\infty f^p(t) d_q t. \qquad (6.2)$$

We first recall some basic notations and definitions. Let $0 < q < 1$ be fixed. In 1910, Jackson defined the general q-integral. For a function $f : [0, b) \to \mathbb{R}$, $0 < b \le \infty$, the q-integral or the q-Jackson integral (see [Jackson (1910a); Kufner and Persson (2003)]) is defined by the following formula:

$$\int_0^x f(t) d_q t = (1 - q) x \sum_{k=0}^\infty q^k f(q^k x), \text{ for } x \in (0, b], \qquad (6.3)$$

and the improper q-integral of a function $f : [0, \infty) \to \mathbb{R}$ is defined by the series

$$\int_0^\infty f(t) d_q t = (1 - q) \sum_{k=-\infty}^\infty q^k f(q^k) \qquad (6.4)$$

provided that the series on the right-hand sides of (6.3) and (6.4) converge absolutely.

For $0 < a < b \le \infty$, the q-integral on $[a, b]$ is defined by

$$\int_a^b f(t) d_q t = \int_0^b f(t) d_q t - \int_0^a f(t) d_q t.$$

In particular, for $x \in (0, \infty)$, it yields that

$$\int_x^\infty f(t) d_q t = \int_0^\infty f(t) d_q t - \int_0^x f(t) d_q t.$$

In the theory of q-analysis, the q-analogue $[\alpha]_q$ of a number $\alpha \in \mathbb{R}$ is defined by

$$[\alpha]_q = \frac{1 - q^\alpha}{1 - q}.$$

6.1.1   *The inequality of Hausdorff operator in q-analysis*

First, we consider the q-integral analogue of the form (6.1):

Theorem 6.1. *Let $\phi \geq 0$. If $1 \leq p < \infty$ and $f \geq 0$ or $p < 0$ and $f > 0$, then the following inequality*

$$\int_0^\infty \left(\int_0^\infty \frac{\phi(t)}{t} f\left(\frac{x}{t}\right) d_q t \right)^p d_q x \leq C_\phi \int_0^\infty f^p(t) d_q t \qquad (6.5)$$

holds with the constant

$$C_\phi = \left(\int_0^\infty \frac{\phi(t)}{t} t^{\frac{1}{p}} d_q t \right)^p. \qquad (6.6)$$

If $0 < p < 1$ and $f \geq 0$, then the inequality (6.5) holds in the reverse direction with the constant C_ϕ. Moreover, when $p < 0$, the constant C_ϕ is the best possible in (6.6).

Remark 6.1. The constant C_ϕ is sharp for the case of $1 \leq p < \infty$, see Section 6.2 later.

Proof. Using the definition (6.4), we have

$$L(f) := \int_0^\infty \left(\int_0^\infty \frac{\phi(t)}{t} f\left(\frac{x}{t}\right) d_q t \right)^p d_q x$$

$$= \int_0^\infty \left((1-q) \sum_{i=-\infty}^\infty \frac{\phi(q^i)}{q^i} f\left(\frac{x}{q^i}\right) q^i \right)^p d_q x$$

$$= (1-q)^{p+1} \sum_{j=-\infty}^\infty \left(\sum_{i=-\infty}^\infty \phi(q^i) f(q^{j-i}) \right)^p q^j$$

$$= (1-q)^{p+1} \sum_{j=-\infty}^\infty \left(\sum_{i=-\infty}^\infty q^{\frac{i}{p}} \phi(q^i) q^{\frac{j-i}{p}} f(q^{j-i}) \right)^p.$$

Let $1 \leq p < \infty$. Applying Minkowski's inequality, we have

$$(L(f))^{1/p} \leq (1-q)^{1+1/p} \sum_{i=-\infty}^\infty q^{\frac{i}{p}} \phi(q^i) \left(\sum_{j=-\infty}^\infty q^{j-i} (f(q^{j-i}))^p \right)^{1/p}$$

$$= (1-q)^{1+1/p} \sum_{i=-\infty}^\infty q^{\frac{i}{p}} \phi(q^i) \left(\sum_{j=-\infty}^\infty q^k (f(q^k))^p \right)^{1/p}$$

$$= \left(\int_0^\infty f^p(t) d_q t \right)^{1/p} \int_0^\infty \frac{\phi(t)}{t} t^{1/p} d_q t$$

$$= C_\phi^{1/p} \left(\int_0^\infty f^p(t) d_q t \right)^{1/p},$$

which means that (6.5) holds with the constant (6.6).

Let $p < 0$ and $f > 0$. Then using Minkowski's inequality, we have

$$L(f) = (1-q)^{p+1} \left(\left(\sum_{j=-\infty}^{\infty} \left(\sum_{i=-\infty}^{\infty} q^{\frac{i}{p}} \phi(q^i) q^{\frac{j-i}{p}} f(q^{j-i}) \right)^p \right)^{1/p} \right)^p$$

$$\leq (1-q)^{p+1} \left(\sum_{i=-\infty}^{\infty} \left(\sum_{j=-\infty}^{\infty} \left(q^{\frac{j-i}{p}} f(q^{j-i}) \right)^p \right)^{1/p} q^{\frac{i}{p}} \phi(q^i) \right)^p$$

$$= (1-q)^{p+1} \left(\sum_{i=-\infty}^{\infty} \left(\sum_{k=-\infty}^{\infty} \left(q^{\frac{k}{p}} f(q^k) \right)^p \right)^{1/p} q^{\frac{i}{p}} \phi(q^i) \right)^p$$

$$= (1-q)^{p+1} \sum_{k=-\infty}^{\infty} q^k \left(f(q^k) \right)^p \left(\sum_{i=-\infty}^{\infty} q^{\frac{i}{p}} \phi(q^i) \right)^p$$

$$= \left(\int_0^{\infty} \frac{\phi(t)}{t} t^{\frac{1}{p}} d_q t \right)^p \int_0^{\infty} f^p(t) d_q t$$

$$= C_\phi \int_0^{\infty} f^p(t) d_q t.$$

This implies that the inequality (6.5) holds with the constant C_ϕ. In this case, we shall show that the constant (6.6) is sharp in (6.5). For $0 < \varepsilon < 1$, let $f_\varepsilon(t) = t^{-1/p-\varepsilon} \chi_{(0,1]}(t)$. Then it follows from $p\varepsilon < 0$ that

$$\int_0^{\infty} f_\varepsilon^p(t) d_q t = (1-q) \sum_{k=-\infty}^{\infty} q^k f_\varepsilon^p(q^k) = (1-q) \sum_{k=0}^{\infty} q^k q^{kp(-1/p-\varepsilon)} = \frac{1-q}{1-q^{-p\varepsilon}}$$

and

$$L(f_\varepsilon) = \int_0^{\infty} \left(\int_0^{\infty} \frac{\phi(t)}{t} f_\varepsilon\left(\frac{x}{t} \right) d_q t \right)^p d_q x$$

$$= (1-q)^{p+1} \sum_{j=-\infty}^{\infty} \left(\sum_{i=-\infty}^{\infty} q^{\frac{i}{p}} \phi(q^i) q^{\frac{j-i}{p}} f_\varepsilon(q^{j-i}) \right)^p$$

$$= (1-q)^{p+1} \sum_{j=-\infty}^{\infty} \left(\sum_{i=-\infty}^{j} \phi(q^i) q^{\frac{i}{p}} q^{-(j-i)(1/p+\varepsilon)} \right)^p$$

$$\geq (1-q)^{p+1} \sum_{j=0}^{\infty} q^{-jp\varepsilon} \left(\sum_{i=-\infty}^{j} q^{i(1/p+\varepsilon)} \phi(q^i) \right)^p$$

$$\geq (1-q)^{p+1} \sum_{j=0}^{\infty} q^{-jp\varepsilon} \left(\sum_{i=-\infty}^{\infty} q^{i(1/p+\varepsilon)} \phi(q^i) \right)^p$$

$$= \frac{1-q}{1-q^{-p\varepsilon}} \cdot \left((1-q) \sum_{i=-\infty}^{\infty} q^{i(1/p+\varepsilon)} \phi(q^i) \right)^p.$$

It implies that

$$C_\phi \geq \frac{L(f_\varepsilon)}{\int_0^\infty f_\varepsilon^p(t) d_q t} \geq \left((1-q) \sum_{i=-\infty}^{\infty} q^{i(1/p+\varepsilon)} \phi(q^i) \right)^p.$$

Letting $\varepsilon \to 0^+$, we obtain

$$C_\phi \geq \left((1-q) \sum_{i=-\infty}^{\infty} q^{i/p} \phi(q^i) \right)^p = \left(\int_0^\infty \frac{\phi(t)}{t} t^{\frac{1}{p}} d_q t \right)^p,$$

which proves that the constant (6.6) is the sharp one.

If $0 < p < 1$, we have

$$L(f) = (1-q)^{p+1} \left(\left(\sum_{j=-\infty}^{\infty} \left(\sum_{i=-\infty}^{\infty} q^{\frac{i}{p}} \phi(q^i) q^{\frac{j-i}{p}} f(q^{j-i}) \right)^p \right)^{1/p} \right)^p$$

$$\geq (1-q)^{p+1} \left(\sum_{i=-\infty}^{\infty} \left(\sum_{j=-\infty}^{\infty} \left(q^{\frac{j-i}{p}} f(q^{j-i}) \right)^p \right)^{1/p} q^{\frac{i}{p}} \phi(q^i) \right)^p$$

$$= (1-q)^{p+1} \left(\sum_{i=-\infty}^{\infty} \left(\sum_{k=-\infty}^{\infty} \left(q^{\frac{k}{p}} f(q^k) \right)^p \right)^{1/p} q^{\frac{i}{p}} \phi(q^i) \right)^p$$

$$= C_\phi \int_0^\infty f^p(t) d_q t.$$

Thus, Theorem 6.1 is proved.   □

6.1.2   *Applications*

In this section, we shall show that the constant in (6.6) is sharp if we take some special functions ϕ. These operators include many classical positive integral operators, such as the Hardy operator and its adjoint operator, the Hilbert operator and the Hardy-Littlewood-Pólya operator.

We have the following statements for $x > 0$.

Theorem 6.2. *For $1 < p < \infty$, $f \geq 0$, the following inequality*

$$\int_0^\infty \left(\int_x^\infty \frac{f(t)}{t} d_q t \right)^p d_q x \leq C_1 \int_0^\infty f^p(t) d_q t \qquad (6.7)$$

holds with the constant

$$C_1 = \frac{q}{[1/p]_q^p}.\tag{6.8}$$

Moreover, the constant C_1 is the best possible in (6.7).

Proof. When $\phi(t) = \frac{\chi_{(0,1)}(t)}{t}$, we can easily see that (6.7) holds with the constant (6.8). Now, we shall show that the constant (6.8) is sharp. For $0 < \varepsilon < 1$, let $f_\varepsilon(t) = t^{-\frac{1}{p}-\varepsilon}\chi_{[1,\infty)}(t)$. Then

$$\int_0^\infty f_\varepsilon^p(t)d_qt = (1-q)\sum_{k=-\infty}^\infty q^k f_\varepsilon^p(q^k) = (1-q)\sum_{k=-\infty}^0 q^k q^{pk(-\frac{1}{p}-\varepsilon)}$$

$$= (1-q)\sum_{k=0}^\infty q^{kp\varepsilon} = \frac{1-q}{1-q^{p\varepsilon}}$$

and

$$L(f_\varepsilon) = \int_0^\infty \left(\int_x^\infty \frac{f_\varepsilon(t)}{t}d_qt\right)^p d_qx$$

$$= \int_0^\infty \left(\int_0^\infty \frac{f_\varepsilon(t)}{t}d_qt - \int_0^x \frac{f_\varepsilon(t)}{t}d_qt\right)^p d_qx$$

$$= \int_0^\infty \left((1-q)\left(\sum_{i=-\infty}^\infty q^i \frac{f(q^i)}{q^i} - x\sum_{j=0}^\infty q^j f(q^j x)\frac{1}{q^j x}\right)\right)^p d_qx$$

$$= (1-q)^{p+1}\sum_{k=-\infty}^\infty q^k \left(\sum_{i=-\infty}^\infty f_\varepsilon(q^i) - \sum_{j=0}^\infty f_\varepsilon(q^{j+k})\right)^p$$

$$= (1-q)^{p+1}\sum_{k=-\infty}^\infty q^k \left(\sum_{i=-\infty}^\infty f_\varepsilon(q^i) - \sum_{j=k}^\infty f_\varepsilon(q^j)\right)^p$$

$$= (1-q)^{p+1}\sum_{k=-\infty}^\infty q^k \left(\sum_{i=-\infty}^{k-1} f_\varepsilon(q^i)\right)^p$$

$$\geq (1-q)^{p+1}\sum_{k=-\infty}^0 q^k \left(\sum_{i=-\infty}^{k-1} q^{i(-\frac{1}{p}-\varepsilon)}\right)^p$$

$$= (1-q)^{p+1}q^{1+p\varepsilon}\sum_{k=0}^\infty q^{kp\varepsilon}\left(\sum_{i=0}^\infty q^{i(\frac{1}{p}+\varepsilon)}\right)^p$$

$$= (1-q)^{p+1}\frac{q^{1+p\varepsilon}}{1-q^{p\varepsilon}}\left(\frac{1}{1-q^{\frac{1}{p}+\varepsilon}}\right)^p.$$

Consequently, it yields that

$$C_1 \geq \frac{L(f_\varepsilon)}{\int_0^\infty f_\varepsilon^p(t)d_qt} = q^{1+p\varepsilon}\left(\frac{1-q}{1-q^{\frac{1}{p}+\varepsilon}}\right)^p.$$

Letting $\varepsilon \to 0^+$, we deduce

$$C_1 \geq q\left(\frac{1-q}{1-q^{\frac{1}{p}}}\right)^p = \frac{q}{[1/p]_q^p},$$

which shows that the constant (6.8) is sharp. $\qquad\square$

Theorem 6.3. *For $1 < p < \infty$, $f \geq 0$, the following inequality*

$$\int_0^\infty \left(\int_0^\infty \frac{f(t)}{x+t}d_qt\right)^p d_qx \leq C_2 \int_0^\infty f^p(t)d_qt \tag{6.9}$$

holds with the best constant

$$C_2 = \left((1-q)\sum_{i=-\infty}^\infty \frac{q^{\frac{i}{p}}}{1+q^i}\right)^p. \tag{6.10}$$

Remark 6.2. It is easy to check that $C_2 < \infty$. In fact, for $1 < p < \infty$ and $0 < q < 1$,

$$\sum_{i=-\infty}^\infty \frac{q^{\frac{i}{p}}}{1+q^i} = \sum_{i=-\infty}^0 \frac{q^{\frac{i}{p}}}{1+q^i} + \sum_{i=1}^\infty \frac{q^{\frac{i}{p}}}{1+q^i}$$

$$\leq \sum_{i=-\infty}^0 q^{(1/p-1)i} + \sum_{i=0}^\infty q^{i/p} = \frac{1}{1-q^{1-1/p}} + \frac{1}{1-q^{1/p}} < \infty.$$

Proof. Let $\phi(t) = (1+t)^{-1}$. The proof of Theorem 6.1 gives that the inequality (6.9) holds with the constant (6.10). We now show that the upper bound in (6.10) is sharp. For $0 < \varepsilon < \min\{1, p-1\}$, let $f_\varepsilon(t) = t^{-(1+\varepsilon)/p}\chi_{[1,\infty)}(t)$. Then by the definition of q-integral,

$$\int_0^\infty f_\varepsilon^p(t)d_qt = (1-q)\sum_{k=-\infty}^\infty q^k f_\varepsilon^p(q^k)$$

$$= (1-q)\sum_{k=-\infty}^0 q^k q^{pk(-\frac{1+\varepsilon}{p})} = (1-q)\sum_{k=0}^\infty q^{k\varepsilon} = \frac{1-q}{1-q^\varepsilon}.$$

The Minkowski inequality leads to

$$L(f_\varepsilon) = \int_0^\infty \left(\int_x^\infty \frac{f_\varepsilon(t)}{x+t}d_qt\right)^p d_qx$$

$$= (1-q)^{p+1} \sum_{j=-\infty}^{\infty} q^j \left(\sum_{i=-\infty}^{0} q^i \frac{q^{i(-\frac{1+\varepsilon}{p})}}{q^j + q^i} \right)^p$$

$$= (1-q)^{p+1} \sum_{j=-\infty}^{\infty} q^j \left(\sum_{i=-\infty}^{0} \frac{q^{i(-\frac{1+\varepsilon}{p})}}{1 + q^{j-i}} \right)^p$$

$$= (1-q)^{p+1} \sum_{j=-\infty}^{\infty} q^{-j\varepsilon} \left(\sum_{i=j}^{\infty} \frac{q^{i(\frac{1+\varepsilon}{p})}}{1 + q^i} \right)^p$$

$$\geq (1-q)^{p+1} \sum_{j=-\infty}^{0} q^{-j\varepsilon} \left(\sum_{i=-\infty}^{\infty} \frac{q^{i(\frac{1+\varepsilon}{p})}}{1 + q^i} - \sum_{i=-\infty}^{j} \frac{q^{i(\frac{1+\varepsilon}{p})}}{1 + q^i} \right)^p$$

$$\geq (1-q)^{p+1} \left(\left(\sum_{j=-\infty}^{0} q^{-j\varepsilon} \left(\sum_{i=-\infty}^{\infty} \frac{q^{i(\frac{1+\varepsilon}{p})}}{1 + q^i} \right)^p \right)^{\frac{1}{p}} \right.$$

$$\left. - \left(\sum_{j=-\infty}^{0} q^{-j\varepsilon} \left(\sum_{i=-\infty}^{j} \frac{q^{i(\frac{1+\varepsilon}{p})}}{1 + q^i} \right)^p \right)^{\frac{1}{p}} \right)^p$$

$$\geq (1-q)^{p+1} \left(\left(\frac{1}{1-q^\varepsilon} \left(\sum_{i=-\infty}^{\infty} \frac{q^{i(\frac{1+\varepsilon}{p})}}{1 + q^i} \right)^p \right)^{\frac{1}{p}} - \sum_{j=-\infty}^{0} q^{-\frac{j\varepsilon}{p}} \sum_{i=-\infty}^{j} \frac{q^{i(\frac{1+\varepsilon}{p})}}{1 + q^i} \right)^p$$

$$= (1-q)^{p+1} \left(\left(\frac{1}{1-q^\varepsilon} \right)^{\frac{1}{p}} \sum_{i=-\infty}^{\infty} \frac{q^{i(\frac{1+\varepsilon}{p})}}{1 + q^i} - \sum_{j=-\infty}^{0} q^{-\frac{j\varepsilon}{p}} \sum_{i=-\infty}^{0} \frac{q^{(i+j)(\frac{1+\varepsilon}{p})}}{1 + q^{i+j}} \right)^p$$

$$\geq (1-q)^{p+1} \left(\left(\frac{1}{1-q^\varepsilon} \right)^{\frac{1}{p}} \sum_{i=-\infty}^{\infty} \frac{q^{i(\frac{1+\varepsilon}{p})}}{1 + q^i} - \sum_{j=-\infty}^{0} q^{j(\frac{1}{p}-1)} \sum_{i=-\infty}^{0} q^{i(\frac{1+\varepsilon}{p}-1)} \right)^p$$

$$= (1-q)^{p+1} \left(\left(\frac{1}{1-q^\varepsilon} \right)^{\frac{1}{p}} \sum_{i=-\infty}^{\infty} \frac{q^{i(\frac{1+\varepsilon}{p})}}{1 + q^i} - \frac{1}{q^{1-\frac{1}{p}}} \frac{1}{1 - q^{1-\frac{1+\varepsilon}{p}}} \right)^p .$$

Consequently, one has

$$C_2 \geq \frac{L(f_\varepsilon)}{\int_0^\infty f_\varepsilon^p(t)d_q t} = (1-q)^p \left(\sum_{i=-\infty}^{\infty} \frac{q^{i(\frac{1+\varepsilon}{p})}}{1 + q^i} - \frac{(1-q^\varepsilon)^{\frac{1}{p}}}{(q^{1-\frac{1}{p}})(1 - q^{1-\frac{1+\varepsilon}{p}})} \right)^p .$$

Letting $\varepsilon \to 0^+$, we have

$$C_2 \geq (1-q)^p \left(\sum_{i=-\infty}^{\infty} \frac{q^{\frac{i}{p}}}{1 + q^i} \right)^p ,$$

which shows that the constant (6.10) is the sharp one in (6.9). Thus, we complete the proof of Theorem 6.3. □

The next result which was proved by Maligranda, Oinarov and Persson in [Maligranda et al. (2014)] can also be directly obtained by Theorem 6.1.

Corollary 6.1. *If either $1 < p < \infty$ and $f \geq 0$, or $p < 0$ and $f \geq 0$, then the following inequality*

$$\int_0^\infty \left(\frac{1}{x} \int_0^x f(t)d_q t \right)^p d_q x \leq C_3 \int_0^\infty f^p(t)d_q t$$

holds with the constant

$$C_3 = \frac{1}{[1 - 1/p]_q^p}.$$

Moreover, the constant C_3 is the best possible.

Considering the q-inequality of the Hardy-Littlewood-Pólya operator, we can obtain the following inequality:

Corollary 6.2. *For $1 < p < \infty$ and $f \geq 0$, the following inequality*

$$\int_0^\infty \left(\int_0^\infty \frac{f(t)}{\max\{t, x\}} d_q t \right)^p d_q x \leq C_4 \int_0^\infty f^p(t)d_q t \qquad (6.11)$$

holds with the constant

$$C_4 = \left(\frac{1}{[1 - 1/p]_q} + \frac{q^{1/p}}{[1/p]_q} \right)^p. \qquad (6.12)$$

Proof. A simple calculation leads to

$$\int_0^\infty \frac{f(t)}{\max\{t, x\}} d_q t = \int_x^\infty \frac{f(t)}{t} d_q t + \frac{1}{x} \int_0^x f(t)d_q t.$$

Then, using the Minkowski inequality, Theorems 6.2 and 6.3, we can easily obtain the inequality (6.11) holds with the constant (6.12). □

6.2 Sharp constants for multivariate Hausdorff q-inequalities

In this section, we focus on the multivariate Hausdorff operator of the form

$$\mathbf{H}_\Phi(f)(x) = \int_{(0,\infty)^n} \frac{\Phi\left(\frac{x_1}{t_1}, \frac{x_2}{t_2}, \ldots, \frac{x_n}{t_n} \right)}{t_1 t_2 \cdots t_n} f(t_1, t_2, \ldots, t_n)\mathbf{dt},$$

where $\mathbf{dt} = dt_1 dt_2 \cdots dt_n$ or $\mathbf{dt} = d_q t_1 d_q t_2 \cdots d_q t_n$ is the discrete measure in q-analysis. The sharp bounds for the multivariate Hausdorff operator on spaces L^p with power weights are calculated, where $p \in \mathbb{R}\backslash\{0\}$.

Let $G = (0, \infty)^n$ and let $\Phi(t_1, t_2, \ldots, t_n)$ be a locally integrable function on G. For any $x = (x_1, x_2, \ldots, x_n) \in G$, the multivariate Hausdorff operator is defined on G by

$$H_\Phi f(x) = \int_G \frac{\Phi\left(\frac{x_1}{t_1}, \frac{x_2}{t_2}, \ldots, \frac{x_n}{t_n}\right)}{t_1 t_2 \cdots t_n} f(t_1, t_2, \ldots, t_n) dt_1 dt_2 \cdots dt_n. \qquad (6.13)$$

Let χ_E be the characteristic function of a set E. If we take $\Phi(t_1, \ldots, t_n) = \prod_{j=1}^{n} \chi_{[1,\infty)}(t_j) t_j^{-1}$, then the Hausdorff operator H_Φ is reduced to the multivariate Hardy operator P_n defined in (1.33). If we take $\Phi(t_1, t_2, \ldots, t_n) = \prod_{j=1}^{n} \chi_{(0,1]}(t_j)$, then the Hausdorff operator H_Φ is reduced to the adjoint of multivariate Hardy operator P_n^*

$$P_n^* f(x) = \int_{x_1}^{\infty} \int_{x_2}^{\infty} \cdots \int_{x_n}^{\infty} \frac{f(t_1, t_2, \ldots, t_n)}{t_1 t_2 \cdots t_n} dt_1 dt_2 \cdots dt_n.$$

For any $x = (x_1, x_2, \ldots, x_n) \in G$, $\alpha = (\alpha_1, \alpha_2, \ldots, \alpha_n)$ with $\alpha_i \in \mathbb{R}$ $(1 \leq i \leq n)$, let $x^\alpha = x_1^{\alpha_1} x_2^{\alpha_2} \cdots x_n^{\alpha_n}$ and $dx = dx_1 dx_2 \cdots dx_n$. Wu and Chen in [Wu and Chen (2014)] showed that the operator H_Φ is bounded on power weighted $L^p(\mathbb{R}^n)$ spaces, that is for $1 \leq p \leq \infty$

$$\left(\int_G |H_\Phi f(x)|^p x^\alpha dx \right)^{1/p} \leq C_0 \left(\int_G |f(x)|^p x^\alpha dx \right)^{1/p}$$

provided that $\Phi(x) \geq 0$ and

$$C_0 = \int_G \Phi(x) \prod_{i=1}^{n} x_i^{(1+\alpha_i)/p - 1} dx < \infty.$$

Moreover, they proved that the constant C_0 is the sharp one. Motivated by their work, a natural question raised is whether the q-analogue of multivariate Hausdorff operator enjoys the same properties as the classical multivariate Hausdorff operator defined in (6.13).

In the following, for simplicity of notation, for $\alpha \in \mathbb{R}$ and $p \in \mathbb{R}\backslash\{0\}$, we shall write $f \in L^p(t^\alpha d_q t)$ if f satisfies

$$\int_0^{\infty} |f(t)|^p t^\alpha d_q t < \infty$$

and let

$$\|f\|_{L^p(t^\alpha d_q t)} = \left(\int_0^{\infty} |f(t)|^p t^\alpha d_q t \right)^{1/p}.$$

Also, we write $f \in L^p(d_q t)$ if $\alpha = 0$, and write $d_q x = d_q x_1 d_q x_2 \cdots d_q x_n$ for any $x = (x_1, x_2, \ldots, x_n) \in G$ and $0 < q < 1$. We now define the q-analogue of multivariate Hausdorff operator by

$$\mathbf{H}_\Phi(f)(x) = \int_G \frac{\Phi\left(\frac{x_1}{t_1}, \frac{x_2}{t_2}, \ldots, \frac{x_n}{t_n}\right)}{t_1 t_2 \cdots t_n} f(t_1, t_2, \ldots, t_n) d_q t_1 d_q t_2 \cdots d_q t_n$$

and the q-analogue of multivariate Hardy operator by

$$\mathbf{H}_n f(x) = \frac{1}{x_1 x_2 \cdots x_n} \int_0^{x_1} \int_0^{x_2} \cdots \int_0^{x_n} f(t_1, t_2, \ldots, t_n) d_q t_1 d_q t_2 \cdots d_q t_n,$$

and the q-analogue of multivariate adjoint Hardy operator by

$$\mathbf{H}_n^* f(x) = \int_{x_1}^\infty \int_{x_2}^\infty \cdots \int_{x_n}^\infty \frac{f(t_1, t_2, \ldots, t_n)}{t_1 t_2 \cdots t_n} d_q t_1 d_q t_2 \cdots d_q t_n.$$

Now we can state the main results.

Theorem 6.4. *Let $\alpha = (\alpha_1, \alpha_2, \ldots, \alpha_n)$ and $\alpha_i \in \mathbb{R}$ for $i = 1, \ldots, n$. Assume that Φ is a nonnegative function and $f \in L^p(x^\alpha d_q x)$. If $1 \le p < \infty$, then the following inequality*

$$\|\mathbf{H}_\Phi f\|_{L^p(x^\alpha d_q x)} \le C_1 \|f\|_{L^p(x^\alpha d_q x)} \tag{6.14}$$

holds, provided that

$$C_1 = \int_G \Phi(t_1, t_2, \ldots, t_n) \prod_{i=1}^n t_i^{\frac{1+\alpha_i}{p} - 1} d_q t_1 d_q t_2 \cdots d_q t_n < \infty. \tag{6.15}$$

If $p < 1$ ($p \ne 0$), then we have the reverse inequality

$$\|\mathbf{H}_\Phi f\|_{L^p(x^\alpha d_q x)} \ge C_1 \|f\|_{L^p(x^\alpha d_q x)}$$

provided that (6.15) holds. Here we assume $f \ne 0$ and $\Phi > 0$ if $p < 0$. Moreover, for $p \in \mathbb{R} \backslash \{0\}$, the constant C_1 is the best possible one.

As applications, we can easily obtain the following results.

Corollary 6.3. *Let $\alpha = (\alpha_1, \alpha_2, \ldots, \alpha_n)$ and $(1 + \alpha_i)/p < 1$ for $i = 1, \ldots, n$. Assume that Φ is a nonnegative function and $f \in L^p(x^\alpha d_q x)$. If $1 \le p < \infty$, then the following inequality*

$$\|\mathbf{H}_n f\|_{L^p(x^\alpha d_q x)} \le C_2 \|f\|_{L^p(x^\alpha d_q x)}$$

holds with

$$C_2 = (1 - q)^n \prod_{i=1}^n \frac{1}{1 - q^{1 - \frac{1}{p} - \frac{\alpha_i}{p}}}.$$

If $p < 1$ ($p \ne 0$), then the following inequality

$$\|\mathbf{H}_n f\|_{L^p(x^\alpha d_q x)} \ge C_2 \|f\|_{L^p(x^\alpha d_q x)}$$

holds. Here we assume that $f \ne 0$ if $p < 0$. Moreover, for $p \in \mathbb{R} \backslash \{0\}$, the constant C_2 is the best possible one.

Corollary 6.4. *Let* $\alpha = (\alpha_1, \alpha_2, \ldots, \alpha_n)$ *and* $(1 + \alpha_i)/p > 0$ *for* $i = 1, \ldots, n$. *Assume that* Φ *is a nonnegative function and* $f \in L^p(x^\alpha d_q x)$. *If* $1 \le p < \infty$, *then the following inequality*

$$\|\mathbf{H}_n^* f\|_{L^p(x^\alpha d_q x)} \le C_3 \|f\|_{L^p(x^\alpha d_q x)}$$

holds with

$$C_3 = (1 - q)^n \prod_{i=1}^{n} \frac{1}{1 - q^{\frac{1 + \alpha_i}{p}}}.$$

If $p < 1$ $(p \ne 0)$, *then we have the reverse inequality*

$$\|\mathbf{H}_n^* f\|_{L^p(x^\alpha d_q x)} \ge C_3 \|f\|_{L^p(x^\alpha d_q x)}.$$

Here we assume that $f \ne 0$ *if* $p < 0$. *Moreover, for* $p \in \mathbb{R} \backslash \{0\}$, *the constant* C_3 *is the best possible one.*

It is interesting to see that the constant C_1 in Theorem 6.4 and the constant C_0 obtained by Wu and Chen in [Wu and Chen (2014)] are in the same integral form, but with different measures, one is continuous and another is discrete. More significantly, we are able to see that C_1 is also the sharp constant in the case $p < 1$ $(p \ne 0)$ with a reverse inequality for the q-analogue multivariate Hausdorff operator $\mathbf{H}_\Phi$. With the same method, we shall in the last section show that C_0 is also the best constant for the reverse inequality

$$\|H_\Phi f\|_{L^p(x^\alpha dx)} \ge C_0 \|f\|_{L^p(x^\alpha dx)}$$

in the case $p < 1$ $(p \ne 0)$.

It should be pointed out that from Theorem 6.4 we can obtain the $L^p(x^\alpha d_q x)$ boundedness for q-analogues of many well-known operators when we take different functions Φ. These operators include the Cesàro operator, the Hardy-Littlewood-Pólya operator, the Riemann-Liouville fractional derivatives, the weighted Hardy operator, among many others.

6.2.1 *Sharp constants for multivariate Hausdorff q-inequalities on weighted Lebesgue spaces*

Proof. Using the definition given in (6.4), we have

$$\mathbf{H}_\Phi f(x) = (1 - q)^n \sum_{j_1 = -\infty}^{\infty} \cdots \sum_{j_n = -\infty}^{\infty} \Phi\left(\frac{x_1}{q^{j_1}}, \ldots, \frac{x_n}{q^{j_n}}\right) f(q^{j_1}, \ldots, q^{j_n}),$$

where $j_1, \ldots, j_n$ are integers. Then for $p \in \mathbb{R}\backslash\{0\}$, by changing variables $k_i = l_i - j_i$ for $1 \le i \le n$, we have that

$$\|\mathbf{H}_\Phi f\|_{L^p(x^\alpha d_q x)} \tag{6.16}$$

$$= \left(\int_G |\mathbf{H}_\Phi f(x)|^p \, x^\alpha d_q x \right)^{1/p}$$

$$= (1-q)^{n(1+\frac{1}{p})} \left(\sum_{l_1=-\infty}^{\infty} \cdots \sum_{l_n=-\infty}^{\infty} \prod_{i=1}^{n} q^{l_i(1+\alpha_i)} \right.$$

$$\times \left| \sum_{j_1=-\infty}^{\infty} \cdots \sum_{j_n=-\infty}^{\infty} \Phi\left(\frac{q^{l_1}}{q^{j_1}}, \ldots, \frac{q^{l_n}}{q^{j_n}} \right) f\left(q^{j_1}, \ldots, q^{j_n} \right) \right|^p \Bigg)^{\frac{1}{p}}$$

$$= (1-q)^{n(1+\frac{1}{p})} \left(\sum_{l_1=-\infty}^{\infty} \cdots \sum_{l_n=-\infty}^{\infty} \prod_{i=1}^{n} q^{l_i(1+\alpha_i)} \right.$$

$$\times \left| \sum_{k_1=-\infty}^{\infty} \cdots \sum_{k_n=-\infty}^{\infty} \Phi\left(q^{k_1}, \ldots, q^{k_n} \right) f\left(q^{l_1-k_1}, \ldots, q^{l_n-k_n} \right) \right|^p \Bigg)^{1/p}.$$

We firstly study the case $1 \le p < \infty$. Assume that (6.15) holds. Using the above expression (6.16) and the Minkowski inequality, we see that

$$\|\mathbf{H}_\Phi f\|_{L^p(x^\alpha d_q x)}$$

$$\le (1-q)^{n(1+\frac{1}{p})} \sum_{k_1=-\infty}^{\infty} \cdots \sum_{k_n=-\infty}^{\infty} \Phi\left(q^{k_1}, \ldots, q^{k_n} \right)$$

$$\times \left(\sum_{l_1=-\infty}^{\infty} \cdots \sum_{l_n=-\infty}^{\infty} \prod_{i=1}^{n} q^{l_i(1+\alpha_i)} \left| f\left(q^{l_1-k_1}, \ldots, q^{l_n-k_n} \right) \right|^p \right)^{1/p}.$$

Changing variables $m_i = l_i - k_i$ for $1 \le i \le n$, the above estimate is

$$(1-q)^{n(1+\frac{1}{p})} \sum_{k_1=-\infty}^{\infty} \cdots \sum_{k_n=-\infty}^{\infty} \Phi\left(q^{k_1}, \ldots, q^{k_n} \right) \prod_{i=1}^{n} q^{k_i(1+\alpha_i)/p}$$

$$\times \left(\sum_{m_1=-\infty}^{\infty} \cdots \sum_{m_n=-\infty}^{\infty} \prod_{i=1}^{n} q^{m_i(1+\alpha_i)} \left| f\left(q^{m_1}, \ldots, q^{m_n} \right) \right|^p \right)^{1/p}$$

$$= \int_G \Phi(t_1, \ldots, t_n) \prod_{i=1}^{n} t_i^{\frac{1+\alpha_i}{p}-1} \, d_q t_1 \cdots d_q t_n \, \|f\|_{L^p(x^\alpha d_q x)},$$

which implies that the inequality (6.14) holds with the constant (6.15).

We need to show that the constant (6.15) is the best one in (6.14). Suppose $N \in \mathbb{Z}_+$ and $0 < \theta < 1$. Let $y = (y_1, y_2, \ldots, y_n) \in G$ and

$$f_{\theta,N}(y) = \prod_{j=1}^{n} y_j^{-(1+\alpha_j)/p} \chi_{\left[q^{N(1+\theta)}, \, q^{-N(1+\theta)}\right]}(y_j).$$

Denote by $\sharp I$ the number of integers in the interval I. A straightforward calculation shows that

$$\|f_{\theta,N}\|_{L^p(x^\alpha d_q x)}^p = (1-q)^n \sum_{j_1=-\infty}^{\infty} \cdots \sum_{j_n=-\infty}^{\infty} \prod_{i=1}^{n} \chi_{\left[q^{N(1+\theta)}, \, q^{-N(1+\theta)}\right]}(q^{j_i})$$

$$= (1-q)^n (\sharp[-N(1+\theta), N(1+\theta)])^n$$

$$= (1-q)^n (2[N(1+\theta)] + 1)^n,$$

where $[N(1+\theta)]$ denotes the integral part of the real number $N(1+\theta)$. Then we have

$$\|\mathbf{H}_\Phi f_{\theta,N}\|_{L^p(x^\alpha d_q x)}^p = \int_G |\mathbf{H}_\Phi f_{\theta,N}(x)|^p x^\alpha d_q x$$

$$= (1-q)^{n(p+1)} \sum_{l_1=-\infty}^{\infty} \cdots \sum_{l_n=-\infty}^{\infty} \prod_{i=1}^{n} q^{l_i(1+\alpha_i)}$$

$$\times \left| \sum_{k_1=-\infty}^{\infty} \cdots \sum_{k_n=-\infty}^{\infty} \Phi\left(q^{k_1}, \ldots, q^{k_n}\right) f_{\theta,N}\left(q^{l_1-k_1}, \ldots, q^{l_n-k_n}\right) \right|^p$$

$$= (1-q)^{n(p+1)} \sum_{l_1=-\infty}^{\infty} \cdots \sum_{l_n=-\infty}^{\infty} \left(\sum_{k_1=-\infty}^{\infty} \cdots \sum_{k_n=-\infty}^{\infty} \Phi\left(q^{k_1}, \ldots, q^{k_n}\right) \right.$$

$$\left. \times \prod_{i=1}^{n} \left(q^{k_i \frac{(1+\alpha_i)}{p}} \chi_{\left[q^{N(1+\theta)}, \, q^{-N(1+\theta)}\right]}(q^{l_i-k_i}) \right) \right)^p$$

$$\geq (1-q)^{n(p+1)} \sum_{l_1=-\infty}^{\infty} \cdots \sum_{l_n=-\infty}^{\infty} \prod_{j=1}^{n} \chi_{[q^N, \, q^{-N}]}(q^{l_j})$$

$$\times \left(\sum_{k_1=-\infty}^{\infty} \cdots \sum_{k_n=-\infty}^{\infty} \Phi\left(q^{k_1}, q^{k_2}, \ldots, q^{k_n}\right) \right.$$

$$\left. \times \prod_{i=1}^{n} \left(q^{k_i \frac{(1+\alpha_i)}{p}} \chi_{\left[q^{N(1+\theta)}, \, q^{-N(1+\theta)}\right]}(q^{l_i-k_i}) \chi_{[q^{\theta N}, \, q^{-\theta N}]}(q^{k_i}) \right) \right)^p$$

$$\geq (1-q)^{n(p+1)} \sum_{l_1=-N}^{N} \cdots \sum_{l_n=-N}^{N} \left(\sum_{k_1=-\infty}^{\infty} \cdots \sum_{k_n=-\infty}^{\infty} \Phi\left(q^{k_1}, \ldots, q^{k_n}\right) \right.$$

$$\times \prod_{i=1}^{n} \left(q^{\frac{k_i(1+\alpha_i)}{p}} \chi_{[q^{\theta N},\, q^{-\theta N}]}(q^{k_i}) \right) \Bigg)^{p}$$

$$= (1-q)^{n(p+1)}(2N+1)^n \left(\sum_{k_1=-\infty}^{\infty} \cdots \sum_{k_n=-\infty}^{\infty} \Phi\left(q^{k_1},\ldots,q^{k_n}\right) \right.$$

$$\left. \times \prod_{i=1}^{n} \left(q^{\frac{k_i(1+\alpha_i)}{p}} \chi_{[q^{\theta N},\, q^{-\theta N}]}(q^{k_i}) \right) \right)^{p}.$$

Thus, we obtain that

$$\frac{\|\mathbf{H}_\Phi f_{\theta,N}\|_{L^p(x^\alpha d_q x)}^p}{\|f_{\theta,N}\|_{L^p(x^\alpha d_q x)}^p}$$

$$\geq \frac{(2N+1)^n(1-q)^{np}}{(2[N(1+\theta)]+1)^n} \left(\sum_{k_1=-\infty}^{\infty} \cdots \sum_{k_n=-\infty}^{\infty} \Phi\left(q^{k_1},\ldots,q^{k_n}\right) \right.$$

$$\left. \times \prod_{i=1}^{n} \left(q^{\frac{k_i(1+\alpha_i)}{p}} \chi_{[q^{\theta N},\, q^{-\theta N}]}(q^{k_i}) \right) \right)^{p}$$

$$\geq \frac{(2N+1)^n(1-q)^{np}}{(2N(1+\theta)+1)^n} \left(\sum_{k_1=-\infty}^{\infty} \cdots \sum_{k_n=-\infty}^{\infty} \Phi\left(q^{k_1},\ldots,q^{k_n}\right) \right.$$

$$\left. \times \prod_{i=1}^{n} \left(q^{\frac{k_i(1+\alpha_i)}{p}} \chi_{[q^{\theta N},\, q^{-\theta N}]}(q^{k_i}) \right) \right)^{p}.$$

Fix θ and let $N \to \infty$. It yields that

$$\lim_{N\to\infty} \frac{\|\mathbf{H}_\Phi f_{\theta,N}\|_{L^p(x^\alpha d_q x)}^p}{\|f_{\theta,N}\|_{L^p(x^\alpha d_q x)}^p}$$

$$\geq \left(\frac{2}{2(1+\theta)} \right)^n \left((1-q)^n \sum_{k_1=-\infty}^{\infty} \cdots \sum_{k_n=-\infty}^{\infty} \Phi\left(q^{k_1},\ldots,q^{k_n}\right) \prod_{i=1}^{n} q^{\frac{k_i(1+\alpha_i)}{p}} \right)^{p}.$$

Now letting $\theta \to 0^+$, we conclude that

$$\lim_{\theta\to 0^+} \lim_{N\to\infty} \frac{\|\mathbf{H}_\Phi f_{\theta,N}\|_{L^p(x^\alpha d_q x)}}{\|f_{\theta,N}\|_{L^p(x^\alpha d_q x)}}$$

$$\geq (1-q)^n \sum_{k_1=-\infty}^{\infty} \cdots \sum_{k_n=-\infty}^{\infty} \Phi\left(q^{k_1},\ldots,q^{k_n}\right) \prod_{i=1}^{n} q^{\frac{k_i(1+\alpha_i)}{p}}.$$

Hence, we finish the proof in the case $1 \leq p < \infty$.

When $p < 1$ and $p \neq 0$, we use the Minkowski inequality to conclude that

$$\|\mathbf{H}_\Phi f\|_{L^p(x^\alpha d_q x)} \geq (1-q)^{n(1+\frac{1}{p})} \sum_{k_1=-\infty}^{\infty} \cdots \sum_{k_n=-\infty}^{\infty} \Phi\left(q^{k_1}, \ldots, q^{k_n}\right)$$

$$\times \left(\sum_{l_1=-\infty}^{\infty} \cdots \sum_{l_n=-\infty}^{\infty} \prod_{i=1}^{n} q^{l_i(1+\alpha_i)} \left| f\left(q^{l_1-k_1}, \ldots, q^{l_n-k_n}\right) \right|^p \right)^{1/p}$$

$$= \int_G \Phi(t_1, \ldots, t_n) \prod_{i=1}^{n} t_i^{\frac{1+\alpha_i}{p}-1} d_q t_1 \cdots d_q t_n \, \|f\|_{L^p(x^\alpha d_q x)}.$$

To show that the constant C_1 is sharp, we need to choose two classes of suitable functions according to the values of p. We divide p into two cases: $0 < p < 1$ and $p < 0$.

(i) $0 < p < 1$. For $N \in \mathbb{Z}_+$, letting $y = (y_1, y_2, \ldots, y_n) \in G$, we take

$$f_N(y) = \prod_{j=1}^{n} y_j^{-(1+\alpha_j)/p} \chi_{[q^N, \, q^{-N}]}(y_j).$$

A direct calculation shows that

$$\int_G |f_N(y)|^p y^\alpha d_q y = (1-q)^n (2N+1)^n.$$

We have also

$$\|\mathbf{H}_\Phi f_N\|_{L^p(x^\alpha d_q x)}^p = \int_G |\mathbf{H}_\Phi f_N(x)|^p x^\alpha d_q x$$

$$= (1-q)^{n(p+1)} \sum_{l_1=-\infty}^{\infty} \cdots \sum_{l_n=-\infty}^{\infty} \prod_{i=1}^{n} q^{l_i(1+\alpha_i)}$$

$$\times \left(\sum_{k_1=-\infty}^{\infty} \cdots \sum_{k_n=-\infty}^{\infty} \Phi(q^{k_1}, \ldots, q^{k_n}) \prod_{i=1}^{n} \left(q^{\frac{(k_i-l_i)(1+\alpha_i)}{p}} \chi_{[q^N, \, q^{-N}]}(q^{l_i-k_i}) \right) \right)^p$$

$$= (1-q)^{n(p+1)} \sum_{l_1=-\infty}^{\infty} \cdots \sum_{l_n=-\infty}^{\infty}$$

$$\left(\sum_{k_1=-\infty}^{\infty} \cdots \sum_{k_n=-\infty}^{\infty} \Phi(q^{k_1}, \ldots, q^{k_n}) \prod_{i=1}^{n} \left(q^{\frac{k_i(1+\alpha_i)}{p}} \chi_{[q^N, \, q^{-N}]}(q^{l_i-k_i}) \right) \right)^p$$

$$= (1-q)^{n(p+1)} \sum_{l_1=-\infty}^{\infty} \cdots \sum_{l_n=-\infty}^{\infty}$$

$$\times \left(\sum_{k_1=l_1-N}^{l_1+N} \cdots \sum_{k_n=l_n-N}^{l_n+N} \Phi(q^{k_1}, \ldots, q^{k_n}) \prod_{i=1}^{n} q^{\frac{k_i(1+\alpha_i)}{p}} \right)^p.$$

It follows from Hölder's inequality that

$$\sum_{l_1=-N}^{N} \cdots \sum_{l_n=-N}^{N} \left(\sum_{k_1=l_1-N}^{l_1+N} \cdots \sum_{k_n=l_n-N}^{l_n+N} \Phi\left(q^{k_1},\ldots,q^{k_n}\right) \prod_{i=1}^{n} q^{\frac{k_i(1+\alpha_i)}{p}} \right)^{p}$$

$$\leq \left(\sum_{l_1=-N}^{N} \cdots \sum_{l_n=-N}^{N} \left(\sum_{k_1=l_1-N}^{l_1+N} \cdots \sum_{k_n=l_n-N}^{l_n+N} \Phi\left(q^{k_1},\ldots,q^{k_n}\right) \prod_{i=1}^{n} q^{\frac{k_i(1+\alpha_i)}{p}} \right) \right)^{p}$$

$$\times \left(\sum_{l_1=-N}^{N} \cdots \sum_{l_n=-N}^{N} 1 \right)^{1-p}$$

$$\leq (2N+1)^n \left(\sum_{k_1=-2N}^{2N} \cdots \sum_{k_n=-2N}^{2N} \Phi\left(q^{k_1},\ldots,q^{k_n}\right) \prod_{i=1}^{n} q^{\frac{k_i(1+\alpha_i)}{p}} \right)^{p}.$$

We now complete the proof in the case $0 < p < 1$, since

$$\lim_{N\to\infty} \frac{\|\mathbf{H}_\Phi f_N\|_{L^p(x^\alpha d_q x)}}{\|f_N\|_{L^p(x^\alpha d_q x)}}$$

$$\leq \lim_{N\to\infty} (1-q)^n \sum_{k_1=-2N}^{2N} \cdots \sum_{k_n=-2N}^{2N} \Phi\left(q^{k_1},\ldots,q^{k_n}\right) \prod_{i=1}^{n} q^{\frac{k_i(1+\alpha_i)}{p}}$$

$$= (1-q)^n \sum_{k_1=-\infty}^{\infty} \cdots \sum_{k_n=-\infty}^{\infty} \Phi\left(q^{k_1},\ldots,q^{k_n}\right) \prod_{i=1}^{n} q^{\frac{k_i(1+\alpha_i)}{p}}.$$

(ii) $p < 0$. Assume $y = (y_1,\ldots,y_n) \in G$. For $\varepsilon > 0$, let

$$f_\varepsilon(y) = \prod_{j=1}^{n} y_j^{-(1+\alpha_j+\varepsilon)/p} \chi_{[1,\infty)}(y_j).$$

Then we have

$$\int_G |f_\varepsilon(y)|^p y^\alpha d_q y = \left(\frac{1-q}{1-q^\varepsilon} \right)^n$$

and that

$$\|\mathbf{H}_\Phi f_\varepsilon\|^p_{L^p(x^\alpha d_q x)}$$

$$= (1-q)^{n(p+1)} \sum_{l_1=-\infty}^{\infty} \cdots \sum_{l_n=-\infty}^{\infty} \prod_{i=1}^{n} q^{l_i(1+\alpha_i)}$$

$$\times \left| \sum_{k_1=-\infty}^{\infty} \cdots \sum_{k_n=-\infty}^{\infty} \Phi\left(q^{k_1},\ldots,q^{k_n}\right) f_\varepsilon\left(q^{l_1-k_1},\ldots,q^{l_n-k_n}\right) \right|^p$$

$$= (1-q)^{n(p+1)} \sum_{l_1=-\infty}^{\infty} \cdots \sum_{l_n=-\infty}^{\infty} \prod_{i=1}^{n} q^{-l_i \varepsilon}$$

$$\times \left(\sum_{k_1=-\infty}^{\infty} \cdots \sum_{k_n=-\infty}^{\infty} \Phi(q^{k_1},\dots,q^{k_n}) \prod_{i=1}^{n} \left(q^{\frac{k_i(1+\alpha_i+\varepsilon)}{p}} \chi_{[1,\infty)}(q^{l_i-k_i}) \right) \right)^{p}$$

$$\geq (1-q)^{n(p+1)} \sum_{l_1=-\infty}^{0} \cdots \sum_{l_n=-\infty}^{0} \prod_{i=1}^{n} q^{-l_i \varepsilon}$$

$$\times \left(\sum_{k_1=l_1}^{\infty} \cdots \sum_{k_n=l_n}^{\infty} \Phi\left(q^{k_1},\dots,q^{k_n}\right) \prod_{i=1}^{n} \left(q^{\frac{k_i(1+\alpha_i+\varepsilon)}{p}} \right) \right)^{p}$$

$$\geq (1-q)^{n(p+1)} \sum_{l_1=-\infty}^{0} \cdots \sum_{l_n=-\infty}^{0} \prod_{i=1}^{n} q^{-l_i \varepsilon}$$

$$\times \left(\sum_{k_1=-\infty}^{\infty} \cdots \sum_{k_n=-\infty}^{\infty} \Phi\left(q^{k_1},\dots,q^{k_n}\right) \prod_{i=1}^{n} \left(q^{\frac{k_i(1+\alpha_i+\varepsilon)}{p}} \right) \right)^{p}$$

$$= \left(\frac{1-q}{1-q^{\varepsilon}} \right)^{n}$$

$$\times \left((1-q)^{n} \sum_{k_1=-\infty}^{\infty} \cdots \sum_{k_n=-\infty}^{\infty} \Phi\left(q^{k_1},\dots,q^{k_n}\right) \prod_{i=1}^{n} \left(q^{\frac{k_i(1+\alpha_i+\varepsilon)}{p}} \right) \right)^{p},$$

where we have used $p < 0$ and $\Phi > 0$. Letting $\varepsilon \to 0^+$, we obtain that

$$\lim_{\varepsilon\to 0^+} \frac{\|\mathbf{H}_\Phi f_\varepsilon\|_{L^p(x^\alpha d_q x)}}{\|f_\varepsilon\|_{L^p(x^\alpha d_q x)}}$$

$$\leq (1-q)^{n} \sum_{k_1=-\infty}^{\infty} \cdots \sum_{k_n=-\infty}^{\infty} \Phi\left(q^{k_1},\dots,q^{k_n}\right) \prod_{i=1}^{n} q^{\frac{k_i(1+\alpha_i)}{p}}.$$

Combining all the estimates, we show that the constant C_1 is sharp and therefore finish the proof of the theorem. $\qquad\square$

6.2.2 *A final remark*

In this section, we can modify the previous argument to yield the best constant for the Hausdorff operator H_Φ given by (6.13) in the L^p spaces with $p < 1$ ($p \neq 0$), ignoring whether these spaces make sense. The following result can be regarded as fixing the gap of Wu and Chen's result in [Wu and Chen (2014)]. Using the similar notation as before, we let $t = (t_1, t_2, \dots, t_n) \in G$ and $dt = dt_1 dt_2 \cdots dt_n$.

Theorem 6.5. *Let* $\alpha = (\alpha_1, \alpha_2, \dots, \alpha_n)$ *and* $\alpha_i \in \mathbb{R}$ *for* $i = 1, \dots, n$.

Assume that Φ is a nonnegative function and $f \in L^p(x^\alpha dx)$. If $p < 1$ ($p \neq 0$), then we have the reverse inequality

$$\|H_\Phi f\|_{L^p(x^\alpha dx)} \geq C_0 \|f\|_{L^p(x^\alpha dx)}$$

provided that

$$C_0 = \int_G \Phi(t) \prod_{i=1}^n t_i^{\frac{\alpha_i+1}{p}-1} dt < \infty. \tag{6.17}$$

Here we assume $f \neq 0$ and $\Phi > 0$ if $p < 0$. Moreover, for $p < 1$ ($p \neq 0$), the constant C_0 is the best possible one.

Proof. By changing variables, we have

$$\begin{aligned}
H_\Phi f(x) &= \int_G \frac{\Phi\left(\frac{x_1}{t_1}, \ldots, \frac{x_n}{t_n}\right)}{t_1 \cdots t_n} f(t_1, \ldots, t_n) dt \\
&= \int_G \frac{\Phi(t_1, \ldots, t_n)}{t_1 \cdots t_n} f\left(\frac{x_1}{t_1}, \ldots, \frac{x_n}{t_n}\right) dt.
\end{aligned} \tag{6.18}$$

Assume that (6.17) holds. Using the equality (6.18) and the Minkowski inequality for the case $p < 1$ ($p \neq 0$), we have

$$\begin{aligned}
\|H_\Phi f\|_{L^p(x^\alpha dx)} &= \left(\int_G \left(\int_G \frac{\Phi(t)}{t_1 \cdots t_n} f\left(\frac{x_1}{t_1}, \ldots, \frac{x_n}{t_n}\right) dt\right)^p x^\alpha dx\right)^{1/p} \\
&\geq \int_G \frac{\Phi(t)}{t_1 \cdots t_n} \left(\int_G f^p\left(\frac{x_1}{t_1}, \ldots, \frac{x_n}{t_n}\right) x^\alpha dx\right)^{1/p} dt \\
&= \int_G \Phi(t) \prod_{i=1}^n t_i^{\frac{\alpha_i+1}{p}-1} dt \, \|f\|_{L^p(x^\alpha dx)}.
\end{aligned}$$

The next step is to show that the constant C_0 is sharp. Similar to the proof of Theorem 6.4, we need to construct two classes of functions for different p.

  Case 1. $0 < p < 1$. For any fixed real number r satisfying $0 < r < 1$, letting $N \in \mathbb{Z}_+$, we take

$$f_N(x) = \prod_{i=1}^n x_i^{-(1+\alpha_i)/p} \chi_{[r^N, r^{-N}]}(x_i), \quad \text{for } x = (x_1, \ldots, x_n) \in G.$$

Let us observe that

$$\|f_N\|_{L^p(x^\alpha dx)}^p = (2N)^n \left(\ln(r^{-1})\right)^n.$$

We have also

$$\|H_\Phi f_N\|_{L^p(x^\alpha dx)}^p$$

$$= \left(\int_G \left(\int_G \frac{\Phi(t)}{t_1 \cdots t_n} f_N \left(\frac{x_1}{t_1}, \ldots, \frac{x_n}{t_n} \right) dt \right)^p x^\alpha dx \right)^{1/p}$$

$$= \left(\int_G \left(\int_G \frac{\Phi(t)}{t_1 \cdots t_n} \prod_{i=1}^n \left(\frac{t_i}{x_i} \right)^{\frac{1+\alpha_i}{p}} \chi_{[r^N, r^{-N}]} \left(\frac{x_i}{t_i} \right) dt \right)^p x^\alpha dx \right)^{1/p}$$

$$= \left(\int_G \left(\int_{x_1 r^N}^{x_1 r^{-N}} \cdots \int_{x_n r^N}^{x_n r^{-N}} \Phi(t) \prod_{i=1}^n t_i^{\frac{1+\alpha_i}{p}-1} dt \right)^p \prod_{i=1}^n x_i^{-1} dx \right)^{1/p}.$$

Considering the last term, we see that

$$\left(\int_{r^N}^{r^{-N}} \cdots \int_{r^N}^{r^{-N}} \left(\int_{x_1 r^N}^{x_1 r^{-N}} \cdots \int_{x_n r^N}^{x_n r^{-N}} \Phi(t) \prod_{i=1}^n t_i^{\frac{1+\alpha_i}{p}-1} dt \right)^p \prod_{i=1}^n x_i^{-1} dx \right)^{\frac{1}{p}}$$

$$\leq \left(\int_{r^N}^{r^{-N}} \cdots \int_{r^N}^{r^{-N}} \left(\int_{r^{2N}}^{r^{-2N}} \cdots \int_{r^{2N}}^{r^{-2N}} \Phi(t) \prod_{i=1}^n t_i^{\frac{1+\alpha_i}{p}-1} dt \right)^p \prod_{i=1}^n x_i^{-1} dx \right)^{\frac{1}{p}}$$

$$= \int_{r^{2N}}^{r^{-2N}} \cdots \int_{r^{2N}}^{r^{-2N}} \Phi(t) \prod_{i=1}^n t_i^{\frac{1+\alpha_i}{p}-1} dt \, \|f_N\|_{L^p(x^\alpha dx)}.$$

Letting $N \to \infty$, we see that

$$\lim_{N \to \infty} \frac{\|H_\Phi f_N\|_{L^p(x^\alpha dx)}}{\|f_N\|_{L^p(x^\alpha dx)}} \leq \lim_{N \to \infty} \int_{r^{2N}}^{r^{-2N}} \cdots \int_{r^{2N}}^{r^{-2N}} \Phi(t) \prod_{i=1}^n t_i^{\frac{1+\alpha_i}{p}-1} dt$$

$$= \int_G \Phi(t) \prod_{i=1}^n t_i^{\frac{1+\alpha_i}{p}-1} dt.$$

Case 2. $p < 0$. For $\varepsilon > 0$, we let

$$f_\varepsilon(x) = \prod_{i=1}^n x_i^{-\frac{1+\alpha_i+\varepsilon}{p}} \chi_{[1,\infty)}(x_i), \text{ for } x = (x_1, \ldots, x_n) \in G.$$

It is easy to see that

$$\|f_\varepsilon\|_{L^p(x^\alpha dx)}^p = \varepsilon^{-n}.$$

We have also

$$\|H_\Phi f_\varepsilon\|_{L^p(x^\alpha dx)}^p$$

$$= \int_G \left(\int_G \frac{\Phi(t)}{t_1 \cdots t_n} f_\varepsilon \left(\frac{x_1}{t_1}, \ldots, \frac{x_n}{t_n} \right) dt \right)^p x^\alpha dx$$

$$= \int_G \left(\int_G \frac{\Phi(t)}{t_1 \cdots t_n} \prod_{i=1}^n \left(\frac{t_i}{x_i} \right)^{\frac{1+\alpha_i+\varepsilon}{p}} \chi_{[1,\infty)} \left(\frac{x_i}{t_i} \right) dt \right)^p x^\alpha dx$$

$$= \int_G \left(\int_0^{x_1} \cdots \int_0^{x_n} \Phi(t) \prod_{i=1}^{n} t_i^{\frac{1+\alpha_i+\varepsilon}{p}-1} dt \right)^p \prod_{i=1}^{n} x_i^{-1-\varepsilon} dx$$

$$\geq \int_1^{\infty} \cdots \int_1^{\infty} \left(\int_0^{x_1} \cdots \int_0^{x_n} \Phi(t) \prod_{i=1}^{n} t_i^{\frac{1+\alpha_i+\varepsilon}{p}-1} dt \right)^p \prod_{i=1}^{n} x_i^{-1-\varepsilon} dx.$$

Since $\Phi(t) > 0$ and $p < 0$, the last term is greater than or equal to

$$\int_1^{\infty} \cdots \int_1^{\infty} \left(\int_0^{\infty} \cdots \int_0^{\infty} \Phi(t) \prod_{i=1}^{n} t_i^{\frac{1+\alpha_i+\varepsilon}{p}-1} dt \right)^p \prod_{i=1}^{n} x_i^{-1-\varepsilon} dx$$

$$= \varepsilon^{-n} \left(\int_G \Phi(t) \prod_{i=1}^{n} t_i^{\frac{1+\alpha_i+\varepsilon}{p}-1} dt \right)^p.$$

Letting $\varepsilon \to 0^+$, we see that

$$\lim_{\varepsilon \to 0^+} \frac{\|H_\Phi f_\varepsilon\|_{L^p(x^\alpha dx)}}{\|f_\varepsilon\|_{L^p(x^\alpha dx)}} \leq \lim_{\varepsilon \to 0^+} \int_G \Phi(t) \prod_{i=1}^{n} t_i^{\frac{1+\alpha_i+\varepsilon}{p}-1} dt$$

$$= \int_G \Phi(t) \prod_{i=1}^{n} t_i^{\frac{1+\alpha_i}{p}-1} dt.$$

This proves the theorem.   □

6.3   Notes

The Hausdorff operators are generalization of Hardy operators. Andersen [Andersen (2003)] studied the n-dimensional Hausdorff operator

$$\mathcal{H}_\Phi f(x) = \int_{\mathbb{R}^n} \frac{\Phi(x/|y|)}{|y|^n} f(y) dy,$$

where Φ is a radial function defined on $\mathbb{R}_+$. A more general Hausdorff operator was studied by Liflyand in [Liflyand (2008)], and by Lerner and Liflyand in [Lerner and Liflyand (2007)]. There are many essential difference between the operator h_Φ in the one-dimensional case and the operator H_Φ in the n-dimensional case. The Hausdorff operator h_Φ in [Liflyand and Móricz (2001)] were considered in dimension two only for the so-called product Hardy space $H^{11}(\mathbb{R} \times \mathbb{R})$:

$$\widetilde{H}_\Phi f(x) = \int_{\mathbb{R}^2} \frac{\Phi(u)}{|u_1 u_2|} f\left(\frac{x_1}{u_1}, \frac{x_2}{u_2} \right) du.$$

In [Weisz (2004)], these and related results were slightly extended. One can find these facts in recent survey articles [Chen et al. (2013)], [Liflyand (2011)] and [Liflyand (2013)].

Maligranda, Oinarov and Persson in [Maligranda et al. (2014)] derived some q-analysis variants of the classical Hardy inequality and obtained their corresponding best constants. Guo and Zhao [Guo and Zhao (2017)] extended the corresponding results of the Hardy operator to the Hausdorff operator (Theorem 6.1 and its applications in Section 6.1.2). Fan and Zhao [Fan and Zhao (2019)] further considered the sharp constants for multivariate Hausdorff q-inequalities (Theorem 6.4 in Section 6.2).

Bibliography

Adams, D. R. and Xiao, J. (2012a). Morrey spaces in harmonic analysis, *Ark. Mat.* **50**, 2, pp. 201–230.

Adams, D. R. and Xiao, J. (2012b). Regularity of Morrey commutators, *Trans. Amer. Math. Soc.* **364**, 9, pp. 4801–4818.

Albeverio, S. and Karwowski, W. (1994). A random walk on p-adics: the generator and its spectrum, *Stochastic Process. Appl.* **53**, 1, pp. 1–22.

Albeverio, S., Khrennikov, A. Y. and Shelkovich, V. M. (2006). Harmonic analysis in the p-adic Lizorkin spaces: fractional operators, pseudo-differential equations, p-adic wavelets, Tauberian Theorems, *J. Fourier Anal. Appl.* **12**, 4, pp. 393–425.

Alvarez, J., Guzmán-Partida, M. and Lakey, J. (2000). Spaces of bounded λ-central mean oscillation, Morrey spaces, and λ-central Carleson measures, *Collect. Math.* **51**, 1, pp. 1–47.

Andersen, K. F. (2003). Boundedness of Hausdorff operators on $L^p(\mathbb{R}^n)$, $H^1(\mathbb{R}^n)$ and $BMO(\mathbb{R}^n)$, *Acta Sci. Math. (Szeged)* **69**, 1–2, pp. 409–418.

Andersen, K. F. and Muckenhoupt, B. (1982). Weighted weak type Hardy inequalities with applications to Hilbert transforms and maximal functions, *Studia Math.* **72**, 1, pp. 9–26.

Avetisov, A. V., Bikulov, A. H., Kozyrev, S. V. and Osipov, V. A. (2002). p-adic models of ultrametric diffusion constrained by hierarchical energy landscapes, *J. Phys. A* **35**, 2, pp. 177–189.

Baiarystanov, A. O., Persson, L. E., Shaimardan, S. and Temirkhanova, A. (2016). Some new Hardy-type inequalities in q-analysis, *J. Math. Inequal.* **10**, 3, pp. 761–781.

Balinsky, A. A., Evans, W. D. and Lewis, R. T. (2015). *The analysis and geometry of Hardy's inequality*, Universitext. (Springer Press, Cham), xv+263 pp.

Bangerezako, G. (2005). Variational calculus on q-nonuniform lattices, *J. Math. Anal. Appl.* **306**, 1, pp. 161–179.

Beatrous, F. and Li, S. (1993). On the boundedness and compactness of operators of Hankel type, *J. Funct. Anal.* **111**, 2, pp. 350–379.

Benedek, A., Calderón, A. P. and Panzone, R. (1962). Convolution operators on Banach space valued funtions, *Proc. Nat. Acad. Sci. U.S.A.* **48**, pp. 256–365.

Bényi, Á and Oh, T. (2006). Best constants for certain multilinear integral operators, *J. Inequal. Appl.* Art. ID 28582, 12 pp.

Bliss, G. A. (1930). An integral inequality, *J. London Math. Soc.* **5**, 1, pp. 40–46.

Boyd, D. W. (1971). Inequalities for positive integral operators, *Pacific J. Math.* **38**, 1, pp. 9–24.

Bramanti, M. and Cerutti, M. (1993). $W_p^{1,2}$-solvability for the Cauchy-Dirichlet problem for parabolic equations with VMO coefficients, *Comm. Partial Differential Equations.* **18**, 9-10, pp. 1735–1763.

Burenkov, V. I., Jain, P. and Tararykova, T. V. (2011). On boundedness of the Hardy operator in Morrey-type spaces, *Eurasian Math. J.* **2**, 1, pp. 52–80.

Campanato, S. (1963). Proprietà di hölderianità di alcune classi di funzioni, *Ann. Scuola Norm. Sup. Pisa Cl. Sci.(3)* **17**, 1–2, pp. 175–188.

Cañestroa, M. I., Salvador, P. O. and Torreblanca, C. R. (2012). Weighted bilinear Hardy inequalities, *J. Math. Anal. Appl.* **387**, 1, pp. 320–334.

Carcía-Cuerva, J. and Rubio de Francia, J. L. (1985). *Weighted norm inequalities and related topics.* North-Holland Mathematics Studies, 116. Notas de Matemática [Mathematical Notes], 104. (North-Holland Publishing Co., Amsterdam), x+604 pp.

Cerdà, J. and Martín, J. (2000). Weighted Hardy inequalities and Hardy transforms of weights, *Studia Math.* **139**, 2, pp. 189–196.

Chanillo, S. (1982). A note on commutators, *Indiana Univ. Math. J.* **31**, 1, pp. 7–16.

Chen, Y. and Ding, Y. (2010). Compactness of the commutators of parabolic singular integrals, *Sci. China Math.* **53**, 10, pp. 2633–2648.

Chen, Y., Ding, Y. and Wang, X. (2009). Compactness of commutators of Riesz potential on Morrey spaces, *Potential Anal.* **30**, 4, pp. 301–313.

Chen, Y., Ding, Y. and Wang, X. (2012). Compactness of commutators for singular integrals on Morrey spaces, *Canad. J. Math.* **64**, 2, pp. 257–281.

Chen, J., Fan, D. and Wang, S. (2013). Hausdorff operators on Euclidean spaces, *Appl. Math. J. Chinese Univ. Ser. B* **28**, 4, pp. 548–564.

Chen, Y. and Lau, K. S. (1989). Some new classes of Hardy spaces, *J. Funct. Anal.* **84**, 2, pp. 255–278.

Chiarenza, F., Frasca, M. and Longo, P. (1993). $W^{2,p}$-solvability of the Dirichlet problem for nondivergence elliptic equations with VMO coefficients, *Trans. Amer. Math. Soc.* **336**, 2, pp. 841–853.

Christ, M. and Grafakos, L. (1995). Best constants for two nonconvolution inequalities, *Proc. Amer. Math. Soc.* **123**, 6, pp. 1687–1693.

Chuong, N. M. and Co, N. V. (2008). The Cauchy problem for a class of pseudodifferential equations over p-adic field, *J. Math. Anal. Appl.* **340**, 1, pp. 629–645.

Chuong, N. M., Egorov, Y. V., Khrennikov, A., Meyer, Y. and Mumford, D. (2007). *Harmonic, wavelet and p-adic analysis.* (World Scientific Publishing Co. Pte. Ltd., Hackensack, N.J.), x+381 pp.

Chuong, N. M., Hong, N. T. and Hung, H. D. (2018). Bounds of weighted multilinear Hardy-Cesàro operators in p-adic functional spaces, *Front. Math.*

China **13**, 1, pp. 1–24.

Coifman, R. R. and Meyer, Y. (1975). On commutators of singular integrals and bilinear singular integrals, *Trans. Amer. Math. Soc.* **212**, pp. 315–331.

Coifman, R. R., Rochberg, R. and Weiss, G. (1976). Factorization theorems for Hardy spaces in several variables, *Ann. of Math.* **103**, 2, pp. 611–635.

Coulhon, T., Müller, D. and Zienkiewicz, J. (1996). About Riesz transforms on the Heisenberg groups, *Math. Ann.* **305**, 2, pp. 369–379.

Cruz-Uribe, D. and Neugebauer, C. J. (1995). The structure of the reverse Hölder classes, *Trans. Amer. Math. Soc.* **347**, 8, pp. 2941–2960.

Deng, D., Duong, X. T. and Yan, L. (2005). A characterization of Morrey-Campanato spaces, *Math. Z.* **250**, 3, pp. 641–655.

Ding, Y. (1997). A characterization of BMO via commutators for some operators, *Northeast Math. J.* **13**, 4, pp. 422–432.

Ding, Y. and Mei, T. (2015). Boundedness and compactness for the commutators of bilinear operators on Morrey spaces, *Potential Anal.* **42**, 3, pp. 717–748.

Drábek, P., Heinig, H. P. and Kufner, A. (1995). Higher dimensional Hardy inequality. General inequalities, 7 (Oberwolfach), pp. 3–16, *Internat. Ser. Numer. Math.* **123**, Birkhäuser, Basel, 1997.

Duong, X. T., Xiao, J. and Yan, L. (2007). Old and new Morrey spaces with heat kernel bounds, *J. Fourier Anal. Appl.* **13**, 1, pp. 87–111.

Ernst, T. (1999). A new notation for q-calculus and a new q-Taylor formula, *Uppsala University Report Depart. Math.* pp. 1–28.

Ernst, T. (2012). *A comprehensive treatment of q-calculus.* (Birkhäuser / Springer Basel AG, Basel), xvi+491 pp.

Euler, L. (1732). Methodus generalis summandi progressiones, *Commentarii academiae scientiarum imperialis Petropolitanae* **6** (1732–1733), pp. 68–97.

Euler, L. (1741). Inventio summae cuiusque seriei ex dato termino generali, *Commentarii academiae scientiarum imperialis Petropolitanae* **8**, pp. 9–22.

Exton, H. (1983). *q-hypergeometric functions and applications.* (Ellis Horwood Ltd., Chichester, Halsted Press, New York), 347 pp.

Fan, D., Lu, S. and Yang, D. (1998). Regularity in Morrey spaces of strong solutions to nondivergence elliptic equations with VMO coefficients, *Georgian Math. J.* **5**, 4, pp. 425–440.

Fan, D. and Zhao, F. (2019). Sharp constants for multivariate Hausdorff q-inequalities, *J. Aust. Math. Soc.* **106**, 2, pp. 274–286.

Faris, W. (1976). Weak Lebesgue spaces and quantum mechanical binding, *Duke Math. J.* **43**, 2, pp. 365–373.

Fazio, G. D. and Ragusa, M. A. (1993). Interior estimates in Morrey spaces for strong solutions to nondivergence form equations with discontinuous coefficients, *J. Funct. Anal.* **112**, 2, pp. 241–256.

Fefferman, C. (1983). The uncertainty principle, *Bull. Amer. Math. Soc. (N.S.)* **9**, 2, pp. 129–206.

Fefferman, C. and Stein, E. M. (1971). Some maximal inequalities, *Amer. J. Math.* **93**, 1, pp. 107–115.

Feichtinger, H. (1987). An elementary approach to Wiener's third Tauberian theorem on Euclidean n-space, *Symposia Mathematica, Vol. XXIX* (Cortona,

1984), pp. 267–301, (Academic Press, New York).

Folland, G. B. and Stein, E. M. (1982). *Hardy Spaces on Homogeneous Groups.* Mathematical Notes, 28. (Princeton University Press, Princeton, N.J.; University of Tokyo Press, Tokyo), xii+285 pp.

Fu, Z. (2008). *Hardy operators, singular integral and related topics*, Ph.D. Dissertation, Beijing Normal University, Beijing, China.

Fu, Z., Grafakos, L., Lu, S. and Zhao, F. (2012). Sharp bounds for m-linear Hardy and Hilbert operators, *Houston J. Math.* **38**, 1, pp. 225–244.

Fu, Z., Liu, Z., Lu, S. and Wang, H. (2007). Characterization for commutators of n-dimensional fractional Hardy operators, *Sci. China Ser. A* **50**, 10, pp. 1418–1426.

Fu, Z., Wu, Q. and Lu, S. (2013). Sharp estimates of p-adic Hardy and Hardy-Littlewood-Pólya operators, *Acta Math. Sin. (Engl. Ser.)* **29**, 1, pp. 137–150.

García-Cuerva, J. (1989). Hardy spaces and Beurling algebras, *J. London Math. Soc.(2)* **39**, 3, pp. 499–513.

Geller, D. (1980). Some results in H^p theory for the Heisenberg group, *Duke Math. J.* **47**, 2, pp. 365–390.

Gilbarg, D. and Trudinger, N. (1983). *Elliptic partial differential equations of second order. Second edition.* (Springer-Verlag, Berlin), xiii+513 pp.

Golubov, B. I. (1997). On the boundedness of the Hardy and the Hardy-Littlewood operators in the spaces ReH^1 and BMO, *Mat. Sb.* **188**, 7, pp. 93–106; translation in *Sb. Math.* **188**, 7, pp. 1041–1054.

Grafakos, L. (2008). *Classical Fourier analysis.* Second edition. Graduate Texts in Mathematics, 249. (Springer, New York), xvi+489 pp.

Grafakos, L. (2009). *Modern Fourier analysis.* Second edition. Graduate Texts in Mathematics, 250. (Springer, New York), xvi+504 pp.

Grafakos, L. and Montgomery-Smith, S. (1997). Best constants for uncentred maximal functions, *Bull. London Math. Soc.* **29**, 1, pp. 60–64.

Guliev, V. S. (1994). Two-weighted L_p-inequalities for singular integral operators on Heisenberg groups, *Georgian Math. J.* **1**, 4, pp. 367–376.

Guo, J. and Zhao, F. (2017). Some q-inequalities for Hausdorff operators, *Front. Math. China.* **12**, 4, pp. 879–889.

Haran, S. (1990). Riesz potentials and explicit sums in arithmetic, *Invent. Math.* **101**, 3, pp. 697–703.

Haran, S. (1993). Analytic potential theory over the p-adics, *Ann. Inst. Fourier(Grenoble)* **43**, 4, pp. 905–944.

Harboure, E., Salinas, O. and Viviani, B. (1998). Reverse Hölder classes in the Orlicz spaces setting, *Studia Math.* **130**, 3, pp. 245–261.

Hardy, G. H. (1920). Note on a theorem of Hilbert, *Math. Z.* **6**, 3–4, pp. 314–317.

Hardy, G. H. (1925). Notes on some points in the integral calculus. LX. An inequality between integrals, *Messenger of Math.* **54**, pp. 150–156.

Hardy, G. H., Littlewood, J. E. and Pólya, G. (1952). *Inequalities,* 2nd ed. (Cambridge, at the University Press), xii+324 pp.

He, Q., Wei, M. and Yan, D. (2019). Characterizations of p-adic central Campanato spaces via commutator of p-adic Hardy type operators, *J. Korean*

Math. Soc. **56**, 3, pp. 767–787.

Heinig, H. P., Kerman, R. and Krbec, M. (2001). Weighted exponential inequalities, *Georgian Math. J.* **8**, 1, pp. 69–86.

Hytönen, T., Pérez, C. and Rela, E. (2012) Sharp reverse Hölder property for A_∞ weights on spaces of homogeneous type, *J. Funct. Anal.* **263**, 12, pp. 3883–3899.

Iwaniec, T. and Sbordone, C. (1998). Riesz transforms and elliptic PDEs with VMO coefficients, *J. Anal. Math.* **74**, pp. 183–212.

Jackson, F. H. (1910a). On q-definite integrals, *Quart. J. Pure Appl. Math.* **41**, pp. 193–203.

Jackson, F. H. (1910b). q-difference equations, *Amer. J. Math.* **32**, 4, pp. 305–314.

Jacobi, G. J. (1829). *Fundamenta nova theoriae functionum ellipticarum*, Königsberg.

Janson, S. (1978). Mean oscillation and commutators of singular integral operators, *Ark. Mat.* **16**, 2, pp. 263–270.

Kac, V. and Cheung, P. (2002). *Quantum calculus.* (Springer-Verlag, N.Y.), x+112 pp.

Khrennikov, A. (1994). *p-adic valued distributions in mathematical physics. Mathematics and its Applications*, Vol. 309, (Kluwer Academic Publishers Group, Dordrecht), xvi+264 pp.

Khrennikov, A. (1997). *Non-Archimedean analysis: quantum paradoxes, dynamical systems and biological models. Mathematics and its Applications*, Vol. 427, (Kluwer Academic Publishers, Dordrecht), xviii+371 pp.

Kim, Y. C. (2009). Carleson measures and the BMO space on the p-adic vector space, *Math. Nachr.* **282**, 9, pp. 1278–1304.

Kim, Y. C. (2008). Weak type estimates of square functions associated with quasiradial Bochner-Riesz means on certain Hardy spaces, *J. Math. Anal. Appl.* **339**, 1, pp. 266–280.

Kochubei, A. N. (2008). A non-Archimedean wave equation, *Pacific J. Math.* **235**, 2, pp. 245–261.

Kolyada, V. I. (2014). Optimal relationships between L^p-norms for the Hardy operator and its dual, *Ann. Mat. Pura Appl.* **193**, 2, pp. 423–430.

Komori, Y. (2003). Notes on commutators of Hardy operators, *Int. J. Pure. Appl. Math.* **7**, 3, pp. 329–334.

Korányi, A. and Reimann, H. M. (1985). Quasiconformal mappings on the Heisenberg group, *Invent. Math.* **80**, 2, pp. 309–338.

Krantz, S. and Li, S. (2001). Boundedness and compactness of integral operators on spaces of homogeneous type and applications II, *J. Math. Anal. Appl.* **258**, 2, pp. 642–657.

Kufner, A., Maligranda, L. and Persson, L. E. (2007). *The Hardy inequality. About its history and some related results.* (Vydavatelský Servis, Plzeň), 162 pp.

Kufner, A. and Persson, L. E. (2003). *Weighted inequalities of Hardy type.* (World Scientific Publishing Co., Inc., River Edge, N.J.), xviii+357 pp.

Kufner, A., Persson, L. E. and Samko, N. (2017). *Weighted inequalities of Hardy type. Second Edition.* (World Scientific Publishing Co., Inc., River Edge, N.J.), xx+459 pp.

Landau, E., Schur, I. and Hardy, G. H. (1926). A Note on a Theorem Concerning Series of Positive Terms: Extract from a Letter, *J. London Math. Soc.* **1**, 1, pp. 38–39.

Lemarié-Rieusset, P. (2007). The Navier-Stokes equations in the critical Morrey-Campanato space, *Rev. Math. Iberoam.* **23**, 3, pp. 897–930.

Lerner, A. K. and Liflyand, E. (2007). Multidimensional Hausdorff operators on the real Hardy spaces, *J. Aust. Math. Soc.* **83**, 1, pp. 79–86.

Lerner, A., Ombrosi, S., Pérez, C., Torres, R. and Trujillo-González, R. (2009). New maximal functions and multiple weights for the multilinear Calderón-Zygmund theory, *Adv. Math.* **220**, 4, pp. 1222–1264.

Li, H. (2009). Fonctions maximales centrées de Hardy-Littlewood sur les groupes de Heisenberg, *Studia Math.* **191**, 1, pp. 89–100.

Li, H. and Qian, B. (2014). Centered Hardy-Littlewood maximal functions on Heisenberg type groups, *Trans. Amer. Math. Soc.* **366**, 3, pp. 1497–1524.

Liflyand, E. (2008). Boundedness of multidimensional Hausdorff operators on $H^1(\mathbb{R}^n)$, *Acta Sci. Math. (Szeged)* **74**, 3-4, pp. 845–851.

Liflyand, E. (2011). Complex and real Hausdorff operators, https://ddd.uab.cat/pub/prepub/2011/hdl\2072\182564/Pr1046.pdf

Liflyand, E. (2007). *Open problems on Hausdorff operators. Complex analysis and potential theory.* (World Scientific Publishing Co., Hackensack, N.J.), pp. 280–285.

Liflyand, E. (2013). Hausdorff operators on Hardy spaces, *Eurasian Math. J.* **4**, 4, pp. 101–141.

Liflyand, E. and Miyachi, A. (2009). Boundedness of the Hausdorff operators in H^p spaces, $0 < p < 1$, *Studia Math.* **194**, 3, pp. 279–292.

Liflyand, E. and Móricz, F. (2001). The multi-parameter Hausdorff operator is bounded on the product Hardy space $H^{11}(\mathbb{R} \times \mathbb{R})$, *Analysis (Munich)*, **21**, 2, pp. 107–118.

Li, X. and Yang, D. (1996). Boundedness of some sublinear operators on Herz spaces, *Illinois J. Math.* **40**, 3, pp. 485–501.

Liu, R. and Zhou, J. (2017). Sharp estimates for the p-adic Hardy type operator on higher-dimensional product spaces, *J. Inequal. Appl.* **2017**, 219, 13 pp.

Long, S. and Wang, J. (2002). Commutators of Hardy operators, *J. Math. Anal. Appl.* **274**, 2, pp. 626–644.

Lu, G. (1995). Embedding theorems on Campanato-Morrey spaces for vector fields and applications, *C. R. Acad. Sci. Paris Sér. I Math.* **320**, 4, pp. 429–434.

Lu, S. (1995). *Four lectures on real H^p spaces.* (World Scientific Publishing Co., Inc., River Edge, N.J.), viii+217 pp.

Lu, S., Ding, Y. and Yan, D. (2007). *Singular integrals and related topics.* (World Scientific Publishing Co. Pte. Ltd., Hackensack, N.J.), viii+272 pp.

Lu, S. and Xu, L. (2005). Boundedness of rough singular integral operators on the homogenous Morrey-Herz spaces, *Hokkaido Math. J.* **34**, 2, pp. 299–314.

Lu, S. and Yang, D. (1992). The Littlewood-Paley function and φ-transform characterization of a new Hardy space HK_2 associated with Herz space, *Studia Math.* **101**, 3, pp. 285–298.

Lu, S. and Yang, D. (1997). Some characterizations of weighted Herz-type Hardy spaces and its applications, *Acta Math. Sinica (N.S.)* **13**, 1, pp. 45–58.

Lu, S. and Yang, D. (1995a). The decomposition of the weighted Herz spaces on $\mathbb{R}^n$ and its applications, *Sci. China Ser. A* **38**, 2, pp. 147–158.

Lu, S. and Yang, D. (1995b). Oscillatory singular integrals on Hardy spaces associated with Herz spaces, *Proc. Amer. Math. Soc.*, **123**, 6, pp. 1695–1701.

Lu, S. and Yang, D. (1995c). The central BMO spaces and Littlewood-Paley operators, *Approx. Theory Appl. (N.S.)* **11**, 3, pp. 72–94.

Lu, S., Yan, D. and Zhao, F. (2013). Sharp bounds for Hardy type operators on higher-dimensional product spaces, *J. Inequal. Appl.* **2013**, 148, 11 pp.

Maligranda, L., Oinarov, R. and Persson, L. E. (2014). On Hardy q-inequalities, *Czechoslovak Math. J.* **64**, 3, pp. 659–682.

Manakov, V. M. (1992). On the best constant in weighted inequalities for Riemann-Liouville integrals, *Bull. London Math. Soc.* **24**, 5, pp. 442–448.

Martín-Reyes, F. J. and Ortega, P. (1998). On weighted weak type inequalities for modified Hardy operators, *Proc. Amer. Math. Soc.* **126**, 6, pp. 1739–1746.

Melas, A. (2003). The best constant for the centered Hardy-Littlewood maximal inequality, *Ann. of Math.* **157**, 2, pp. 647–688.

Miao, Y. and Qi, F. (2009). Several q-integral inequalities, *J. Math. Inequal.* **3**, 1, pp. 115–121.

Mitrinović, D. S., Pečarić J. E. and Fink, A. M. (1993). *Classical and new inequalities in analysis.* (Kluwer Academic Publishers Group, Dordrecht), xviii+740 pp.

Morrey, C. (1938). On the solutions of quasi-linear elliptic partial differential equations, *Trans. Amer. Math. Soc.* **43**, 1, pp. 126–166.

Muckenhoupt, B. (1978). *Weighted norm inequalities for classical operators.* Harmonic analysis in Euclidean spaces (Proc. Sympos. *Pure Math., Williams Coll., Williamstown, Mass.), Part 1, pp. 69–83, Proc. Sympos. Pure Math., XXXV, Part,* Amer. Math. Soc., Providence, R.I., 1979.

Muckenhoupt, B. (1972). Hardy's inequality with weights, *Studia Math.* **44**, 1, pp. 31–38.

Neri, U. (1975). Fractional integration on the space H^1 and its dual, *Studia Math.* **53**, 2, pp. 175–189.

Niu, P., Zhang, H. and Wang, Y. (2001). Hardy type and Rellich type inequalities on the Heisenberg group, *Proc. Amer. Math. Soc.* **129**, 12, pp. 3623–3630.

Opic, B. and Kufner, A. (1990). *Hardy-type inequalities.* Pitman Research Notes in Mathematics Series, vol. 219, (Longman Scientific, Technical, Harlow), xii+333 pp.

Pachpatte, B. G. (1992) On multivariate Hardy type inequalities, *An. Ştiinţ. Univ. Al. I. Cuza Iaşi Secţ. I a Mat.* **38**, 2, pp. 355–361.

Palagachev, D. and Softova, L. (2004). Singular integral operators, Morrey spaces and fine regularity of solutions to PDE's, *Potential Anal.* **20**, 3, pp. 237–263.

Paluszyński, M. (1995). Characterization of the Besov spaces via the commutator operator of Coifman, Rochberg and Weiss, *Indiana Univ. Math. J.* **44**, 1, pp. 1–17.

Rim, K. S. and Lee, J. (2006). Estimates of weighted Hardy-Littlewood averages on the p-adic vector space, *J. Math. Anal. Appl.* **324**, 2, pp. 1470–1477.

Rogers, K. M. (2005). A van der Corput lemma for the p-adic numbers, *Proc. Amer. Math. Soc.* **133**, 12, pp. 3525–3534.

Rogers, K. M. (2004). Maximal averages along curves over the p-adic numbers, *Bull. Austral. Math. Soc.* **70**, 3, pp. 357–375.

Rosenblum, M. (1962). Summability of Fourier Series in $L^p(d\mu)$, *Trans. Amer. Math. Soc.* **105**, 1, pp. 32–42.

Sawyer, E. (1985). Weighted inequalities for the two-dimensional Hardy operator, *Studia Math.* **82**, 1, pp. 1–16.

Sawyer, E. (1984). Weighted Lebesgue and Lorentz norm inequalities for the Hardy operator, *Trans. Amer. Math. Soc.* **1**, 1, pp. 329–337.

Schur, I. (1911). Bemerkungen zur Theorie der beschränkten Bilinearformen mit unendlich vielen Veränderlichen, *J. Reine Angew. Math.* **140**, pp. 1–28.

Shi, S., Fu, Z. and Lu, S. (2020). On the compactness of commutators of Hardy operators, *Pacific J. Math.* **307**, 1, pp. 239–256.

Shi, S. and Lu, S. (2013). Some characterizations of Campanato spaces via commutators on Morrey spaces, *Pacific J. Math.* **264**, 1, pp. 221–234.

Shi, S. and Lu, S. (2014). A characterization of Campanato space via commutator of fractional integral, *J. Math. Anal. Appl.* **419**, 1, pp. 123–137.

Shi, S. and Lu, S. (2015). Characterization of the central Campanato space via the commutator operator of Hardy type, *J. Math. Anal. Appl.* **429**, 2, pp. 713–732.

Siskakis, A. (1987). Composition semigroups and the Cesàro operator on H^p, *J. London Math. Soc.* **36**, 1, pp. 153–164.

Siskakis, A. (1990). The Cesàro operator is bounded on H^1, *Proc. Amer. Math. Soc.* **110**, 2, pp. 461–462.

Stein, E. M. (1993). *Harmonic analysis: real-variable methods, orthogonality, and oscillatory integrals.* Princeton Mathematical Series, 43. Monographs in Harmonic Analysis, III. (Princeton University Press, Princeton, N.J.), xiv+695 pp.

Stein, E. M. and Weiss, G. (1971). *Introduction to Fourier analysis on Euclidean spaces.* Princeton Mathematical Series, No. 32. (Princeton University Press, Princeton, N.J.), x+297 pp.

Stempak, K. (1994). Cesàro averaging operators, *Proc. Roy. Soc. Edinburgh (Sect. A)* **124**, 1, pp. 121–126.

Sulaiman, W. T. (2011). New types of q-integral inequalities, *Adv. Pure Appl. Math.* **1**, 3, pp. 77–80.

Talenti, G. (1969). Osservazioni sopra una classe di disuguaglianze, *Rend. Sem. Mat. Fis. Milano* **39**, pp. 171–185.

Tomaselli, G. (1969). A class of inequalities. *Boll. Un. Mat. Ital.* **2**, 4, pp. 622–631.

Uchiyama, A. (1978). On the compactness of operators of Hankel type, *Tohoku Math. J.* **30**, 1, pp. 163–171.

Varadarajan, V. S. (1997). Path integrals for a class of p-adic Schrödinger equations, *Lett. Math. Phys.* **39**, 2, pp. 97–106.

Vladimirov, V. S., Volovich, I. V. andZelenov, E. I. (1994). *p-adic analysis and mathematical physics. Series on Soviet and East European Mathematics, 1.*

(World Scientific Publishing Co., Inc., River Edge, N.J.), xx+319 pp.

Weisz, F. (2004). The boundedness of the Hausdorff operator on multi-dimensional Hardy spaces, *Analysis (Munich)* **24**, 2, pp. 183–195.

Wiener, N. (1930). Generalized harmonic analysis, *Acta Math.* **55**, 1, pp. 117–258.

Wiener, N. (1932). Tauberian theorems, *Ann. of Math.* **33**, 1, pp. 1–100.

Wu, Q. and Fu, Z. (2016). Sharp estimates for Hardy operators on Heisenberg group, *Front. Math. China* **11**, 1, pp. 155–172.

Wu, Q., Mi, L. and Fu, Z. (2013). Boundedness of p-adic Hardy operators and their commutators on p-adic central Morrey and BMO spaces, *J. Funct. Spaces Appl.* **2013**, Art. ID 359193, 10 pp.

Wu, S. and Debnath, L. (2007). Inequalities for convex sequences and their applications, *Comput. Math. Appl.* **54**, 4, pp. 525–534.

Wu, X. and Chen, J. (2014). Best constants for Hausdorff operators on n-dimensional product spaces, *Sci. China Math.* **57**, 3, pp. 569–578.

Xiao, J. (2001). L^p and BMO bounds of weighted Hardy-Littlewood averages, *J. Math. Anal. Appl.* **262**, 2, pp. 660–666.

Yang, D., Yang, Do. and Zhou, Y. (2010). Localized Morrey-Campanato spaces on metric measure spaces and applications to Schrödinger operators, *Nagoya Math. J.* **198**, pp. 77–119.

Yosida, K. (1995). *Functional analysis.* Reprint of the sixth (1980) edition. Classics in Mathematics. (Springer-Verlag, Berlin), xii+501 pp.

Yuan, W., Sickel, W. and Yang, D. (2010). *Morrey and Campanato meet Besov, Lizorkin and Triebel.* Lecture Notes in Mathematics, 2005. (Springer-Verlag, Berlin), xii+281 pp.

Zhao, F., Fu, Z. and Lu, S. (2014). M^p weights for bilinear Hardy operators on $\mathbb{R}^n$, *Collect. Math.* **65**, pp. 87–102.

Zhao, F., Fu, Z. and Lu, S. (2012). Endpoint estimates for n-dimensional Hardy operators and their commutators, *Sci. China Math.* **55**, 10, pp. 1977–1990.

Zhao, F. and Liu, R. (2021). Sharp inequalities between L^p-norms for the higher dimensional Hardy operator and its dual, *Math. Inequal. Appl.* **24**, 1, pp. 275–289.

Zhao, F. and Lu, S. (2013). A characterization of λ-central BMO space, *Front. Math. China* **8**, 1, pp. 229–238.

Zhao, F. and Lu, S. (2015). The best bound for n-dimensional fractional Hardy operators, *Math. Inequal. Appl.* **18**, 1, pp. 233–240.

Zhao, F. and Ma, L. (2019). Higher-dimensional weighted Knopp type inequalities, *Math. Inequal. Appl.* **22**, 2, pp. 619–629.

Zhu, Y. and Zheng, W. (1998). Besov spaces and Herz spaces on local fields, *Sci. China Ser. A* **41**, 10, pp. 1051–1060.

Zienkiewicz, J. (2005). Estimates for the Hardy-Littlewood maximal function on the Heisenberg group, *Colloq. Math.* **103**, 2, pp. 199–205.

Zuniga-Galindo, W. A. (2004). Pseudo-differential equations connected with p-adic forms and local zeta functions, *Bull. Austral. Math. Soc.* **70**, 1, pp. 73–86.

Index

Printed in the United States
by Baker & Taylor Publisher Services